Agricultural Science: Conventional and Organic Farming

Agricultural Science: Conventional and Organic Farming

Editor: Magen Kensington

Callisto Reference
www.callistoreference.com

Callisto Reference,
118-35 Queens Blvd., Suite 400,
Forest Hills, NY 11375, USA

Visit us on the World Wide Web at:
www.callistoreference.com

ISBN: 978-1-64116-740-6 (Hardback)

Cataloging-in-publication Data

Agricultural science : conventional and organic farming / edited by Magen Kensington.
p. cm.
Includes bibliographical references and index.
ISBN 978-1-64116-740-6
1. Agriculture. 2. Organic farming. 3. Land use, Rural. 4. Crop science. I. Kensington, Magen.
S419 .A37 2023
630.9--dc23

Table of Contents

Preface

Agricultural science is a multidisciplinary field of biology, which includes components of economic, natural and social sciences. These components are utilized in the study and practice of agriculture. The two types of farming which are studied within agricultural science are conventional farming and organic farming. Conventional farming is a method which involves the usage of pesticides and other chemicals for obtaining highest production of crops. Organic farming is an agricultural approach which emphasizes on methods of companion planting and crop rotation along with the utilization of organic fertilizers including bone meal, compost manure and green manure. There is a significant distinction between organic and conventional farming in terms of tools and methods used to cultivate crops, but they also have varied ethics and principles. Organic farming strives to maintain the quality of ecology, water and land, whereas conventional farming solely looks for increased profits and yields. The aim of this book is to present researches that have transformed the study of conventional and organic farming and aided their advancement. It is appropriate for students seeking detailed information in this area of agricultural science as well as for experts.

Various studies have approached the subject by analyzing it with a single perspective, but the present book provides diverse methodologies and techniques to address this field. This book contains theories and applications needed for understanding the subject from different perspectives. The aim is to keep the readers informed about the progresses in the field; therefore, the contributions were carefully examined to compile novel researches by specialists from across the globe.

Indeed, the job of the editor is the most crucial and challenging in compiling all chapters into a single book. In the end, I would extend my sincere thanks to the chapter authors for their profound work. I am also thankful for the support provided by my family and colleagues during the compilation of this book.

Editor

1

Organic Farming Practices among Livestock and Fish Farmers in Southern Nigeria

Charity N. Atoma, Olufemi M. Adesope and Linda C. Familusi

Abstract

To provide solutions to the issues of climate change, food insecurities, environmental degradation, food safety and sustainability in production, organic farming was developed as opposed to the conventional method of farming which involved the use of synthetic fertilizers and other agro-chemicals. The objectives of the study was to determine the level of awareness of organic farming practices and to determine the use and level of use of organic farming practices among livestock and fish farmers in the study area. Multistage sampling techniques were employed. Primary data was collected through structured questionnaire. Sample size comprised of 115 fish farmers and 99 livestock farmers which made a total of 214 rural farmers. Results showed that fish farmers were more aware of organic farming practices rela-tive to livestock farmers. Also, fish farmers used more of organic farming methods for fishing activities than livestock farmers. However, the level of awareness and level of use of organic farming practices was low for both farmers. Therefore, it was recommended that extension campaign on organic farming sensitization should be increased.

Keywords: organic agriculture, awareness, use level

1. Introduction

Agriculture is the mainstay of the Nigerian economy. Agriculture in its nature is multifunctional. This multi-functionality nature relates to food production, security and safety, environment and landscape, water management and social and economic focus [1]. Agriculture has the capability to feed its population, serve as a source of revenue to the nation, provide employment and employment opportunities and serve as source of raw materials to agro-allied industries [2]. In recent times however, these functions could not be met given that food production in the nation could no longer meet up with the rapid population growth and thus reduction in exports [2]. In order to address the issues of insufficient food supply to meet the need of the nation's ever growing population, the use of agro-chemicals was adopted. This led to an increase in crop and animal production through the use of synthetic fertilizers and other agro-chemicals. Nevertheless, a number of side effects from the use of synthetic fertilizers and other agro-chemicals have been recorded and hence has led to the development of organic farming.

Inorganic fertilizers usually cause water pollution due to the leaching and washing away of such agro-chemicals by erosion in rivers [3]. It argued that most inorganic fertilizers and chemicals used in agriculture are manufactured using non-renewable resources such as fossil fuel which usually contributes to pollution and environmental degradation and hence unsustainable agricultural production [4]. Organic farming is a multifunctional system with benefits which cuts across economic, environmental and social functions. The multifunctional benefits of organic farming includes its contributions to the improvement of livelihoods, food security, resilience to climate change, increase in yields in a long run bases, reducing financial risks, creating market opportunities, improving health and the environment, combating desertification among other numerous benefits [5].

Organic farming represents a deliberate attempt to make the best use of local natural resources and is an environmental friendly system of farming. Organic farming is a production system that excludes the use of synthetically manufactured chemicals like fertilizers, pesticides, growth enhancers/regulators, food additives, fungicides and herbicides [6–8]. Organic farming practices include crop rotation, biological pest control, crop residues, cover crop, legumes, organic fertilizers, animal manures and green manures among others [9]. Organic farming technology is frequently regarded as the solution to environmental problems that are related to agriculture as well as food safety [10]. It usually has "zero impact on the environment" [4, 9, 11].

Organic farming practices has been shown to affects soil microbiological and chemical properties by increasing soil nutrient availability, microbial biomass and microbial activity, which represent a set of sensitive indicators of soil qual-ity [12, 13]. In addition to other benefits especially as it relates to health and food safety, organic farming has shown to be safer and healthy [14, 15]. It also results to increased levels of flavonoids[1] when used for the production of fruits and vegetables [14] and protects against cancer and other age related diseases to a lesser extent [16]. Organic farming increases insect pollination and overall specie richness [17]. The major goal of organic farming activities is a sustainable production of food with little or no effect on the environment. This goal and many others have not been achieved by conventional farming hence the need to encourage organic farming which is capable of providing solutions to the current environmental challenges like the climate change and environmental hazards and also help to achieve maximal production of quality food sustainably [18]. Food and Agriculture Organization clearly states that organic agriculture promotes ecological resilience, improves biodiversity, healthy management of farm and surrounding environment and building community knowledge and strength [19].

Keynote address emphasized that in Nigeria and many developing countries, organic agriculture is just developing [15]. As at the ending of 2016, report showed that Africa as a whole contributes to only 3% (i.e., 1.8 million hectares) of the total organic farmlands of the world [20, 21]. This contribution is mostly accounted for from East African countries with little or nothing from Nigeria. These points to the need to fast track organic farming development in Nigeria. Hence, the International Federation of Organic Agriculture Movement (IFOAM) and their members were charged with the responsibilities of developing organic farming in Nigeria during the second national conference on organic agriculture which held in Nigeria [5]. However, literatures argued that the yield from organic agriculture is lower than the yield from the conventional methods [22]. This may be the case

[1] Flavonoids helps to protect against cardiovascular disease

but notwithstanding, the net farm income of organic farmers was reported more profitable than that of conventional farmers [23]. Part of the (better) profits from certified OA resulted from the premium paid by contracting companies. The revenue generated from organic farming is higher than that of conventional methods because of the higher product prices generated from it [9].

Thus, the general objectives of the study are to ascertain the household level of awareness and use of organic farming practices in South-South Nigeria. Specifically, the study determined the level of awareness of organic farming practices among livestock and fish farmers, identified the use of organic farming practices among livestock and fish farmers and determined the level of use of organic farming practices among livestock and fish farmers in the study area.

2. Methodology

The area of study is South-South Nigeria which comprises six (6) states namely: Delta, Bayelsa, Edo, Rivers, Cross River and Akwa Ibom as shown in **Figure 1**.

The study population comprises of rural households engaged in livestock and fish farmers. Multistage random selection technique was employed. The first stage was the random selection of three states—Bayelsa, Delta and Akwa-Ibom. The study population is the livestock and fish farmers in Bayelsa, Delta and Akwa-Ibom. Delta state is divided into three agricultural zones—Delta North, Delta South and Delta Central out of which Delta central was selected. Bayelsa state is also classified into three agricultural zones—Brass, Yenagoa and Sagbama out of which brass zone was sampled. Akwa-Ibom is divided into six agricultural zones—Abak, Eket, Etinan, Ikot Ekpene, Oron and Uyo zones. Two zones Uyo and Ikot Ekpene were samples. A total of 99 livestock farmers and 115 fish farmers were sampled. The lists of farmers were gotten from the zonal managers in charge of each zone. Primary data were collected through the use of a questionnaire and interview schedule. Data obtained were analyzed using descriptive statistics as mean and percentages. Four point Likert scale was used to measure the response of famers in terms of their awareness level of organic agriculture and their use and use level of organic farming practices.

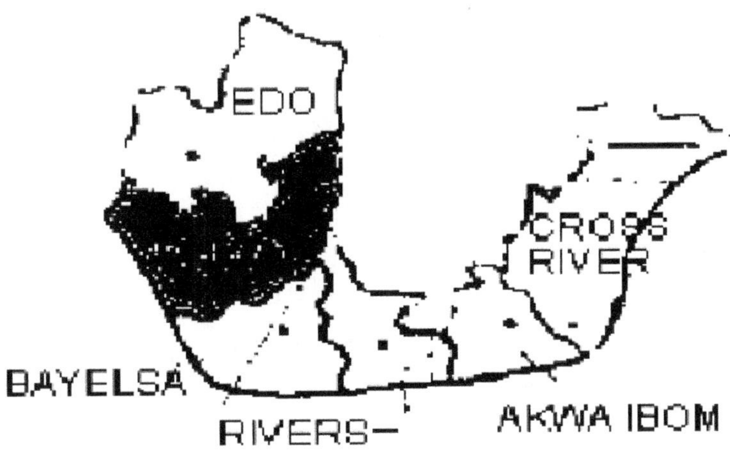

Figure 1.
Map of South-South region of Nigeria. Source: [26].

3. Results and discussions

Table 1 shows the result for the level of awareness of organic farming practices among livestock farmers. The mean and percentages of the response were clearly shown.

Organic Farming Practices for livestock	Not at all	Low	Moderate	High	Mean	Decision
Adequate land holding	33 (33.3)	40 (40.4)	14 (14.1)	12 (12.1)	2.05	Not aware
Farm diversification	44 (44.4)	29 (29.3)	25 (25.5)	1 (1.0)	1.82	Not aware
Free movement of animals/ Provision of fresh air and natural day light	61 (61.6)	6 (6.1)	25 (25.3)	7 (7.1)	1.77	Not aware
Protection against adverse weather condition	57 (57.6)	2 (2.0)	30 (30.3)	10 (10.1)	1.92	Not aware
Resting areas	57 (57.6)	16 (16.2)	13 (13.1)	13 (13.1)	1.81	Not aware
Clean and dry beddings	55 (55.6)	3 (3.0)	29 (29.3)	15 (15.2)	2.09	Not aware
Enough space for exercise	52 (52.5)	3 (3.0)	29 (29.3)	15 (15.2)	2.07	Not aware
Access to fresh drinking water by livestock	12 (12.1)	1 (1.0)	59 (59.6)	27 (27.3)	3.02	Aware
Allowing livestock to express natural behaviour	29 (29.3)	11 (11.1)	40 (40.1)	19 (19.2)	2.50	Aware
Use of local breed	59 (59.6)	2 (2.0)	26 (26.3)	12 (12.1)	1.90	Not aware
Natural reproduction technique	53 (53.6)	11 (11.1)	20 (20.2)	15 (15.2)	1.96	Not aware
Produce without genetic engineering , ionising radiation or sewage sludge	51 (51.5)	25 (25.3)	15 (15.2)	8 (8.1)	1.79	Not aware
Adequate feeding	16 (16.2)	64 (64.6)	0 (0.0)	19 (19.2)	2.86	Aware
Animal feeding is 100% organic	32 (32.3)	3 (3.0)	24 (24.2)	40 (40.4)	2.72	Aware
Prompt treatment of sick animals	27 (27.3)	10 (10.1)	51 (51.5)	11 (11.1)	1.74	Not aware
Manage animals without antibiotics	74 (74.7)	10 (10.1)	7 (7.1)	8 (8.1)	1.48	Not aware
Traditional/natural treatment of sick animals	62 (62.6)	11 (11.1)	15 (15.2)	11 (11.1)	1.74	Not aware
Vaccinate only during disease outbreak	69 (69.7)	16 (16.2)	8 (8.1)	6 (6.1)	1.50	Not aware
Manage without added growth hormones	45 (45.5)	29 (29.3)	15 (15.2)	10 (10.1)	1.89	Not aware
Accurate record keeping	33 (33.3)	1 (1.0)	40 (40.4)	25 (25.3)	2.57	Aware

Table 1.
Distribution of livestock farmers by level of awareness of organic farming practice.

Table 1 showed that farmers were adequately aware of allowing livestock access to fresh drinking water ($\bar{x}$ = 3.02), adequate feeding ($\bar{x}$ = 2.86), animal feeding of 100% organic ($\bar{x}$ = 2.72) and accurate record keeping ($\bar{x}$ = 2.57). The four practices above had mean scores above the discriminating index. The other practices were below the discriminating index of 2.50. The grand mean on the level of adoption for livestock farmers was 2.06. This shows that farmer's level of awareness of organic farming practices for livestock production is low. This is in-line with [24]. They implied that low awareness of organic agriculture was as a result of low coverage. Therefore, this study suggested that the low farmer's awareness could be as a result of poor extension campaign in organic livestock practices.

Table 2 shows the level of awareness of organic farming practices among fish farmers in the study area. The mean and percentages of the response were clearly shown.

Table 2 revealed that farmers were aware of such organic farming practices as eco-friendly design ($\bar{x}$ = 2.91), high quality water source ($\bar{x}$ = 2.90), pond

Organic Farming Practices	Not at all	Low	Moderate	High	Mean	Decision
Eco-friendly design	22 (19.1)	4 (3.5)	51 (44.3)	38 (33.0)	2.91	Aware
Manage without growth Hormone	31 (27.0)	3 (2.6)	61 (53.0)	20 (17.4)	2.60	Aware
Antibiotics is only used in clinical cases where no other treatment would work	65 (56.5)	2 (1.7)	19 (16.5)	29 (25.2)	2.10	Not Aware
Cultivate without genetic engineering.	44 (38.3)	5 (4.3)	21 (18.3)	45 (39.1)	2.58	Aware
Site is far from polluting substances	63 (54.8)	4 (3.5)	20 (17.4)	28 (24.3)	2.11	Not Aware
High quality water source (stream, river)	31 (27.0)	1 (0.9)	31 (27.0)	52 (45.2)	2.90	Aware
Organic fertilizer	54 (47.0)	4 (3.5)	20 (17.4)	31 (27.0)	2.24	Not Aware
Low stock density 10kg/m	39 (39.9)	6 (5.2)	47 (40.9)	23 (20.0)	2.46	Not Aware
Manage without synthetic appetizer and colouring	40 (34.8)	15 (13.0)	44 (38.3)	16 (13.9)	2.31	Not Aware
Polyculture	70 (60.9)	13 (11.3)	18 (15.7)	14 (12.2)	1.79	Not Aware
Proper record keeping	65 (56.5)	3 (2.6)	18 (15.7)	29 (25.2)	2.09	Not Aware
Pond protection from predators	13 (11.3)	2 (1.7)	36 (31.3)	64 (55.7)	3.31	Aware
Use of resistant species	29 (25.2)	1 (0.9)	34 (29.6)	51 (44.3)	2.95	Aware
Natural treatment (homeopathy)	43 (37.4)	8 (7.0)	11 (9.6)	53 (46.1)	2.64	Aware

Source: Field survey, 2015

Table 2.
Distribution of fish farmers by level of awareness of organic farming practices.

Organic farming practices	Use	%	Non use	%
Organic Farming Practices for livestock				
Adequate land holding	44	44.5	55	55.5
Farm diversification	39	39.4	60	60.5
Free movement of animals Provision of fresh air and natural day light	50	50.5	49	49.5
Protection against adverse weather condition	29	29.9	70	70.1
Resting areas	25	25.3	74	74.7
Clean and dry beddings	41	41.4	58	58.6
Enough space for exercise	46	46.6	53	53.4
Access to Fresh drinking water	75	75.8	24	24.2
Allowing livestock to Express natural behaviour	63	63.6	36	36.4
Use of local breed	49	49.5	50	50.5
Natural reproduction technique	57	57.6	42	42.4
Produce without genetic engineering , ionizing radiation or sewage sludge	40	40.4	59	59.6
Adequate feeding	73	73.3	26	26.3
Animal feeding is 100% organic	52	52.5	47	47.5
Prompt treatment of sick animals	60	60.6	39	39.4
Manage animals without antibiotics	21	21.2	78	78.8
Traditional/natural treatment of sick animals	36	36.4	63	63.6
Vaccinate only during disease outbreak	23	23.2	76	76.8
Manage without added growth hormones	31	31.3	68	68.7
Accurate record keeping.	54	54.5	45	45.5

Table 3.
Distribution of livestock farmers by use of organic farming practices.

protection from predators ($\bar{x}$ = 3.36), use of resistant species ($\bar{x}$ = 2.95, natural treatment ($\bar{x}$ = 2.64), cultivation without genetic engineering ($\bar{x}$ = 2.58) and man-agement without growth hormones ($\bar{x}$ = 2.60). Other practices were below mean score of ($\bar{x}$ = 2.50). The grand mean was 2.49. This implies a moderate awareness level which could be as a result of organic fish farming practices being in line with the traditional method of fish farming.

Table 3 shows the use of organic farming practices by livestock farmers in the study area.

According to **Table 3**, organic farming practices commonly used by livestock farmers includes fresh drinking water (76%), adequate feeding (73%), allowing livestock to express natural behavior (64%), prompt treatment of sick animals (61%), natural reproduction technique (58%), accurate record keeping (55%), animal feed is 100% organic (53%), free movement of animals/provision of fresh air and natural day light (51%) and use of local breed (50%). Out of 20 organic livestock practices, only 9 were above average and this is not up to 50% rating. This is not surprising since most livestock farmers are yet to be abreast with what organic livestock entails hence the low awareness level.

Table 4 shows the result for the use of organic farming practices by fish farmers in the study area.

Organic Farming Practices	Use	%	Non use	%
Eco-friendly design	91	79.1	24	20.9
Manage without growth Hormone	84	73.0	31	27.0
Antibiotics is only used in clinical cases where no other treatment would work	70	60.9	45	39.1
Cultivate without genetic engineering.	64	55.7	51	44.3
Site is far from polluting substances	87	75.7	28	24.3
High quality water source (stream, river)	63	54.8	52	45.2
Organic fertilizer	56	48.7	59	51.3
Low stock density 10k/m	35	30.4	80	69.6
Manage without synthetic appetizer and colouring	46	40.0	69	60.0
Poly-culture	59	51.3	56	48.7
Proper record keeping	53	46.1	62	53.9
Pond protection from predators	93	80.9	22	19.1
Use of resistant species	80	69.6	35	30.4
Natural treatment (homeopathy)	65	66.5	50	43.5

Table 4.
Distribution of fish farmers by use of organic farming practices.

The use of organic farming practices among fish farmers varied slightly in percentages as shown in **Table 4**. However, the commonly used organic farming practices includes pond protection from predators (81%), eco-friendly design (79%), site protection far from polluting substances (76%), manage without growth hormones (73%), use of resistant varieties had (70%), natural treatment (67%), antibiotics is used in clinical cases where no other treatment would work (61%), cultivated without genetic engineering (56%), high quality water source (55%) and poly-culture (51%). The use of organic farming practices by fish farmers was relatively high compared to organic farming practices by livestock farm-ers. This could be attributed to the fact that most of the organic practices are in line with the traditional practices of the people.

Table 5 shows the result for the level of use of organic farming practices among livestock farmers.

From **Table 5**, organic livestock production practices' in South-South Nigeria is low (grand mean = 1.93) as only 2 (10%) out of 20 outlined practices had mean score of 2.50 (discriminating index) and above. That is access to fresh drinking water (mean score = 2.77) and adequate feeding (mean score = 2.65). This result implied that the level of use is rare.

Table 6 shows the level of use of organic farming practices among fish farmers in the study area.

The results from **Table 6** revealed that out of the fourteen (14) practices out-lined, fish farmers regularly engaged in the use of three of such practices which are the use of eco-friendly design (x^{-} = 2.56), site being far from polluting substances (x^{-} = 2.57) and pond protection from predators (x^{-} = 2.70). From the result, the others were considered not being used. The grand mean of 1.99 indicates that the level of use of organic farming practices by fish farmers in the study area is low.

This could be associated with some challenges or difficulties in carrying out such practices and lack of awareness of the dangers associated with the conventional

Organic Farming Practices for livestock	Never	Rarely	Regularly	Very regularly	Mean	Decision
Adequate land holding	55 (55.5)	9 (9.1)	25 (25.3)	10 (10.1)	1.89	NU
Farm diversification	60 (60.5)	9 (9.1)	20 (20.2)	10 (10.1)	1.79	NU
Free movement of animals/Provision of fresh air and natural day light	49 (49.5)	16 (16.2)	34 (34.3)	0 (0.00)	1.84	NU
Protection against adverse weather condition	70 (70.1)	1 (1.0)	28 (28.3)	0 (0.00)	1.57	NU
Resting areas	74 (74.7)	5 (5.1)	20 (20.2)	0 (0.00)	1.45	NU
Clean and dry beddings	58 (58.6)	1 (1.0)	39 (39.4)	1 (1.0)	1.83	NU
Enough space for exercise	53 (53.4)	15 (15.2)	20 (20.2)	11 (11.1)	1.89	NU
Access to fresh drinking water	24 (24.2)	1 (1.0)	48 (48.5)	26 (26.3)	2.77	U
Allowing livestock to Express natural behaviour	36 (36.4)	3 (3.0)	57 (57.6)	3 (3.0)	2.27	NU
Use of local breed	50 (50.5)	1 (1.0)	37 (37.4)	11 (11.1)	2.09	NU
Natural reproduction technique	42 (42.4)	1 (1.0)	46 (46.5)	10 (10.1)	2.24	NU
Produce without genetic engineering , ionizing radiation or sewage sludge	59 (59.6)	4 (4.0)	32 (32.3)	4 (4.0)	1.81	NU
Adequate feeding	26 (26.3)	1 (1.0)	54 (54.5)	18 (18.2)	2.65	U
Animal feeding is 100% organic	47 (47.5)	3 (3.0)	34 (34.3)	15 (15.2	2.17	NU
Prompt treatment of sick animals	39 (39.4)	1 (1.0)	57 (57.6)	2 (2.0)	2.22	NU
Manage animals without antibiotics	78 (78.8)	3 (3.0)	17 (17.2)	1 (1.0)	1.40	NU
Traditional/natural treatment of sick animals	63 (63.6)	8 (8.1)	27 (27.3)	1 (1.0)	1.65	NU
Vaccinate only during disease outbreak	76 (76.8)	11 (11.1)	11 (11.1)	1 (1.0)	1.36	NU
Manage without added growth hormones	68 (68.7)	2 (2.0)	29 (29.3)	0 (0.00)	1.60	NU
Accurate record keeping	45 (45.5)	3 (3.0)	50 (50.5)	1 (1.0)	2.07	NU

Source: Field survey, 2015

Table 5.
Distribution of livestock farmers by level of use of organic practices.

Organic Farming Practices	Never	Rarely	Regularly	Very. regularly	Mean	Decision
Eco-friendly design	24 (20.9)	6 (5.2)	81 (70.4)	4 (3.5)	2.56	U
Manage without growth hormone	31 (27.0)	5 (4.3)	79 (68.7)	0 (0.0)	2 41	NU
Antibiotics is only used in clinical cases where no other treatment would work	45 (39.1)	29 (25.2)	40 (34.8)	1 (0.9)	1.97	NU
Cultivate without genetic engineering.	51 (44.3)	7 (6.1)	38 (33.0)	19 (16.5)	2.21	NU
Site is far from polluting substances	28 (24.3)	9 (7.8)	62 (53.9)	16 (13.9)	2.57	U
High quality water source (stream, river,	52 (45.2)	2 (1.7)	53 (46.1)	8 (7.0)	2.14	NU
Organic fertilizer	59 (51.3)	11 (9.6)	43 (37.4)	2 (1.7)	1.89	NU
Low stock density 10k/m	80 (69.6)	2 (1.7)	29 (25.2)	4 (3.5)	1.62	NU
Manage without synthetic appetizer and colouring	69 (60.0)	1 (0.9)	30 (26.1)	15 (13.0)	1.92	NU
Polyculture	56 (48.7)	7 (6.1)	50 (43.5)	2 (1.7)	1.98	NU
Proper record keeping	62 (53.9)	1 (0.9)	44 (38.3)	8 (7.0)	1.98	NU
Pond protection from predators	22 (19.1)	1 (0.9)	81 (70.4)	11 (9.6)	2.70	U
Use of resistant species	35 (30.4)	3 (2.6)	69 (60.0)	8 (7.0)	2.43	NU
Natural treatment (homeopathy)	50 (43.5)	35 (30.4)	26 (22.6)	4 (3.5)	1.86	NU

Table 6.
Distribution of fish farmers by level of use of organic farming practices.

practices. This does not augur well for the quest for healthy living. There was a positive relationship between knowledge of agricultural practice and innovativeness of farmers [25]. Thus the need to improved awareness of such practices to farmers.

4. Conclusion

The study concludes that the level of awareness of farmers to organic farming practices is low. However, fish farmers are better aware of such practices than livestock farmers. The justification between these major differences in the level of awareness of organic practices in farming activities could reflect on the livelihood of the southern communities in Nigeria. It is known that the major occupation is fishing. Therefore, it is only normal to be better aware of existing and improved techniques to improve fish farming other than the livestock counterpart. The study recommends that in the campaign for increased awareness of organic agriculture, special attention should be taken to create awareness to farmers on how organic farming practices can be applied for livestock production as well since this sector had shown a lower level of awareness.

Conclusion is also drawn on the use of organic farming practices among farmers. The use of organic farming practices is higher for fish farmers relative to livestock farmers. The rationale to this difference is drawn from the observation that some of the indigenous knowledge and traditional practices of fish farmers were similar to identified organic practices. However for livestock farmers, the opposite was observed and thus the low use of organic practices for production. It is recommended that extension personnel should educate farmers on the adoption of organic farming methods in production with special interest on livestock farmers. This is believed to contribute to the improvement in the use of organic farming methods for fish and livestock production. Particularly, livestock farmers should be educated on the use of such practices as: no antibiotics used, vaccination only during disease outbreak, protection of animals from adverse weather conditions and farm diversification. Likewise, fish fishers should be educated on such practices as: low stock density, no synthetic appetizer and coloring, proper record keeping and use of organic fertilizers.

In spite of the observation that fish farmers used more of organic farming methods relative to livestock farmers, the level of use of organic farming practices among both groups of farmers is low, despite the importance drawn from using organic methods for agricultural production. This draws to the need for increased extension campaign to sensitize farmers and sustain the interest in organic agriculture. It is also recommended that policymakers should create more windows of opportunities and incentives as well as the enabling environment to encourage more farmers to partici-pate in organic farm production. This is believed to contribute to the increase in the level of awareness, use and practices of organic agriculture in South-South Nigeria.

Author details

Charity N. Atoma[1], Olufemi M. Adesope[2*] and Linda C. Familusi[2]

1 Department of Agricultural Extension and Management, Delta State Polytechnic, Ozoro, Nigeria

2 Department of Agricultural Economics and Extension, Faculty of Agriculture, University of Port Harcourt, Nigeria

*Address all correspondence to: olufemi.adesope@uniport.edu.ng

References

[1] Todorova S, Ikova J. Multifunctional agriculture: Social and ecological impacts on the organic farms in Bulgaria. Procedia Economics and Finance. 2014;**9**:310-320. DOI: 10.1016/S2212-5671(14)00032-X

[2] Oluwaseyi AB. The prospects of agriculture in Nigeria: How our fathers lost their way—A review. Asian Journal of Economics, Business and Accounting. 2017;**4**(2):1-30

[3] Smil V. Enriching the Earth: Fritz Haber, Cartosch, and the Transformation of World Food Production. Boston: MIT Press; 2001

[4] Atoma CN, Atoma JO. Analysis of organic farming practices amongst crop farmers in Delta State, Nigeria. Journal of Information and Knowledge Management. 2015;**6**(3):213-220

[5] IFOAM. Organic Agriculture—Role in Countering Climate Change. Germany: IFOAM; 2007. Available from: www. ifoam.org/en/benefits-organic agriculture

[6] Anderson JB, Jolly D, Green R. Determinants of farmer adoption of organic production methods in the fresh-market produce sector in California. In: A Logistic Regression Analysis. San Francisco, California: Western Agricultural Economics Association; 2005. Available from: http://ageconsearch. umn.edu/bitstream/36319/1/sp05an01. pdf [Accessed: July 2011]

[7] Adesope OM, Mattews-Njoku EC, Oguzor NS, Ugwuja VC. Effect of socio-economic characteristics of farmers on their adoption of organic farming practices. In: Peeyush S, Vikas A, editors. Crop Production Technologies. UK: IntechOpen; 2012. pp. 211-220. DOI: 10.5772/30712

[8] Kutama AS, Abdullahi MA, Umar S, Binta UB, Ahmad MK. Organic farming in Nigeria: Problems and future prospects. Global Advanced Research Journal of Agricultural Science. 2013;**2**(10):256-262

[9] Mgbenka RN, Onwubuya EA, Ezeano CI. Organic farming in Nigeria: Need for popularization and policy. World Journal of Agricultural Sciences. 2015;**11**(6):346-355

[10] Agbamu JU. Agricultural research extension farmer linkages in Japan: "Policy issues for sustainable agricultural development in developing countries". International Journal of Social and Policy Issues. 2002;**1**:252-263

[11] Fess TL, Benedito VA. Organic versus conventional cropping sustainability: A comparative system analysis. Sustainability. 2018;**10**(1):1-42. https://doi.org/10.3390/su10010272

[12] Marinari S, Mancinelli R, Campiglia E, Grego S. Chemical and biological indicators of soil quality in organic and conventional farming systems in Central Italy. Ecological Indicators. 2006;**6**(2006):701-711

[13] Adebayo S, Oladele IO. Organic Agricultural Practices among Small Holder Farmers in South Western Nigeria. UK: IntechOpen; 2014. http://dx.doi.org/10.5772/57598

[14] Mitchell AE, Hong YJ, Koh E, Barrett DM, Bryant DE, Denison RF, et al. Ten-year comparison of the influence of organic and conventional crop management practices on the content of flavonoids in tomatoes. Journal of Agricultural and Food Chemistry. 2007;**2007**(55):6154-6159

[15] Gbadegesin AS. A Keynote Address presented at the 9th National Conference on Organic Agriculture Abeokuta, Nigeria; 11-15 November 2013; 2013. pp 1-11

[16] Commenges D, Scotet V, Renaud S, Jacqmin-Gadda H, Barberger-Gateau P, Dartigues JF. Intake of flavonoids and risk of dementia. European Journal of Epidemiology. 2000;**16**:357-363

[17] Fuller RJ, Norton LR, Feber RE, Johnson PJ, Chamberlain DE. Benefits of organic farming to biodiversity vary among taxa. Biology Letters. 2005;**1**:431-434. DOI: 10.1098/rsbl.2005.0357

[18] IFOAM. Principles of Organic Agriculture Preamble. 2005. 4 pp. Available from: http//www.ifoam.org/organicfacts/principles/pdfs/IFOAM_FS_principles for website pdf

[19] Food and Agriculture Organization of the United Nations (FAO). Organic Agriculture and Climate Change. Rome: Food and Agriculture Organization; 2008

[20] Willer H, Lernoud J, Kemper L. The world of organic agriculture 2018: Summary. In: Wliier H, Lernoud J, editors. FiBL & IFOAM-Organics International: The Workd of Organic Agriculture Statistics and Emerging Trends 2018. Germany: Medienhaus Plump; 2018. pp. 22-33

[21] Lernoud J, Willer H, Schlatter B. Africa: Current statistics. In: Willer H, Lernoud J, editors. FiBL & IFOAM-Organics International. Germany: Medienhaus Plump; 2018. pp. 177-185

[22] Seufert V, Ramankutty N, Foley JA. Comparing the yields of organic and conventional agriculture. LETTER. Nature. 2012;**485**:229-234. DOI: 10.1038/nature11069

[23] International Federation of Organic Agriculture Movements (IFOAM). Productivity and Profitability of Organic Farming Systems in East Africa. Bonn, Germany: IFOAM; 2013

[24] Oyewole MF, Oloyede F, Meludu N. Coverage of organic agricultural news in Nigerian newspapers. In: R. G. U, editor. Proceedings of the 4th ISOFAR Scientific Conference "Building Organic Bridges," at the Organic World Congress 2014; 13-15 October; Istanbul, Turkey; pp. 863-866

[25] Sakib H, Alfred SI, Prodhan FA. Farmers knowledge on aquaculture practices in Bogra District of Bangladesh. International Journal of Agricultural Extension. 2014;**2**(2):121-127

[26] Kperogi FA. Daily Trust. Nigeria: Why the Nigerian English Phrase 'South- South' is Bad English. 8 June 2014. https://allafrica.com/stories/201406092316.html

2

Economic Approach to Risk Analysis of Naturally Occurring Radioactive Materials (NORMs) in Dairy Milk Products Consumed in Nigeria

Ezekiel O. Agbalagba and Hannah O. Agbalagba

Abstract

One of the UN SDGs is for sustainable food for all; thus the need for safety in milk consumption in Nigeria which is also consonant with *the NAFDAC* regulatory plan of safety of food in Nigeria. This research work examined the radioactivity content in milks (powdered and liquid) consumed in Nigeria using sodium iodide (7.6 cm × 7.6 cm NaI(Tl)) detector. The estimated total cost of health detriment of consumption of the investigated milk products shows that the children age group has the highest cost health detriment per-caput dose with an estimated total cost of health detriment of US $17.26 million, followed by the adult age group with an estimated cost implication of US $11.86 million, and infants with the least computed cost implication of US $10.192 million. The overall results show that the milks consumed in Nigeria are radiologically safe and may not constitute any direct radiological health burden to consumers of these milk brands. Optimization of radiation protection mechanism for cost-benefit analysis is recommended.

Keywords: gamma spectroscopy, milk samples, natural radioactivity, economic-benefit analysis

1. Introduction

Radionuclide consequence recognizes no boundaries and therefore it is trans-border in nature. It can migrate through food, air, and soil and be transported to faraway countries from where its pollution or contamination occurs. An assessment of any release of radioactivity to the environment is important for the protection of public health, especially if the released radionuclides can enter the food chain. Milk as a staple food may naturally represent a comprehensive radioactive food chain because cows consume grass and are exposed to the same radioactive elements as food crops and water supply.

Contamination pathway: emitted radionuclides go into the human body through multifaceted mechanisms which include the intake of foodstuffs via food chain from natural sources. Vegetables and green leafy are susceptible to exterior contamination during the growing season, whereas roots and tubers get contaminated through the ingestion of nutrient from the soil [1, 2]. Grains are subjected to contamination mostly during storage or fallout may occur during the growing season as in the case of Fukushima Daiichi nuclear fallout. These liberated radionuclides may be transported into the grains and grasses through the plant growth process and find their way into the food chain when grazed by a cow [3]. Pollution of dairy products like milk is largely due to animal grazing and consumption of contaminated grass and drinking waters. Thus grass is essentially a direct source or pathway of radionuclides to animals and to man through meat and milk consumption. If dairy milk starts testing positive for high level of radioactive elements, this is indicative of radioactive contamination of the total food supply [4]. Milk samples that contain high levels of radioactivity when ingested by man could accumulate in certain parts of the body, for example, uranium-238 and radium-226 accumulates in human kidney and lungs, and thorium-232 accumulates in human liver, skeleton, tissue, and lungs, while potassium-40 accumulates in the muscles [5]. The accumulation of these radionuclides in any vital organs of the human body will affect the health condition that may cause various forms of diseases and weakening of the immune system and contribute to the increase in mortality rate [6].

1.1 Statement of the problem

Since the end of Second World War in 1948, research works on radionuclide contamination of food in the environment, and its transfer mechanism and pathway to animals and human population have been reported with vigor (ICPR, 1993 & 2000, [2, 7–11]); milk has been one of the staple foodstuff products that was featured prominently in the food items studied [9]. This may be due to its vital position in a family's daily food consumption plan, thus a reliable indicator of natural radionuclides to man for its high consumption rate globally. It is one of the essential food for human nutrition and contains most of the macronutrients, namely, protein, carbohydrates, fat, vitamins (A, B, and D groups), and trace elements such as calcium, phosphate, magnesium, zinc, and selenium [7, 12]. Milk is a rudimentary foodstuff for the infants compared to adults, on apparent body weight basis. Thus, milk consumed in Nigeria need to be assessed for radiological risk level for proper economic benefit analysis. It is therefore pertinent to set a radionuclide regulatory framework necessary in establishing guidelines relating to radiation protection in milk as a staple foodstuff. To the best of our knowledge, comprehensive data base on levels of radionuclides in staple foodstuffs and standard radionuclide regulatory framework for food imported and consumed in the country are inadequate, these lay credence to this research work.

1.2 Aim and objectives of the study

This research work was designed to examine the level of natural radionuclides present in liquid and powdered milk products consumed in Nigeria with a view to establishing their specific activities and compare same with values reported in other parts of the world. Assessment of the annual internal dose from the intake of the milk product will be determined, while the doses to the different sensitive human organs were estimated to establish their radiological risk to man. The result would be a contribution to the creation of a standard catalogue of natural radioactivity in foodstuff (milk) in Nigeria. This can serve as a baseline data for possible evaluation of future change in activity levels of these milk products due to environmental factors, ingredient composition, etc. The result obtained in this study will also be useful to the country food and drug regulatory body which has the obligation to protect the public health by ensuring that only the right quality of food and drugs are imported and consumed in the country.

2. Materials and methods

2.1 Sample collection

To collect milk samples that represent a fair proportion of milk products consumed in Nigeria, a survey was carried out, which involves visiting homes, major supermarkets, fast food outlets, hotels, and major milk distributors throughout the six geopolitical zones in Nigeria. The survey revealed that the acceptability of brands of the milk (powdered or liquid) is dictated by its availability, cost, and social class of consumers. The availability of specific products in a region is closely related to the proximity to the manufacturer or key distributor and the awareness due to advertisement of the products in the area. This in situ assessment has shown that 10 leading powdered milk and 11 liquid milk brands were consumed by a large sector of the populace in the six geopolitical zones of the country in the following order: Peak > Cowbell > Coast > Loya > Nunu > Miksi > Dano > real milk for powdered milks and peak > Three Crown > Coast > Hollandia > Olympic > Nunu > Nutri milk > vital milk > Lady liberty > Bridel in liquid milks, of these Dano milk (Demark), Bridel milk (France), and Lady liberty (USA) are imported milk brands. These 10 powdered milk brands and 11 liquid milk brands were then sourced from various shopping malls and local markets. Two hundred (200) grams of the powdered milk collected was put in a cylindrical polystyrene container and sealed with tapes to prevent radon permeability, while for liquid milk, 200 cl of the homogenous samples was filled into a Marinelli beaker which was hitherto washed, rinsed with diluted H_2SO_4 acid to prevent the samples from being contaminated, and sealed and weighed as samples for gamma spectroscopy analysis.

2.2 Gamma spectroscopy analysis

The collected sealed samples were left for a minimum of 4 weeks to allow secular equilibrium prior to the counting of the samples for radioactivity concentration. The counting for radioactivity was carried out using a $7.6cm \times 7.6cm$ NaI(Tl) detector for 10 h (Model Bircom, USA) housed in a 10-cm-thick lead shield to reduce background gamma radiation. The detector with energy resolution (FWHM) of 7.5% at 662 keV was coupled to a set of electronics which consist of pre-amplifier, main amplifier, analog-to-digital converter (ADC), and a Canberra Multichannel Analyzing (MCA) computer system. The integrated spectroscopy system was used for the power supply and the data acquisition of the energy spectra and utilized SAMPO S100 software package from Canberra (MAESTRO window USA). The energy calibration of the detector was performed between the gamma energy range of 83 keV and 1875 keV using International Atomic Energy Agency standard point sources (^{109}Cd, ^{57}Co, ^{137}Cs, ^{54}Mn, and ^{22}Na), the energy range of the radionuclide to be identified. To simulate the milk samples, 100 g of IAEA-375 reference sample was used. The radioactivity concentrations of $^{226}Ra/^{238}U$ were determined from the photopeaks of 609.32 keV (^{214}Bi), 1120.20 keV (^{214}Bi), and 352.6 keV (^{214}Pb) and that of ^{232}Th from 969.3 keV (^{228}Ac) and 583.78 keV (^{208}Tl), while the radioactivity of ^{40}K was evaluated from 1460.3 keV photopeak following the decay of ^{40}K. The background spectrum measured under the same settings for both the standard and sample measurement was used to correct the computed sample activity concentration in agreement with Arogunjo et al. [13].

3. Results and discussion

3.1 Results

The measured radioactivity levels of the three naturally occurring radionuclides, ^{226}Ra, ^{232}Th, and ^{40}K, identified in the sampled milk under investigation including their uncertainty are presented in **Table 1**.

The obtained results of the radioactivity concentration indicate that ^{226}Ra activity in powdered milk varied from 14.2 ± 5.9 Bq kg^{-1} to 26.8 ± 8.3 Bq kg^{-1} with a mean activity concentration of 19.3 ± 7.2 Bq kg^{-1}, and ^{232}Th activity varied from 8.8 ± 3.6 Bq kg^{-1} to 15.0 ± 5.9 Bq kg^{-1} with a mean activity concentration of 12.1 ± 4.8 Bq kg^{-1}, while ^{40}K varied from 317.5 ± 77.6 Bq kg^{-1} to 589.8 ± 94.6 Bq kg^{-1} with a mean activity level of 468.0 ± 72.7 Bq kg^{-1}. The ^{226}Ra radioactivity concentration in liquid milk varied from 12.2 ± 4.7 Bq kg^{-1} to 21.2 ± 8.3 Bq kg^{-1} with a mean activity concentration of 16.6 ± 6.3 Bq kg^{-1}, and ^{232}Th activity concentration varied from 6.8 ± 3.0 Bq kg^{-1} to 13.2 ± 6.1 Bq kg^{-1} with a mean activity concentration of 10.6 ± 4.3 Bq kg^{-1}, while ^{40}K varied from 218.6 ± 39.4 Bq kg^{-1} to 484.2 ± 67.9 Bq kg^{-1} with a mean activity level of 317.6 ± 58.5 Bq kg^{-1}. It was observed that ^{40}K samples have the highest activity concentration trailed by ^{226}Ra samples, while ^{232}The samples had the least activity level. Evaluation of the three naturally occurring radionuclide concentrations in the milk samples with the UNSEAR 2000 permissible limit for powdered milks and liquid milk samples shows that all ^{226}Ra and ^{232}Th activity levels in both samples are well below the permissible limits, while ^{40}K exceeded the permissible limit in Peak milk, Peak Chocolate, Cowbell Chocolate, Loya, Miksi, Coast, Real, and Nunu milk for powdered milk and exceeded the permissible limit for liquid milk in Vital and Nutric milk. It is pertinent to note that the activity levels determined in the current study are above the values gotten for milk consumed in some countries like Saudi Arabia [7]; Iran/France [14]; Jordan [15]; Egypt [16]; and Syria [17], while the values are comparable to those milk consumed in other countries like New Zealand [14] and Brazil [18]. Conversely, this result shows that the mean activity concen-trations in the various milk brands sampled are well within their international permissible limits. The difference in the radioactivity levels in the various brands of milk sampled suggests that the source of raw materials used for the production of the milks is a contributory factor.

The estimated Radium equivalent (Ra_{eq}), annual gonadal dose equivalent (AGED), internal hazard (H_{in}) indices, annual effective dose equivalent (AEDE) received, and excess lifetime cancer risk (ELCR) for the various milk products sampled are presented in **Table 2**.

3.2 Radium equivalent activity (Ra_{eq})

The model of the radium equivalent activity establishes the use of a single index to define the gamma output or compare the specific activities of materials containing ^{226}Ra, ^{232}Th, and ^{40}K by a single quantity, which takes into consideration the radiation risk associated with these NORMs [19–22]. The radium equivalent activity represents a weighted factor of activities of the three natural radionuclides

S/N	Powdered milk sample	Radioactivity concentration (Bq kg^{-1})		
		^{226}Ra	^{232}Th	^{40}K
1	Peak milk	26.8 ± 8.3	9.5 ± 2.2	536.1 ± 84.3
2	Peak Chocó milk	17.6 ± 6.9	12.3 ± 4.1	424.7 ± 67.9
3	Cowbell milk	17.4 ± 7.1	13.1 ± 5.4	317.5 ± 77.6
4	Cowbell Chocó milk	16.4 ± 6.7	8.8 ± 3.6	544.2 ± 83.4
5	Nunu milk	26.5 ± 6.5	10.1 ± 4.9	513.5 ± 74.3
6	Loya milk	14.2 ± 5.9	14.7 ± 6.1	413.3 ± 83.3
7	Miksi milk	15.1 ± 7.2	12.8 ± 5.5	589.8 ± 94.6
8	Coast	18.6 ± 6.0	13.3 ± 6.3	422.4 ± 64.4
9	Real milk	21.0 ± 8.5	15.0 ± 5.9	519.9 ± 95.6
10	Dano milk (Demark)	19.5 ± 8.4	10.9 ± 4.2	398.4 ± 66.3
	Mean activity conc.	**19.3** ± **7.2**	**12.1** ± **4.8**	**468.0** ± 72.7
S/N	**Liquid milk sample**	**Radioactivity concentration (Bq l^{-1})**		
		^{226}Ra	^{232}Th	^{40}K
1	Hollandia milk	12.2 ± 4.7	12.6 ± 5.8	321.4 ± 66.3
2	Vital milk	21.2 ± 8.4	9.9 ± 3.8	476.1 ± 84.1
3	Nutric milk	15.0 ± 5.9	8.9 ± 3.2	484.2 ± 67.9
4	Peak milk	13.2 ± 4.6	8.0 ± 3.1	218.7 ± 49.0
5	Three Crown milk	17.2 ± 8.3	13.2 ± 5.7	324.2 ± 51.9
6	Olympic milk	16.3 ± 4.5	12.5 ± 4.6	268.9 ± 58.6
7	Coast milk	18.1 ± 5.5	13.2 ± 6.1	312.8 ± 49.3
8	Nunu milk	17.5 ± 6.3	9.9 ± 3.6	218.6 ± 39.4
9	Condensed Peak milk	19.2 ± 8.4	10.6 ± 4.2	272.9 ± 50.1
10	Bridel milk (France)	15.8 ± 6.5	11.3 ± 4.6	282.2 ± 73.3
11	Lady Liberty (USA)	14.2 ± 6.1	6.8 ± 3.0	313.3 ± 53.3
	Mean activity conc.	**16.6** ± **6.3**	**10.6** ± **4.3**	**317.6** ± **58.5**

Table 1.
Specific activity concentration of the powdered and liquid milk products commonly consumed in Nigeria.

measured and is based on the estimation that 1.0 Bqkg^{-1} of ^{226}Ra, 0.7 Bq kg^{-1} of ^{232}Th, and 13.0 Bq kg^{-1} of ^{40}K produce equal radiation dose rates [23, 24].

The radium equivalent activity concentration is given as [23]:

$$Ra_{eq} = C_{Ra} + 1.43\ C_{Th} + 0.0770\ C_K \tag{1}$$

where C_{Ra}, C_{Th}, and C_K are the radioactivity concentrations in Bq kg^{-1} of ^{226} Ra, ^{232}Th, and ^{40}K, respectively.

The consumption of any foodstuff whose Ra_{eq} concentration exceeds 370 Bq kg^{-1} should be discouraged to avoid radiation hazards. The result of the estimated radium equivalent dose rates in **Table 2** shows a value range of 60.6 ± 20.8 Bq kg^{-1} in Cowbell milk to 82.5 ± 24.3 Bq kg^{-1} in real milk with a mean value of 72.6 ± 19.7 Bq kg^{-1} in powdered milk brands, while in liquid milk brand, the range of value is 41.5 ± 12.9 Bq l^{-1} in Peak milk to 72.0 ± 20.3 Bq l^{-1} in Nutric milk with a

Milk samples	Ra_{eq} ($Bqkg^{-1}/Bq\,l^{-1}$)	AGED (μSvy^{-1})	H_{in}	AEDE (μSvy^{-1})	ELCR × 10^{-3} ($mS\ vy^{-1}$)
Powder milk ($Bq\ kg^{-1}$)					
Peak milk	81.7 ± 17.9	290.6	0.3	200.2	0.7
Peak Chocó milk	67.9 ± 18.0	239.1	0.2	166.3	0.6
Cowbell milk	60.6 ± 20.8	208.3	0.2	146.2	0.5
Cowbell Chocó milk	70.8 ± 18.3	258.2	0.2	177.2	0.2
Nunu milk	65.7 ± 19.4	230.8	0.2	196.7	0.6
Loya milk	67.0 ± 21.1	235.0	0.2	164.8	0.6
Miksi milk	78.8 ± 22.3	285.3	0.3	197.7	0.2
Coast milk	70.2 ± 19.9	245.8	0.2	171.2	0.7
Real milk	82.4 ± 24.3	290.7	0.3	202.6	0.7
Dano milk (foreign)	80.4 ± 19.2	285.0	0.3	160.4	0.6
Mean value	72.6 ± 20.1	256.9	0.25	178.4	0.6
Global standard	370	300	≤1.0	450	0.29 ($mS\ vy^{-1}$)
Liquid milk ($Bq\ L^{-1}$) × 10^{-3}					
Hollandia milk	54.9 ± 18.1	191.0	0.2	134.4	0.5
Vital milk	53.7 ± 18.8	184.8	0.2	177.1	0.6
Nutric milk	72.0 ± 20.3	256.4	0.3	163.8	0.6
Peak milk	65.0 ± 15.8	235.6	0.2	99.6	0.3
Three Crown milk	48.0 ± 14.5	170.6	0.2	147.2	0.5
Olympic milk	41.5 ± 12.7	142.9	0.2	131.5	0.5
Coast milk	61.1 ± 20.5	210.2	0.2	146.7	0.5
Nunu milk	54.9 ± 15.6	187.1	0.2	114.8	0.4
Condensed Peak milk	61.1 ± 18.0	209.3	0.2	132.5	0.5
Bridel milk (foreign)	48.4 ± 14.6	163.9	0.2	129.5	0.5
Lady Liberty (foreign)	55.3 ± 18.3	189.1	0.2	117.7	0.4
Mean value		173.2	0.20	135.9	0.5
World standard	370	300	≤1.0	450	0.29 ($mS\ vy^{-1}$)

Table 2.
Mean radium equivalent and summary of computed radiological risk parameter of milk samples.

mean value of 56.2 ± 17.0 $Bq\,l^{-1}$. All the values obtained are within the international acceptable limit for Ra_{eq} and therefore comply with the radium equivalent standard for radioactivity concentration. The percentage contributions of the three naturally occurring radionuclides in the powdered and liquid milk samples are shown in **Figures 1** and **2**, respectively. These percentages were calculated based on the estimation that 1 Bq kg^{-1} of ^{226}Ra, 0.7 Bq kg^{-1} of ^{232}Th, and 13 Bq kg^{-1} of ^{40}K produce the same radiation dose rates in the radium equivalent [24]; the average percentage contribution of ^{40}K is 42%, and for ^{226}Ra the percentage contribution to the entire milk content is 20%, while ^{232}Th percentage contribution to the powdered milk samples is 38%. The percentage contribution of the three natural radionuclides (^{226}Ra, ^{232}Th, and ^{40}K) to the liquid milk samples were computed to be 24, 36, and 40%, respectively, which shows that ^{232}Th contributes the highest radioactivity dosage to both the powdered and liquid milk activity concentration.

3.3 Annual gonad equivalent dose (AGED)

The gonads, the active bone marrow, and the bone surface cells are classified as organs of interest by UNSCEAR (2003). The annual gonadal dose equivalent (AGED, mS vy^{-1}) owing to the specific activities of ^{226}Ra, ^{232}Th, and ^{40}K was computed using [25]:

$$\text{AGED}\ \left(\mu\text{S vy}^{-1}\right) = 3.09\text{C}_{\text{Ra}} + 4.18\text{C}_{\text{Th}} + 0.314\text{C}_{\text{K}} \qquad (2)$$

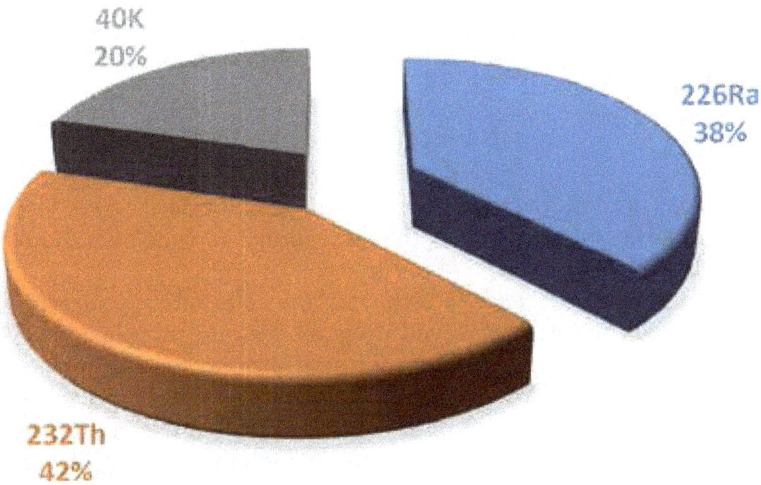

Figure 1.
Percentage contribution of the three natural radionuclides measured in powdered milk.

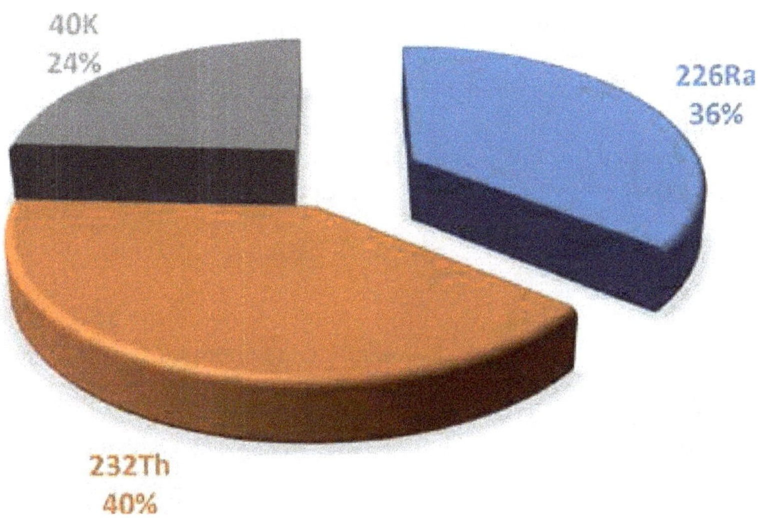

Figure 2.
Percentage contribution of the three natural radionuclides measured in liquid milk.

where C_{Ra}, C_{Th}, and C_K represent the radioactivity levels of ^{226}Ra, ^{232}Th, and ^{40}K, respectively.

If the radioactivity of the food source is higher than the world permissible value of 0.30 mSvy^{-1}, the model suggests that the food product is a potential source of radiological health risk to the consumer. In the studied powdered milk samples, AGED activity concentration varied from 208.3 μS vy^{-1} to 290.7 μS vy^{-1}, whereas in liquid milk samples analyzed, the estimated activity levels varied from 170.6 μS vy^{-1} to 256.4 μSvy^{-1}. Although the obtained 290.7 μS vy^{-1} activity concentration value in *Real* powdered milk is approximately the maximum permissible limit value, all other obtained estimated values in both liquid and powdered brands of milk sampled are well within the global maximum permissible values of 300 μS vy^{-1} [26]; thus the different milk samples examined may not cause any immediate radiological health side effects in terms of AGED index.

3.4 Internal gamma indices

The internal hazard (H_{in}) index is defined according to Zarie and Al-Mugren [27] as:

$$H_{in} = C_{Ra}/185 + C_{Th}/259 + C_K/4810 \tag{3}$$

where C_{Ra}, C_{Th}, and C_K represent the radioactivity levels of ^{226}Ra, ^{232}Th, and ^{40}K correspondingly. H_{in} should be less than the unity for the radiation risk to be insignificant.

The internal hazard index (H_{in}) estimates the internal exposure rate to oncogenic radon nuclides and its short-lived progeny. The mean radioactivity values of the computed H_{in} for powdered and liquid milk products are 0.3 and 0.2, respectively, which are well below the unity value recommended as the permissible limit. It is observed from the values obtained that powdered milks contain more of the natural radionuclides than liquid milk samples, which may be a result of the activity concentration through evaporation to dryness of the raw liquid milk. But all the milk samples analyzed met the minimum internal gamma index requirement for consumption.

3.5 Annual effective dose equivalent (AEDE)

The annual effective dose equivalent received was computed from absorbed dose rate by applying a dose conversion factor of 0.7 Sv Gy^{-1} and the occupancy of 0.8 (19/24) recommended by UNSCEAR [26] and Veiga et al. [28]. Therefore, the annual effective dose equivalent (μS vy^{-1}) was calculated using the formula [26]:

$$\text{AEDE}\left(\mu\text{S vy}^{-1}\right) = \text{absorbed dose}\left(\text{nG yh}^{-1}\right) \times 8760\ \text{h} \times 0.7\ \text{SvGy}^{-1} \times 0.8 \times 10^{-3} \tag{4}$$

The estimated annual effective dose equivalent obtained for the powdered and liquid milk brands analyzed is shown in **Table 2**. The values obtained for the powdered milk varied from 146.2 μS vy^{-1} to 202.6 μS vy^{-1} with a mean activity dose level of 178.4 μS vy^{-1}, while in the liquid milk, the activity dose levels ranged from 99.6 μS vy^{-1} to 177.1 μS vy^{-1} with a mean activity concentration of 135.9 μS vy^{-1}. The obtained result shows that the milk samples surveyed have their annual effective dose equivalent values lower than the world average values of 450 μS vy^{-1} [26, 29–31]. This result shows that the various sampled milk brands are radiologically safe going by the global recommended permissible limit of annual effective dose rate.

3.6 Excess lifetime cancer risk

Excess life cancer risk predicts the likelihood of developing cancer over a lifetime at a certain exposure rate. It is a value representing the number of extra cancers expected in a given number of people on exposure to a carcinogen at a given dose.

Excess lifetime cancer risk is given as Taskin et al. [32]:

$$\text{ELCR} = \text{AEDE} \times \text{DL} \times \text{RF} \tag{5}$$

The parameters used are defined; thus AEDE is the annual effective dose equivalent, DL is average duration of life (estimated to be 70 years), and RF is the risk factor (S/v), i.e., fatal cancer risk per Sievert. ICRP uses a RF of 0.05 for the public for stochastic effects [32]. The intake of milk containing an elevated level of radionuclide may increase the chance of cancer risk. If the radioactivity in the milk is higher than the world average, it could be a source of radiation to the human body and some specific organs, in that their ELCR would be greater than the world average of 0.29 mS vy^{-1} in such body. The estimated ELCR obtained in all the measured samples is lower than the international standard limit.

Table 3 presents the estimated values of the annual effective dose of the three age groups of infants, children, and adult represented in this evaluation.

Table 4 shows the dose conversion factor for the three different age groups used in estimating the annual effective dose rate to individual consumption of the various milk samples in Nigeria.

3.7 Annual effective dose for different age groups

"The annual effective dose E_D to individuals due to the ingestion of the radionuclides in powder and liquid milk is estimated using the equation.

$$E_D = A_c \times I_{in} \times E \tag{6}$$

where the parameters; E_D is the annual effective dose in (Svy^{-1}), A_c is the activity concentration of the radionuclides in milk; E is the dose conversion factor and I_{in} is the annual intake of milk with respect to the age group" [15].

Table 3 presents the result of the computed annual ingestion dose to the three age groups (infants, children, and adults), with an estimated annual milk intake of 15 kg, 14 kg, and 10 kg, respectively, and using the conversion factor values in **Table 4**. The results obtained indicate that the infants' accumulated maximum and minimum effective dose values are 1743.0 μS vy^{-1} and 892.7 μS vy^{-1}, respectively, with an average dose rate of 1345.4 μS vy^{-1}. For the children, the obtained values for the annual accumulated maximum and minimum effective ingestion doses are 1048.9 μS vy^{-1} and 546.8 μS vy^{-1}, respectively, with an average dose rate of 822.5 μS vy^{-1}. The accumulated maximum and minimum annual effective ingestion doses obtained for adults are 235.1 μS vy^{-1} and 122.3 μS vy^{-1}, respectively, with an average annual dose rate of 181.1μSvy^{-1}. The obtained dose rates in the three different age groups show that the annual ingestion dose exceeded the UNSCEAR [26], maximum permissible limit by 68% for the infants, and 2.8% for the children; however, the adult value is 22.6% below the recommended maximum permissible limit. This obtained result indicates that the ingestion rate of milk by infants and

Radionuclides	Powdered milk	Annual effective dose(μSv)		
		Infants	Children	Adult
^{40}K	Maximum	371.6	107.4	47.5
	Minimum	200.0	57.8	25.6
	Average	294.8	85.2	37.7
^{226}Ra	Maximum	386.4	300.5	97.7
	Minimum	204.3	158.9	51.7
	Average	277.9	216.2	70.3
^{232}Th	Maximum	1279.9	817.4	134.3
	Minimum	748.1	477.8	78.5
	Average	1034.6	660.7	108.5
	Accumulated mean	1607.3	962	186.5
Radionuclides	**Liquid milk**	**Annual effective dose(μSv)**		
		Infants	**Children**	**Adult**
^{40}K	Maximum	305.1	88.1	39.0
	Minimum	137.7	39.8	17.6
	Average	200.1	57.8	25.6
^{226}Ra	Maximum	305.9	237.9	77.3
	Minimum	175.4	136.4	44.3
	Average	239.0	185.9	60.4
^{232}Th	Maximum	1132.0	722.9	118.8
	Minimum	579.7	370.2	60.8
	Average	906.3	578.8	95.1
	Accumulated maximum	1743.0	1048.9	235.1
	Accumulate minimum	892.8	546.4	122.8
	Accumulated mean	1345.4	822.5	181.1
	UNSCEAR 2000 standard	200–800	200–800	200–800

Table 3.
Annual effective dose to infant, children, and adult age groups due to intake of natural radionuclides in powdered and liquid milk samples of quantities 15 kg y^{-1}, 1 4 k g y^{-1}, and 8 kg y^{-1}, respectively.

	Dose conversion factors (nSv/Bq)		
	^{40}K	^{226}Ra	^{232}Th
Infant (1–2 y)	42	4700	4600
Children (7–12 y)	13	800	290
Adult (>17 y)	6.2	280	230

Table 4.
The dose conversion factors of ^{40}K, ^{226}Ra, and ^{232}Th for the infant, children, and adult age groups [33].

children should be reduced to avoid any future amassed radiological health side effect [34]; accordingly an optimized amount of milk is required to be recommended as a yearly permissible limit for consumption.

Table 5 shows the optimized annual consumption rate to stay within the international recommended range.

Following the exceeding of the UNSCEAR [26] μSvy^{-1} permissible limit from the recommended quantity of milk to be consumed by the three age groups, **Table 5**

		Annual effective dose (μSv)	
		Infants	**Children**
Radionuclides in powdered milk			
^{40}K	Average	137.59	66.92
^{238}U	Average	129.70	169.84
^{232}Th	Average	482.79	519.09
	Accumulated average	750.1	755.9
Radionuclides in liquid milk			
^{40}K	Average	106.71	53.67
^{238}U	Average	127.49	172.64
^{232}Th	Average	483.36	537.42
	Accumulated average	717.6	763.7
	UNSCEAR 2000 standard	800	800

Table 5.
Annual effective dose to infants and children with a proposed annual consumption of 7 and 11 kg, respectively, of powder milk samples and 8 and 13 kg, respectively, for liquid milk.

Organs	Effective dose rate to organs ($mS\ vy^{-1}$): powdered milk		
	Infants	**Children**	**Adult**
Lungs	0.8	0.5	0.1
Ovaries	0.8	0.5	0.1
Bone marrow	0.9	0.5	0.1
Testes	1.1	0.6	0.1
Kidneys	0.8	0.5	0.1
Liver	0.6	0.4	0.1
Whole body	0.9	0.5	0.1

Organs	Effective dose rate to organs ($mSvy^{-1}$): liquid milk		
	Infants	**Children**	**Adult**
Lungs	0.7	0.4	0.1
Ovaries	0.6	0.4	0.1
Bone marrow	0.7	0.5	0.1
Testes	0.9	0.5	0.1
Kidneys	0.7	0.4	0.1
Liver	0.5	0.3	0.1
Whole body	0.7	0.5	0.1

Table 6.
Dose rate to different organs of the body due to radionuclides in powder and liquid milk.

Organ or tissue	Conversion factor (F)
Lungs	0.64
Ovaries	0.58
Bone marrow	0.69
Testes	0.82
Whole body	0.68
Kidney	0.62
Liver	0.46

Table 7.
Average values of F for different organs or tissues [33].

presents an optimized annual quantity of powdered milk to be consumed as 7 kg and 11 kg for children and infants, respectively, and 8 kg and 13 kg were recommended for children and infants, respectively, for liquid milk, while adult was not optimized because the value obtained was within the international permis-sible limit. Using the optimized quantity, the annual accrued values obtained for infants and children for powdered milk were 750.1 μSv and 755.9 μSv, respectively, while the accumulated values recorded for liquid milk were 717.6 μSv and 763.7 μSv for infants and children, respectively. The values recorded are well below the 800 μSv [26] yearly recommended permissible limit.

Table 6 presents the result of calculated effective dose rate to the different human organs for consuming powdered and liquid milk.

Table 7 presents the average values of F, for different organ or tissue uses in the computation of the effective dose rate of these organs.

3.8 The effective dose rate (D_{organ}) in mS vyr^{-1} to different body organs or tissues

The annual effective dose to organ models evaluates the total amount of radionuclides consumed by man over a period of 1 year that goes to and accumulate in the different sensitive organs and tissues of the human body. The effective dose rate transported to a particular organ was calculated using the relation:

$$D_{organ}\left(mSvy^{-1}\right) = O \times E_D \times F \tag{7}$$

The parameter E_D represents the annual effective dose, O represents the occupancy factor with a value of 0.8, and F is the conversion factor of organ dose from consumption of the food.

Table 6 presents the obtained computed values of the effective dose rate assimilated by the various organs evaluated, while **Table 7** presents the conversion factors and F values for the seven organs/ tissues. The computed dose values obtained in powdered milks revealed that the human testes (organ) received the greatest dose of average values of 1.1 mS vy^{-1}, 0.6 mS vy^{-1}, and 0.1 mS vy^{-1} for infants, children, and adults, respectively; meanwhile, the dose received by the liver was established to be the least with average dose values of 0.6 mS vy^{-1}, 0.4 mS vy^{-1}, and 0.1 mS vy^{-1}, respectively. The computed results obtained from liquid milk follow the same trend with testes recording the highest radionuclide dose ingestion with average estimated dose levels of 0.9 mS vy^{-1}, 0.5 mS vy^{-1}, and 0.1 mS vy^{-1} for infants, children, and adults, respectively, and the least dose intake values of 0.5 mS vy^{-1}; 0.3 mS vy^{-1}, and 0.1 mS vy^{-1} were detected in liver for infants, children, and adults, respectively. It was observed from the results that the dose ingestion to infants' testes exceeded the world acceptable limit of 1.0 mSv annually to the human-sensitive organs/tissues. The relatively higher dose to the testes and low-dose intake to the liver are justifiable from food nutrient absorption rate [15, 31]. The high radiological dose to the testes may justify the rampant rate of prostate cancer among young men which may be linked to accumulated effects of the dose intake over time, which need further clinical investigation [34].

4. Risk factors and cost-benefit analysis

Cost-benefit analysis in radiation protection is the projection of radionuclide injection effects of consumption of food product versus the nutritional benefits derived from the intake of the food at low individual doses, below the dose limits. Applying the direct relationship between dose and effect, the health impairment is directly proportional to the effective dose-equivalent occasioned by the consumption of milk product. The proportionality element is termed the risk factors [35, 36].

4.1 Committed effective dose (C_D)

The committed effective dose to an individual assessed for three age groups (0–1 yr infant, 1–7 yrs children, and >17 yrs age group for adults) over a typical lifetime of 50 years was computed using the formula:

$$C_D = 50 \times E_D \tag{8}$$

The calculated committed effective doses to the various age groups are presented in **Table 8**. The obtained doses to infants and children were futuristic and a forecast of the probable dose to be committed to the individuals in the period of 50 years. The obtained dose values for adult over an average lifetime of 50 years for powdered milk and liquid milk were 9.3 mS vy^{-1} and 9.1 mS vy^{-1} doses, respec-tively. The committed doses obtained for children were 48.1 mS vy^{-1} and 41.1 mS vy^{-1} for powdered and liquid milk, respectively, while for infants it is 80.4 mS vy^{-1} and 67.3 mS vy^{-1}, respectively. The result obtained points to the fact that constant ingestion of the milk products may result to accumulation of radio-nuclides in some organs of the human, with more radionuclide dose intake in powdered milk than consuming liquid milk. Moreover, the values gotten are well within international standard.

Table 8 presents the summary of the result of the risk analysis and health detriment effect values that may arise from the consumption of milk product samples investigated in this study.

Table 9 presents the population of the three age groups under examination in this study, obtained from the National Population Commission report [37], and the computed two-thirds (2/3) of this population used for the evaluation of the collective effective dose equivalent.

4.2 Collective effective dose equivalent

It is insufficient that the risk to individuals is set at an adequately low level in radiation protection, but the total detriment to the public resulting from exposure to radiation should be kept as low as is reasonably achievable (ALARA) for health, economic, and social factors. The evaluation of the collective/total detriment to health for the public is the sum of detriments to the individuals making up the

Age group	Annual Effective dose E_D (μSv)		Committed effective dose C_D (mS vy^{-1})		Collective effective dose equivalent S_E (man-Sv)		Total health detriment (G) (man)		Mean THD (G) (man)	Total cost of detriment (man-Sv) $million
	Powdered milk	Liquid milk	Powdered milk	Liquid milk	Powdered milk	Liquid milk	Powdered milk	Liquid milk		
Infant (0–1 yr)	1607.3	1345.4	80.4	67.3	5,551.5	4,640.9	91.6	76.6	168	10.2
Children (7–12 yrs)	962.0	822.5	48.1	41.1	9,305.8	7,950.3	153.5	131.2	285	17.3
Adult (>17 yrs)	186.5	181.1	09.3	9.1	6,023.1	5,840.7	99.4	96.4	196	11.9

Table 8.
Summary of risk analysis of radiation dose from milk products.

Age	Population (P_i)	Two-thirds of the population ($\frac{2}{3}P_i$)
Infant (0–1 yr)	7,771,348	5,180,899
Children (7–12 yrs)	21,763,942	14,509,295
Adult (>17 yrs)	72.660,755	48,440,503

Table 9.
Nigeria population in different age groups [37].

public due to a level of radiation exposure. The statement of proportionality between stochastic biological effects and dose equivalent also applies to the collective detriment to health being directly proportional to the collective effective dose equivalent [38]. Hence the collective effective dose equivalent, S_E in a population comprising of N_i individuals, is evaluated as [39]:

$$S_E = \sum N_i H_{Ei} \tag{9}$$

where S_E represents the collective effective dose equivalent (person, Sv) and N_i is the number of persons in a population that are exposed to the radiation, while H_{Ei} represents the mean effective dose equivalent ($\mu S\ vy^{-1}$).

The Nigerian Population Commission [37] puts the population figure of people living in Nigeria as 140,431,790 with the age groups of 0–1 year, 7–12 years, and >17 years having population figures of 7,771,348; 21,763,942; and 72,660,755, respectively.

In Nigeria, approximately two-thirds of the population are expected or projected to make one brand of milk or the other as staple food; consequently two-thirds of the population of the different age ranges/groups are probable to have a radiation dose intake from milk products.

The formula for collective effective dose equivalent was accordingly modified to read:

$$S_E = \sum \frac{2}{3}(N_i H_{Ei}) \tag{10}$$

Table 9 presents the population of three age group brackets being studied in Nigeria. Since two-thirds of the estimated population of Nigerian projected to consume one brand of milk or the other, it is estimated that the 5,180,899 infants, 14,509,295 children, and 48,440,503 adults signifying 48.5% of the total population of Nigeria consume milk products. The estimated collective effective dose equivalent S_E obtained revealed that the value for infants is 5551.5man-Sv in powdered milk and 4640.9man-Sv in liquid milk. In the children age bracket, the values obtained are 9305.8man-Sv for powdered milk and 7950.3man-Sv for liquid milk. Similarly, in the adult age bracket, the collective effective dose equivalent values obtained are 6023.1man-Sv for powdered milk and 5840.7man-Sv for liquid milk. The values obtained revealed that the children population is probable to have the highest radionuclide dose from milk intake.

4.3 Total health detriment

The objective gross or total health detriment also known as collective health detriment "G" (man), resulting from exposure to gamma radiation in an environment or ingestion of irradiated products by man, is evaluated using the formula [38]:

$$G = R_T S_E \quad (11)$$

From Eq. (11), R_T represents the total risk factor the body organs are exposed to, as recommended by the International Commission on Radiological Protection, ICRP [39], where these risk factors are used in assessing the fatal radiation-induced cancers and severe hereditary effects in the first two generations. It has a gross value of $1.65 \times 10^{-2}\ Sv^{-1}$, with $1.25 \times 10^{-2}\ Sv^{-1}$ representing the value for fatal radiation-induced cancers and $0.4 \times 10^{-2}\ Sv^{-1}$ representing the value for severe hereditary effects, for the first two generations, while S_E is the collective effective dose equivalent (man–Sv) [38].

The estimation of the total health detriment is vital and necessary because any health detriment on this populace will impact negatively on the entire population.

The total health detriment for the three age brackets computed in the different milk products examined is presented in **Table 8**. The total health detriment to man obtained for powdered milk are 91.6 for infants, 153.5 for children, and 99.4 for adults. In the liquid milks, the values obtained for the age groups are 76.6 for infants, 131.2 for children, and 96.4 for adults. This calculated total health detriment indicates that for every 5,180,899 Nigerian infants consuming milk products, 168 of them have the probability to have radiological health-related side effects from the intake of milk products. Similarly, the evaluation of the gross health detriment indicates that for every 14,509,295 Nigerian children consuming milk products, 285 are likely to have a radiological health risk from the intake of milk, while of the estimated 48,440,503 Nigerian adults that consume milk products, 196 are likely to have a radiological health hazard, with fatal radiation-induced cancers the most probable, going by the risk factor of $1.25 \times 10^{-2}\ Sv^{-1}$. The radiological index obtained shows a ratio of 1:30,839 for infants, 1:50,910 for children, and 1:247,145 for adults, with infants' radiological index ratio being the highest. This indicates that the infants are most vulnerable radiologically in milk intake.

4.4 Cost of detriment

The correlation between the cost of the health detriment and collective effective dose equivalent is a linear one. It is expressed as [38]:

$$Y_C = \alpha S_E \quad (12)$$

where Y_C is the cost of health detriment, SE is the collective effective dose equivalent (man-Sv), and α is the collective dose equivalent constant. If it were possible to arrive at a common monetary value for the cost of radiation harmful stochastic health effects, then α would have a unique value. But it is not practically possible due to socioeconomic considerations that vary from country to country and from time to time. A review of literatures indicates a wide range of α values ranging from 1000 to 100,000 US dollars [38]. Considering the low per capita income of African countries including Nigeria which is below US $100 per day, the cost of detriment analysis value of US $1000 was assigned to human life. This value does not necessarily mean a real monetary value of life but rather is proposed to provide measures by which fair and consistent resources are allocated to radiation protection [21]. From the computed result of the total cost of health detriment from the collective effective dose equivalent to the different age group in **Table 8**, it was observed that children age group has the highest cost health detriment per-caput dose with an estimated total cost of health detriment of US $17.256 million, followed by adults with an estimated cost implication of US $11.864 million, while infants have the least with an estimated cost implication of US $10.192.

5. Conclusion

This investigation presents the gamma spectrometry evaluation of the natural radioactivity in powdered and liquid milk consumed in Nigeria. The radionuclide concentration of the milk samples was found to be dictated by the source the milks products were derived. Although the specific activity concentration of milk samples varied, their mean activities were less than the world permissible dose limit for the public. The radium equivalent activities obtained for all the milk samples (powdered or liquid) considered were all below the criterion limit of radiation dose (1.0 Sv y^{-1}). All the calculated radiological risk parameters show that none of the milk samples exceeded their recommended allowable level. It was however found from the annual effective dose calculation that the consumption of powder and liquid milk by infants and children at the rate of 14 kg y^{-1} and 15 kg y^{-1}, respectively, may lead to a radiation dose to vital organs of the body above normal recommended values, but optimized quantity was suggested for the group to stay within the recommended permissible limit. On the cost-benefit analysis, the estimated collective effective equivalent dose values obtained show that the children population receives the highest dose. The total health detriment values obtained revealed a low detrimental effect to consumers of these milk brands. The calculated values of the total cost of health detriment revealed that the children age group has the highest cost health detriment per-caput dose, followed by adults, while infants have the least; these values obtained are observed to be low. The overall result therefore shows that the powdered and liquid milks consumed in Nigeria are radiologically safe and may not cause immediate or significant radiation health hazard to consumers of the examined milk brands. However, optimizing radiation protection by means of this cost-benefit analysis is recommended.

Author details

Ezekiel O. Agbalagba* and Hannah O. Agbalagba
Federal University of Petroleum Resources, Nigeria

*Address all correspondence to: agbalagba.ezekiel@fupre.edu.ng; ezek64@yahoo.com

References

[1] Badran MM, Sharshar T, Elinimer T. Levels of ^{137}Cs and ^{40}K in edible parts of some vegetables consumed in Egypt. Journal of Environmental Radioactivity. 2003;**67**:181-190

[2] Tchokossa P, Olomo JB, Balogun FA, Adesanmi CA. Assessment of radioactivity contents of food in the oil and gas producing areas in Delta State, Nigeria. International Journal of Science and Technology. 2013;**3**(4):245-250

[3] Albrecht A, Schultze U, Liedgens M, Fluhler H, Frossard E. Incorporating soil structure and root distribution into plant uptake models for radionuclides: Toward a more physically based transfer model. Journal of Environmental Radioactivity. 2002;**59**:329-350

[4] Huff EA. Fukushima radiation treats US milk supplies at levels 2000 percent higher than EPA maximum. Update on Fukushima nuclear plant disaster 2011 online report. 2011

[5] Tawalbeh AA, Samat SB, Yasir MS, Omar M. Radiological impact of drinks intakes of naturally occurring radionuclides on adults of Central Zone of Malaysia. The Malaysian Journal of Analytical Sciences. 2012;**16**(2): 187-193

[6] Adeniji AE, Alatise OO, Nwanya AC. Radionuclide concentrations in some fruit juices produced and consumed in Lagos, Nigeria. American Journal of Environmental Protection. 2013;**2**(2): 37-41

[7] Al-Zahrani JH. Natural radioactivity and heavy metals in milk consumed in Saudi Arabia and population dose rate estimate. Life Science Journal. 2012; **9**(2):651-656

[8] Emumejaye K. Determination of potassium-40 concentration in some powdered milk samples consumed in Delta State, Nigeria. Journal of Applied Physics. 2012;**2**:08-12

[9] Gaso MI, Segovia N, Cervantes ML, Herrera T, Perez-Silva E. Internal radiation dose from ^{137}Cs due to the consumption of mushrooms from a Mexican temperate mixed forest. Radiation Protection Dosimetry. 2000; **87**:213-216

[10] Jibiri NN, Ajao AO. Natural activities of 40K, 238U and 232Th in elephant grass (*Pennisetum purpureum*) in Ibadan metropolis, Nigeria. Journal of Environmental Radioactivity. 2005;**78**: 105-111

[11] Melquiades FL, Appoloni CR. Natural radiation levels in powered milk samples. Food Science and Technology. 2004;**24**(4):501-504

[12] Buldini PL, Cavalli S, Sharma JL. Matrix removal for the ion chromatographic determination of some trace elements in milk. Microchemical Journal. 2002;**72**:277-284

[13] Arogunjo AM, Ofuga EE, Afolabi MA. Levels of natural radionuclides in some Nigerian cereals and tubers. Journal of Environmental Radioactivity. 2005;**82**:1-6

[14] Hosseni T, Fathivan AA, Barati H, Karimi M. Assessment of radionuclides in imported foodstuffs in Iran. Journal of Radiation Research. 2006;**4**(3): 149-153

[15] Zaid QA, Khled MA, Anas MA, Abdalmajeid MA. Measurement of natural and artificial radioactivity in powder milk corresponding annual effective dose. Radiation Protection Dosimetry. 2010;**138**(3):278-283

[16] Ibrahim HS, Abdelfatah FH, Nadia HE, Hussen AM, Mohammed AN.

Radiolo-icalon solids, foodstuff and fertilizers in the Alexandria region Egypt. Turkish Journal of Engineering and Environmental Sciences. 2007;**31**: 9-17

[17] Al-Masri MS, Mukallati H, Al-Hamwi A, Khalili H, Hassan M, Assaf H, et al. Natural radionuclides in Syrian diet and their daily intake. Journal of Radioanalytical and Nuclear Chemistry. 2004;**260**(2):405-412

[18] Melquiades FL, Appoloni CR. ^{40}K, ^{137}Cs and ^{232}Th activities in Brazilian milk samples measured by gamma ray spectrometry. Indian Journal of Pure and Applied Physics. 2002;**40**:5-11

[19] Baratta EJ. Radium, Radon and Uranium in Drinking Water. Washington, DC: Lewis Publisher; 1990. pp. 203-213

[20] Frame P. Radium Equivalent Health Physics Society. 2006. Available from: http://hps.org/documents/background_ radiation_fact_sheet.pdf

[21] Roy S, Alam MS, Miah FK, Alam B. Concentration of naturally occurring radionuclides and fission products in bricks samples fabricated and used in and around Great Dhaka City. Radiation Protection Dosimetry. 2000;**88**:255-260

[22] Sam AK, Abbas N. Assessment of radioactivity and the associated hazards in local and imported cement types used in Sudan. Radiation Protection Dosimetry. 2001;**93**:275-277

[23] Diab HM, Nouh SA, Hamdy A, El-Fiki SA. Evaluation of natural radioactivity in a cultivated area around a fertilizer factory. Journal of Nuclear and Radiation Physics. 2008;**3**(1):53-62

[24] Gang S, Diyun C, Zeping T, Zhiquang Z, Wenbiao X. Natural radioactivity levels in topsoil from the Pearl river delta zone, Guangdong, China. Journal of Environmental Radioactivity. 2012;**103**:48-53

[25] Al-Jundi J, Salah W, Bawa'aneh MS, Afaneh F. Exposure to radiation from the natural radioactivity in Jordanian building materials. Radiation Protection Dosimetry. 2006;**118**:93-96

[26] UNSCEAR (United Nation Scientific Committee on the Effects of Atomic Radiation). Sources and effect of ionizing radiation. In: Report to the General Assembly with Scientific Annaxes. New York: United Nations; 2000

[27] Zarie KA, Al Mugren KS. Measurement of natural radioactivity and assessment of radiation hazard in soil samples from Tayma area (KSA). Isotope and Radiation Research. 2010; **42**(1):1-9

[28] Veiga RG, Sanches N, Anjos RM, Macario K, Bastos J, Iguatemy M, et al. Measurement of natural radioactivity in Brazilian beach sands. Radiation Measurement. 2006;**41**:189-196

[29] IAEA. International Aomic Energy Agency. Measurement of radiation in food and the environment. In: Technical Reports Series 295. Vienna; 1989

[30] UNSCEAR (United Nations Scientific Committee on the Effects of Atomic Radiation). Sources and effects of ionizing radiation. In: Report to General Assembly. New York: United Nations; 1993

[31] World Health Organization, WHO. Guideline for drinking water quality; measurement of natural and artificial radioactivity in powder milk corresponding annual effective dose radiation protection. In: Recommendations. Vol. 1. Geneva: WHO; 1993

[32] Taskin HM, Karavus P, Ay A, Touzogh S, Hindiroglu S, Karaham G.

Radionuclide concentration in soil and lifetime cancer risk due to the gamma radioactivity in Kirklareli, Turkey. Journal of Environmental Radioactivity. 2009;**100**:49-53

[33] ICRP (International Commission on Radiological Protection). Age-dependent doses to members of the public from intake of radionuclides. Part 5: Compilation of ingestion and inhalation coefficients. In: ICR Publication 72. Oxford: Pergamon Press; 1996

[34] Marmuleva NI, Barinov EY, Petukhov VL. Radionuclides accumulation in milk and its products. Journal de Physique IV. 2003;**107**(2): 827-829

[35] ICRP, International Commission on Radiological Protection. Age-dependent dose to member of the public from intake of radionuclides. Part II. In: Publication-67. Oxford: Pergamon Press; 1993

[36] IAEA (International Atomic Energy Agency). International Basic Safety Standards for Protection against Ionizing Radiation and for Safety Radiation Sources. No. 115. Vienna: IAEA; 2003

[37] Nigeria Population Commission, NPC. Population distribution by age and sex. In: Priority Table. Volume IV. Population and Housing Census 2006. Federal Republic of Nigeria; 2010

[38] Ahmed JV, Daw HT. Cost-benefit analysis and radiation protection. A technical presentation on nuclear safety and environmental protection. IAEA Bulletin. 1991;**22**(5/6):13-22

[39] ICRP (International Commission on Radiological Protection). The 1990–91 recommendation of the International Commission on Radiological Protection. Publication 60. Annals of ICRP. 1991;**21**: 1-3

3

Insect Pest Management in Organic Farming System

Hamadttu Abdel Farag El-Shafie

Abstract

Due to the regulations of organic farming, few options remain for organic farmers to manage pests and diseases in their crops compared to conventional farming. However, major pests could still be managed through manipulation of the agroecosystem processes in advantage of the crops and disadvantage of pests. The limited number of active plant protection substances authorized for use in organic farming can provide support to natural and biological control agents in suppression of pests and diseases. This chapter highlights the principles and strategies of crop protection in organic farming, the cultural practices adopted, the active substances allowed for use to suppress pests, and the impacts on faunal and floral biodiversity. A case study of organic date palm cultivation is discussed.

Keywords: organic farming, holistic approach, biopesticides, pest management

1. Introduction

Organic agriculture is a holistic production system that sustains the health of soils, ecosystems, and people. It relies on ecological processes, biodiversity, and cycles adapted to local conditions rather than the use of inputs with adverse effects. Organic agriculture combines tradition, innovation, and science to benefit the shared environment and promote fair relationships and a good quality of life for all involved [1]. Holistic means near-closed nutrient and energy cycle system consider-ing the whole farm as one organism [2]. Organic agriculture relies on a number of farming practices based on ecological cycles and aims at minimizing the environ-mental impact of the food industry, preserving the long-term sustainability of soil and reducing to a minimum use of nonrenewable resources [3]. Organic agriculture is both a philosophy and a system of farming aiming to produce food that is nutri-tious and uncontaminated with substances that could harm human health [4]. Organic farming benefits to the ecosystem include conservation of soil fertility, carbon dioxide storage, fossil fuel reduction, preserving landscape, and preserva-tion of biodiversity [3].

Pest management in organic farming is achieved by using appropriate cropping techniques, biological control, and natural pesticides (mainly extracted from plant or animal origins). Weed control, the main problem for organic growers, can be managed through cultural practices including mechanic cultivation, mulching, and flaming. Organic farming is characterized by higher diversity of arthropod fauna and conservation of natural enemies than conventional agriculture [3, 5].

According to the IFOAM [1], organic agriculture is guided by four principles: health (soil, plant, animal, and human), ecology (living ecological systems and cycles), fairness (environment and life opportunities), and care (protect the health and well-being of current and future generations as well as the environment). The US Congress passed the organic food product act in 1990, while the European Union (EU) set up the first regulations on organic farming in 1991, and in the same year, the Codex Alimentarius Commission officially recognized organic agriculture. Gomiero et al. [3] gave more details on history of organic farming, total global areas, organic standards, and impact on the environment. The chapter deals with pest management in organic farming system with an example of organic date production as case study.

2. Principles and strategies of crop protection in organic farming system

Pest management in organic farming is a holistic (whole-farm) approach that largely depends on the ecological processes and biodiversity in the agroecosystem. Accordingly, most IPM tactics, principles, and components match with organic farming systems [6]. The goal of this strategy is to prevent pests from reaching economically damaging levels without causing risk to the environment. Successful IPM programs in organic farming may have the following components: (1) monitoring crops for pests, (2) accurately identifying pests, (3) developing economic thresholds, (4) implementing integrated pest control tactics, and (5) record keeping and evaluation.

The factors that render crop habitat unsuitable for pests and diseases include limitation of resources, competition, parasitism, and predation [7]. These fac-tors play an important role in maintaining equilibrium of the agroecosystem and suppression of harmful pests. Faunal and floral diversities play a substantial role in pest and disease management in organic farming system [8, 9]. The four principles of pest management in organic farming system, namely, prevention, avoidance, monitoring, and suppression, will be discussed in this chapter with special reference to date palm as case study.

3. Differences between organic and conventional farming with respect to plant protection

Few options of plant protection substances are available for certified organic growers compared to conventional ones. Thus, they should capitalize on the natural processes and management of the ecosystem to control harmful organisms. Organic farms had a more diverse arthropod fauna, on average, than conventional farms. The average for five 30-second vacuum samples per farm was approximately 40 arthropod species in conventional tomato compared to 66 species in organic tomato fields. Additionally, natural enemies (parasitoids plus predators) were more abundant on organic farms [10]. Arthropod biodiversity, as measured by species richness, was, on average, one-third greater on organic farms than on conventional farms [10].

Under organic farming systems, the fundamental components and natural processes of ecosystems, such as soil organism activities, nutrient cycling, and species distribution and competition, are used directly and indirectly as farm management tools to prevent pest populations from reaching economically damaging levels. Soil fertility and crop nutrients are managed through tillage and cultivation practices, crop rotations, and cover crops and supplemented with manure, composts, crop waste material, and other allowed substances.

Organic farming (OF)	Conventional farming (CF)
Synthetic fertilizers and synthetic pesticides are not permitted	Synthetic fertilizers and synthetic pesticides are allowed
Genetically modified organisms (GMOs) are not allowed	GMOs can be used
Soils have higher water holding capacity than CF	Soils have less water holding capacity than OF
OF has larger floral and faunal biodiversity than CF (complex crop pattern)	CF has smaller biodiversity than OF (simple crop pattern)
The agricultural landscape is characterized by heterogeneity (multicultural system)	The agricultural landscape is characterized by homogeneity (monocultural system)
Minimizing the use of nonrenewable resources by recycling plant and animal waste into the soils (on-farm inputs)	Depends largely on nonrenewable resources (off-farm inputs)
OF is more sustainable than CF	CF is less sustainable compared to OF
Strictly regulated by international and national institutional bodies such as Codex Alimentarius and IFOAM	Not strictly regulated
Crop protection depends mainly on natural processes such as soil fertility, crop cycle, and biodiversity (more preventive)	Crop protection relies mainly on human intervention with synthetic chemicals (more curative)

Table 1.
Fundamental differences between organic and conventional farming.

Soil-borne and root pathogens are usually found in low levels in organic farming as compared to conventional farming [11]. Pathogens such as *Pythium* spp., *Sclerotium rolfsii*, *Phytophthora* spp., and some *Fusarium* can survive on organic matter of the soil, in the absence of their hosts for long periods, and are thus difficult to be controlled with crop rotation. Additionally, airborne pathogens cannot be controlled with cultural practice such as crop rotation [12]. Powdery mildew and rust diseases (airborne) and insect pests such as aphids and whiteflies (sucking insects) are less serious in organic farming than in conventional farming due to lower nitrogen concentrations in foliar tissues or phloem of plants in the former compared with the latter [11]. Almost all pesticides available for organic farming have short residual effects and work through direct contact mode of action as compared to the persistent systemic pesticides used in conventional farming. **Table 1** gives the main differences between organic and conventional farming with respect to soil fertility, biodiversity, and other criteria.

4. Crop protection practices in organic farming

Practices and tactics used in organic farming are based on the three management strategies, which include prevention, monitoring, and suppression. These practices will be intensively discussed in the following paragraphs:

4.1 Identification and monitoring of crop pests

Crop pests include insects, weed, plant pathogens, invertebrate, and vertebrate animals. Identification of insect pests and their natural enemies is an important step in any pest management program. Insect pests and natural enemies could be identified using keys and field guides or otherwise consulting an official identification bodies. Unlike insect pests, plant pathogens including fungi, bacteria, virus, and nematodes are difficult to identify in the field and may need laboratory diagnosis. However, signs of insect damage and symptoms of plant diseases may be easily distinguished in the field. Weeds could be easily identified using key and field guides.

Monitoring is the regular inspection or scouting of field crops for pests, including insects, pathogens, nematodes, and weeds, to determine their abundance and level of damage. It serves as an early warning system for the presence of pests and diseases providing information for decision-making regarding management action and evaluation of control methods. Insect pests can be monitored through visual observation, pheromone and light traps, sticky traps, water traps, yellow traps, sweep nets, beating trays, and pitfall traps. Scouting data are used to develop economic thresholds, a useful decision-making tool to start control action when a pest population reaches or exceeds the specified economic threshold.

4.2 Tactics used for pest prevention and suppression in organic farming

A successful integrated pest management (IPM) program in organic farming incorporates a variety of pest management tactics such as cultural, mechanical/ physical, biological, and biopesticide (allowed for organic use) tactics individually or in combination. Each control tactic, discussed below, employs a different set of mechanisms for preventing and suppressing pest populations.

4.2.1 Cultural pest control

The goal of cultural control is to alter the environment, the condition of the host, or the behavior of the pest to prevent or suppress an infestation. It disrupts the normal relationship between the pest and the host and makes the pest less likely to survive, grow, or reproduce [13]. In agricultural crops, crop rotation, selection of crop plant varieties, timing of planting and harvesting, irrigation management, crop rotation, and use of trap crops help reduce populations of weeds, microorganisms, insects, mites, and other pests. These cultural practices are more preventive than curative and thus may require planning in advance [13–15]. The diversified habitat provides these parasites and predators with alternative food sources, shelter, and breeding sites [16]. Tillage can cause destruction of the insect or its overwinter-ing chamber, removal of the protective cover, elimination of food plants, and dis-ruption of the insect life cycle generally killing many of the insects through direct contact, starvation or exposure to predators, and weather [13]. The use of trap strip crops can control insect damage at the field edges and at the same time avail refuge and food for beneficial insects. Insect resistance is an important component of pest and disease management. Quality-based resistance can be induced in plants through management of nutrients and irrigation. Intercropping and biodiversity play an important role in pest management in organic farming [13].

4.2.2 Mechanical and physical pest control

One of the simplest methods of physical or mechanical pest control is handpicking insects or hand-pulling weeds. This method works best in those situations where the pests are visible and easily accessible [17]. Physical or mechanical disruption of pests also includes such methods as mowing, hoeing, flaming, soil solarization, tilling or cultivation, and washing [17]. Animals such as kangaroos cause damage by eating yellow dates; hence, fruit bunches are covered to protect them from such damage [18].

Devices that can be used to exclude insect pests from reaching crops in organic farming include, but not limited to, row covers, protective nets with varying mesh size according to the pest in question, and sticky paper collars that prevent crawling insects from climbing the trunks of trees. Water pressure sprays can be employed to dislodge insect pests such as aphids and mites from the plant surface. Insect vacuums, on the other hand, could be used to remove insects from plant surface and collect them into a collection box.

4.2.3 Biological pest control

Biological methods are the use of beneficial organisms that can be used in the field to reduce insect pest populations. Biological control is grouped into three categories: importation or classical biological control, which introduces pest's natural enemies to the locations where they do not occur naturally, augmentation involves the supplemental release of natural enemies, boosting the naturally occurring population, and conservation, which involves the conservation of existing natural enemies in the environment [19]. The role of beneficial species on pests is of relatively greater importance in organic agriculture than in conventional agriculture, because organic growers do not have recourse to highly potent insecticides (such as synthetic pyrethroids) with which to tackle major pest problems [13].

4.2.4 Biopesticide control

Biopesticides are characterized by having minimal or no risk to the environment, natural enemies, and nontarget organisms due to their mode of action, rapid degradation, and the small amounts applied to control pests. They are slow acting, have a relatively critical application times, and suppress rather than eliminate a pest population [20]. Biopesticides have limited field persistence and shorter shelf life and present no residue problems. Thus, they are approved for pest management in organic crops.

5. Plant protection products (PPPs) authorized in organic farming

The crop protection in organic farming is holistic, and, hence, it is extremely difficult to separate inputs as plant nutrients (fertilizers) and plant protectants (pesticides) [6]. Plant protection products authorized for use in organic farming differ among countries depending on the differences in crops, pests, and crop-ping systems, as well as regulations and standards adopted by these countries [21]. Organically approved pesticides fall into the following groups: biorational, inorganics, botanicals, microbial, oils, and soaps. The most widely used as insecticides are microorganisms, natural pyrethrins, rapeseed oil, and paraffin; the most widely used as fungicides are copper compounds, sulfur, and microorganisms. The rules of organic agriculture allow the use of unregistered products such as nettle slurry, which is used against aphids. It can be prepared on the farm or shared among farmers [21, 22].

The basic substance concept was introduced by the EU regulation 1107 in 2009. It was defined as substance not intendedly used for plant protection purposes; however, it can still be used in protection of plants either directly or as a diluent. According to this definition, substances used as foodstuff such as vinegar and sunflower oil can be used as plant protection [23]. The basic substances of plant and animal origin, which are used as foodstuff, can be legally used in crop protec-tion in organic farming with the exception of being used as herbicides. These basic substances include chitosan hydrochloride, fructose, sucrose, *Salix* spp. cortex, and *Equisetum arvense* L. (field horsetail) which are used as elicitors of the plant self-defense mechanism. Sunflower oil, whey, and lecithins are used as fungicides, while vinegar is used as fungicide and bactericide, and *Urtica* sp. is used as insecticide, fungicide, and acaricide [21]. In organic farming, only active substances listed in the Commission Regulation (EC) No. 889/2008 (**Table 2**) can be used. New update is frequently being made by the EC to add or remove PPPs from the list.

Name of product	Purpose and specifications of use
Azadirachtin from the neem tree (*Azadirachta indica*)	
Beeswax	Used as protectant for treatment of cuts and wounds after pruning or in grafting
Plant oils	Used for control of small-bodied insects such as thrips, aphids, and whiteflies
Laminarin (from *Laminaria digitata*) or kelp or brown algae seaweed	A polysaccharide from the group of the glucans, used to protect plants against fungi and bacteria. Kelp should be grown according to the organic standards
Pheromones	Used only in traps and dispensers
Pyrethrins from the leaves of *Chrysanthemum cinerariaefolium*	Used as insecticide
Pyrethroids (only deltamethrin or lambdacyhalothrin)	Used only in traps with attractants or pheromones
Quassia from the plant *Quassia amara*	Only insecticide and repellent
Microorganisms, e.g., *Bacillus thuringiensis*, *Beauveria bassiana*, and *Metarhizium anisopliae*	Origin should not be GMOs
Spinosad from the soil bacterium *Saccharopolyspora spinosa*	Used as insecticide
Ethylene	Insecticidal fumigant against fruit flies
Paraffin oil	Used as insecticide against small-bodied insects
Fatty acids (soft soaps)	Insecticide against mite, thrips, and aphids
Lime sulfur (mixture of calcium hydroxide and sulfur)	Used as fungicide
Kieselgur (diatomaceous earth) from the hard-shelled diatom protist (chrysophytes)	Used as mechanical insecticide
Naturally occurring aluminum silicate (kaolin)	As insect repellent against a wide range of insects at a rate of 50 kg/ha
Calcium hydroxide	Used as fungicide
Sodium hypochlorite (bleach or as javel water). It is a disinfectant with numerous uses, and its effect is due to the chlorine	Used in seed treatment as viricide and bactericide
Sulfur	Used as broad-spectrum inorganic contact fungicide and acaricide
Copper compounds such as: copper hydroxide, copper oxychloride, copper oxide, tribasic copper sulfate, and Bordeaux mixture (copper sulfate and calcium hydroxide)	Used as fungicide and bactericide maximum of 6 kg copper per ha annually
Sheep fat (obtained from fatty sheep tissues by heat extraction and mixed with water to obtain an oily water emulsion)	A triglyceride consisting predominantly of glycerine esters of palmitic acid, stearic acid, and oleic acid. A repellent by smell against vertebrate pests such as deer and other game animals. It should not be applied to the edible parts of the crop
Quartz sand	Used as repellent against vertebrate pests

Table 2.
Plant protection products approved by the European Union (EU) for use in organic farming [24].

6. A case study of organic date palms

There are about 100 million date palms in the world mostly distributed in Asia and North Africa, producing 7.78 million ton of dates annually [25]. The interna-tional famous date palm cultivars include Medjool, Deglet Noor, Barhee, Halawy, Khalas, and Khadrawy. Organic dates are now produced in many countries around the world including Tunisia, Israel, Saudi Arabia, Egypt, Sudan, Iran, Algeria, and the USA. Date palm, whether grown conventionally or organically, has numerous pests and diseases including 132 species of arthropod (insects and mites), 52 verte-brate pests (birds, rodents, bats), and 28 non-arthropod pests (slugs and snails, para-sitic nematode) [26, 27]. Additionally, more than 16 important fungal, phytoplasma, and unidentified diseases attack the date palm. The major ones include Bayoud, black scorch, *Diplodia*, Khamedj inflorescence rot, Belaat, graphiola leaf spot, Al-Wijam disease, brittle leaf disease, and Faroun disease [28]. These pests and diseases may cause substantial losses in date palm groves if left unmanaged. Therefore, a well-planned and supervised pest management program is important to maintain a sustainable date palm production in organic farming system. Some examples of injuries inflicted by pests on date palm and dates are shown in **Figures 1** and **2**.

Date palm pests of economic important in organic farming could be prevented through an IPM program comprising the following components: selection of plan-ning materials, pest monitoring, cultural management, and conservation of natural enemies of pests.

6.1 Selection of planning materials

To a healthy vigorous palm that yield good quality date fruits, one should start with good planting materials whether tissue culture seedlings, offshoots, or mature palms. Planting materials should be adapted to the area where to be grown, in addition of being healthy and free from pests and diseases. Such planting materi-als should be obtained from nurseries certified for organic date palm production, where strict quarantine measures and protocols are applied. Many serious pests and diseases of date palm including the invasive red palm weevil spread rapidly through movement of infested planting materials [29]. Dubas bug, scale insects, longhorn beetle, and rhinoceros beetle also invade new areas through transportation of

Figure 1.
Symptoms of damage on the fruit bunch stalk (left) due to Oryctes elegans and on the trunk (right) due to Jebusaea hammerschmidti.

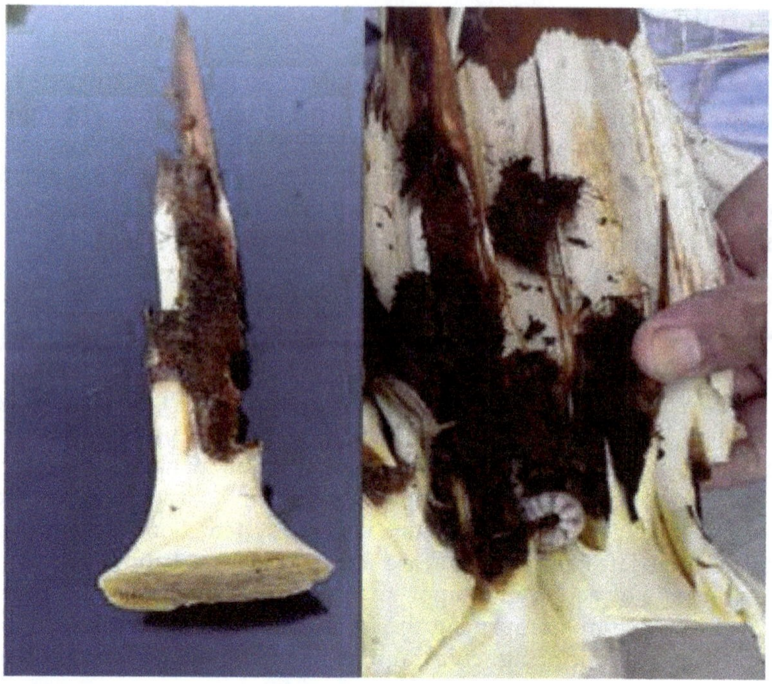

Figure 2.
Fatal damage caused by the larvae of the longhorn beetle Jebusaea hammerschmidti on the apical meristem (Goumara) of a date palm.

infested offshoots and mature palms (**Figure 3**). Thus, application of preventive and protective controls through strict implementation of agricultural quarantine controls, as well as non-trading of any offshoots or infected palms, are essential for the establishment of new date palm plantation.

6.1.1 Characteristics of a good date palm offshoot

a. Make sure that the offshoot belongs to the cultivar that is intended to be grown. Selection should be made during harvesting time of the mother palm, because it is easy to identify the date palm cultivar from the characteristics of its fruit.

b. The offshoot should be 3–4 years old, with length of approximately 1–1.5 m and diameter of 25–35 cm with an average weight of 20–30 kg.

c. The offshoot should contain numerous undamaged roots.

d. The offshoot should be free of insect pests and diseases.

e. The offshoot should be mature and hence will have a better chance of survival after transplanting. Bearing fruits and having daughter offshoots indicate the maturity of the offshoot.

f. Care must be taken not to wound the offshoot during detachment from the mother palm, as the wounds would predispose the offshoot for bacterial and fungal diseases, as well as for opportunistic insect pests such as the dynastic beetles, termites, and red palm weevil.

Figure 3.
Many important pests and diseases of date palm can be introduced into new areas through transporting unhealthy planting materials.

6.2 Pest monitoring and mass trapping

Monitoring of major date palm pests is essential for decision-making such as determination of economic threshold that largely help in starting control actions and avoidance of routine preventive treatments. Pheromone trapping could be used to determine population cycles and prediction of pest outbreaks. Pheromones can also be employed in mating disruption, attack and kill, and male inhalation tech-niques to reduce pest populations [30]. The same devices of pheromone and light traps can also be used for mass trapping of adult insect pests, particularly gravid females that lead to drastic reduction in pest population (**Figure 4**) [31].

6.3 Cultural management

Services of date palm that are important in the management of pests and diseases include irrigation management, field sanitation, removal of weeds, organic fertilization, old frond pruning, frond base cutting, offshoots removal, pollination, fruit thinning, spines removal, fruit bagging, and harvesting. Each one of the abovementioned opera-tions is carried out at specific time of the year with specific purpose; however, each operation can control palm pests and diseases in one way or another. Thus, adoption of date palm calendar for each locality will provide control of date palm pests and diseases.

6.3.1 Organic fertilization and irrigation (soil condition)

Management of irrigation to avoid conditions that are congenial to the development of pests and diseases (e.g., red palm weevil) is an important soil conditioning practice in organic farming. Another important practice is maintaining soil health and nutrients to increase palm immunity against pests' attack, such as the longhorn beetle, which is known to inflict serious damage on weak unattended undernourished date palms. Healthy palms with balanced nutrients and irrigation withstand attack by this opportunistic insect pest. High humidity, which is conducive to the buildup of Dubas, is expected to prevail in densely planted orchards. High soil moisture (flood irrigation and basin irrigation) increases the infestation by the red palm weevil in date palm groves [32]. Care has to be taken when applying organic manure to newly transplanted date palm offshoots, because it may contain eggs and different stages of the rhinoceros beetles, which are considered serious pests of date palm. However, the organic manure can be disinfested from these grubs

Figure 4.
Solar light trap (top left), pheromone-baited trap (top right), adult borers collected by the light trap (bottom right), and adult of red palm weevils mass trapped through pheromone trap (bottom right).

and other insect pests using physical methods or chemicals permitted in organic farming system. In this respect, the farm wastes including eradicated palms can be pulverized and used for production of compost (**Figure 5**). Organic fertilizers are added to date palm during the end of October–December. This is to promote date palm growth and increase its immunity against pests and diseases. About 5–50 kg of organic fertilizer is required per palm, depending on age.

6.3.2 Palm spacing

Well-spaced date palms (8 × 8 m) have no problem of dub bug insect which represents a real problem in narrowly spaced plantations [33]. Densely spaced palms facilitate the spread of crawling mites and scale insects from one palm to another. Sallam et al. [32] reported high incidence of red palm weevil infestation in closely spaced date palms. He attributed the high infestation to the high in-grove humidity caused by densely planted farms.

6.3.3 Pruning of date palm

Pruning is the most important practice that contributes significantly in management of pests and diseases, and it includes the removal of old dry fronds (leaves),

Figure 5.
Shredding machine for pulverizing date palms severely infested by the red palm weevil, Rhynchophorus ferrugineus.

offshoots, aerial offshoots, fibers, and spines (**Figure 6**). Frond removal has two parts: cutting of fronds from the lower whorls of the canopy (Tagleem) and cutting the rachis base (petioles) 1–2 years after frond cutting (Takreeb) [34, 35]. The advantages of frond pruning are listed below:

a. Facilitates climbing of the date palm by the farmers.

b. Reduces fire hazards in date palm plantations, particularly during dry seasons.

c. Improves aeration around the palm trunk and thus reduces humidity and discourage hiding and oviposition by trunk borers.

d. Reduces transpiration rate of newly transplanted palms and hence increases the chance of palm survival.

e. Reduces hiding places for unwanted arthropods such as cockroaches, scorpions and non-arthropods such as snail, slugs, as well as vertebrate pests (birds and rats)

f. Facilitates handpicking of large-sized grubs and adults of trunk borers.

The following precautions are recommended to be taken during pruning process:

a. Prune only fronds after 3–7 years (old dry fronds) on only palms that are 7 years old or above.

b. Curry out pruning during December–January, when temperatures are low to avoid infestation by the red palm weevil where activity of the weevils is at the lowest level.

c. Treatment of wounds and pruned surfaces immediately with bee wax or any other substance allowed in organic farming to obscure the kairomones (odor emitted by the palm) which attract the red palm weevil and other palm pests.

d. Avoidance of palm overpruning as fronds protects the palm's heart from excessive heat as well as from cold during winter.

e. Cutting frond base should be inclined outward with downsloping 45° to avoid accumulation of rainwater in the area between the base of the frond and trunk.

f. Disinfection of pruning equipment such as saws, sheers, and sickles to avoid the spread of fungal diseases such as black scorch and *Fusarium* wilts.

It has been stated that tillage practices and leaf pruning had the greatest effect in reducing termite, long antennae, and horned beetles, respectively. On the other hand, sucker removal operations had the greatest effect in reducing the severity of injuries of horned and long antennae beetles in date palm trees [15]. In addition, larvae of long antennae beetles can complete overwintering in the petioles of damaged leaves. Therefore, pruning the dry, damaged, and old leaves can reduce the severity of injuries of borer pests. Termites attack the dry and damaged parts of date palm tree, so pruning the petiole is very effective in reducing nutrient availability, population growth, and severity of injury [15].

6.3.4 Pollination, fruit, and bunch thinning

For good quality date fruits, pollen grain should be obtained from certified bodies to be sure that they are free of pests and diseases such as the inflorescence beetle *Macrocoma* sp. and the fungal pathogen *Mauginiella scaettae* and *Thielaviopsis paradoxa* , which cause Khamedj inflorescence rot and black scorch diseases,

Figure 6.
Pruned palm trunk showing cut frond (A), fibers (B), and cut frond base or petiole pruning (C).

respectively [36]. In this respect, the author stated that extracting pollen and mixing with talc/flour or with water for mechanical pollination proved to be cost-effective and more efficient in prevention of inflorescence pests and disease than traditional pollination methods.

Fruit thinning has two types: strand thinning either made by cutting the end of the strands or removal 30% of the strands from the center of the spathes [35]. It is carried out in February–March 2–3 days after female spathes opening and before pollination. Bunch thinning, on the other hand, involves the removal of the whole bunch and is usually done after pollination. It is carried out in a way that 6–8 bunches are left in each mature date palm. The number of bunches per palm should corresponds to the number of green functioning fronds, i.e., 9–12 green fronds per bunch to ensure high yield of date fruits with high quality [35]. The bunch thinning should be made even on all sides of the palm taking into account the distribution of bunch loads. This is essential to avoid curving of palm head as the case with the cultivar Barhi. Weak infested or infected bunches with undersized fruits and incomplete pollination should be removed first during thinning process. Latifian [37] reported that bunch pruning helped in decreasing the lesser moth, *Batrachedra amydraula* infestation.

6.3.5 Fruit bunch bagging, harvesting, and sorting

The use of insect-proof fruit bunch covers, made of woven monofilament polyethylene yarn (40 mesh), excludes all insect pests including beetles, ants, flies, rats, and birds (**Figure 7**). These bags are more expensive than the loose net bags. Bunch covering and bunch-remained pruning had suitable effects in decreasing the date spider mite, *Oligonychus afrasiaticus*, raisin moth *Cadra figulilella*, and the lesser date moth, *Batrachedra amydraula* infestation [38, 39]. Early harvesting of cultivars such as Barhee, Deglet Noor, and Medjool provides satisfactory control against ripening dates including date moth, raisin moth, carob moth, greater date moth, and sap beetles [40, 41]. Fruit bagging and early harvesting provide effective control against fruit depredation by frugivorous birds [42]. Culling of infected/infested date fruit during harvesting and field drying is considered as an important step in the management of pests and diseases during transit and storage [36].

6.3.6 Phytosanitation in date palm groves

Both field and palm sanitation can have a profound effect in reducing the population of pests and diseases of date palm. The removal of fallen date fruits on the basin of the palm and in the leaf axil of unpruned palms helps provide control for the nitidulid beetles, lesser date moth, and other insect pests [40]. The fallen fruits provide suitable breeding site for these insect pests as well as for rats and birds. Thus, all dried litter around palms should be carefully removed. In organic farms, grazing animals such as goats, horses, and donkeys may be used to clean weeds, fallen fruits, and other farm wastes [40]. Neglected date palm farms represent suitable breeding sites for serious date palm pests including the red palm weevils, longhorn beetle, and rhinoceros beetle [29, 35]; thus, infested old neglected palms should be eradicated.

6.4 Conservation and enhancement of natural enemies of pests

The date palm agroecosystem comprises diverse groups of natural enemies including insect predators, parasitoids, spiders, predatory mites, birds, entomopathogenic nematodes, and microorganisms. In this respect, El-Shafie et al. [26]

Figure 7.
The white-eared bulbul Pycnonotus leucotis (top), damage on dates due to bulbul (bottom left), and bunch covering to control birds (bottom right).

listed 90 species of predators and parasitoids from 9 orders and 23 families. Out of the listed species, the most important are the general predator *Chrysoperla carnea* and the braconid wasp *Bracon* spp. that is highly associated with the date moth *Cadra cautella*. Predatory mites from the family Phytoseiidae such as *Phytoseiulus persimilis* and *Neoseiulus* sp. and *Trichogramma* parasitoids are common. Al-Khatri [43] reported more than 70% parasitism of Dubas bug in Oman by the specialist egg parasitoid, *Pseudoligosita babylonica*. He also mentioned other species of Dubas natural enemies including the hymenopterous *Bocchus hyalinus*, *Aprostocetus* sp., and *Aphanogmus* sp. as well as the coccinellid *Cheliomenes sexmaculata*.

Several measures taken in date palm plantation can enhance survival and biodiversity of natural enemies. For example, the exclusion of synthetic pesticides by rules of organic farming is the cornerstone in conservation of natural enemies of pests. Intercropping of date palm with annual plants may avail new habitats for predators of pest such as the lacewing. Soils with high population of diversified beneficial organisms such as ground beetles (carabids) and earwigs, which are commonly to be encountered in the date palm agroecosystem (El-Shafie, unpublished data), are expected to maintain low levels of harmful pests. On the other hand, cultural control techniques create a balance between pests and their natural

Pest	Time of appearance	Possible control measures
Red palm weevil, *Rhynchophorus ferrugineus*	All the year round with adult peaks in March–May and October–November	Pheromone trapping of adults, removal and destruction of infested palm, strict quarantine measures to prevent entry of the weevil in date grooves, application of azadirachtin, the *Beauveria bassiana*, and other biological control agents
Termites (*Microcerotermes diversus*, *Odontotermes smeathmani*)	All the year round	Keeping palm healthy palms, removal of dry fronds and litters from around palm basin, application of azadirachtin as curative measures
Green pit scale insect (*Palmaspis phoenicis*) and white scale (*Parlatoria blanchardi*)	All the year round	Pruning and removal of infested fronds, adequate fertilization and irrigation, application of mineral oils (96%) at a rate of 10/1000 liters of water, application of azadirachtin
Weeds	All the year round	Mechanical weeding, grazing by farm animals, use of covers to smother weeds
Rodents	All the year round	Use of mechanical traps, provision of nesting sites for predatory birds, such as owls, that can effectively control rodents in date palm grooves
Inflorescence weevil (*Derelomus* sp.), inflorescence beetle (*Macrocoma* sp.)	With beginning of inflorescence February–March	Use of uninfested pollen, dusting with microfine sulfur at a rate of 50 g/ palm
Bayoud disease, *Fusarium* wilt caused by *F. oxysporum* f. sp. *albedinis*	All the year round	Cultivation of resistant date palms, removal and incineration of infested palms, avoidance of the spread of the disease pathogen through irrigation, use of organic fertilizer rich in chitin to enhance the development of actinomycetes which antagonize the pathogen
Inflorescence rot (Khamedj disease) caused by *Mauginiella scaettae*	February–March	Avoid the use of infected pollen, treatment of the palm with Bordeaux mixture (0.3–0.5%) after harvest and before inflorescence of the next year as preventive measures Treatment (dusting) with microfine sulfur at a rate of 50 g/palm
Black scorch disease caused by *Thielaviopsis paradoxa*	All the year round	Avoid making wound on the palm, sanitation measures such as removal and destruction of badly infected palms, application of Bordeaux mixture, and use of microfine sulfur (80%) at a rate of 2.5 g/1000 liters of water after harvest
Diplodia disease (basal leaf rot) caused by the fungus *Diplodia phoenicum*	All the year round	Use of healthy uninfected offshoots, avoidance of making wounds in palms, disinfection of pruning equipment, application of copper sulfate or copper carbonate
Lesser date moth (Humeira) (*Batrachedra amydraula* Meyer)	February–March	Field sanitation including removal of fallen fruits, use of pheromone or light traps, use of *Bacillus thuringiensis*, biological control using egg parasitoid *Trichogramma* and the larval parasitoid *Bracon* sp.
The old world dust mite (*Oligonychus afrasiaticus*)	April–July	Removal of weeds around palms, which may act as alternative host for the mite, use of windbreak to reduce dust storms, spraying, bunches with a strong stream of water to dislodge mites and destroy webbing; use of predatory mites and coccinellids, dusting bunches with sulfur
The longhorn beetle (*Jebusaea hammerschmidti*), the bunch borers (*Oryctes agamemnon arabicus*, *Oryctes elegans*), and the frond borer (*Phonopate frontalis*)	April–July Larvae of the longhorn beetle are found inside the palm all year round	Pruning of old dry fronds, avoid using uncured farm manure as organic fertilizer, handpicking of larvae during frond base cutting, light trapping of adult beetles, maintaining healthy palms, application of the fungi *Beauveria bassiana*, *Metarhizium anisopliae*, and the entomopathogenic nematode *Rhabditis blumi*

Pest	Time of appearance	Possible control measures
Date palm Dubas bug (*Ommatissus lybicus*)	March–April September–October	Pruning of infested lower fronds to remove Dubas eggs, spraying with azadirachtin (2–3 ml/per liter of water), application of agricultural soaps, biological control with fungi such as *Beauveria* and the egg parasitoid *Oligosita* sp.
Fruit rots	June–July	Bunch covering and avoidance of fruit injuries by insects and birds
Birds	July–October	Covering of bunches during Khalal stage with bird-proof nets
Pests of stored dates	September–November	Bunch bagging to exclude pests that start infestation in the field, sanitation and disinfestation storehouses before use, freezing dates at −18°C

Table 3.
Calendar of major pests and diseases in organically grown date palms and their management in the Gulf region.

enemies, and they are more effective in the prevention of outbreaks of date palm borer pests [15]. The growing of hedgerows, strip crops, and windbreaks provides suitable habitats and source of pollen and nectar for beneficial organisms [3, 16]. Provision of nesting boxes for owls in date palm groves has a noticeable reduction in the population of field rats [27]. In addition to the abovementioned measures to conserve natural enemies, repeated release of purchased predators and parasitoids can maintain their numbers, which cause substantial reduction in pest populations. In this context, Ali and Hama [33] reported that the release of *Trichogramma* sp. twice a year at a rate of 300–500 individuals/palm contributed significantly in the integrated management of the lesser date moth, *C. cautella*.

6.5 Synopsis

The major date palm pests and diseases prevailing in organic date palm plantation, which cause economic damage, are listed in **Table 3**, with possible measures to control them.

7. Impact of pest management in organic farming on the environment

As mentioned earlier in this chapter, pest management in organic farming depends mainly on crop husbandry and biological control. The prohibition of synthetic fertilizers and pesticides leads to conservation of natural enemies including predators and parasitoids. The absence of harmful pesticides also increases diversity of pollinators of crops and minimizes pesticide residues in food products [13, 16, 19]. The community of microorganisms flourishes well in organically managed farms leading to increased organic matter decomposition, soil fertility, and sustainability of the ecosystem. Organic farming enhances the biodiversity of the ecosystem through multicropping and growing of hedges and refuges for beneficial insects as well as wildlife [3]. Preserving biodiversity contributes much in reducing the initial invasion and subsequent establishment of organic farms by pests and diseases [3, 8, 9, 44].

8. Conclusions

Crop protection in organic farming is more preventive than curative. Thus, husbandry practices such as crop rotation, fertilization, cultivation, use of resistant varieties, and preservation of natural enemies play an essential role in pest management. Plant protection products (PPPs) permitted in organic farming should only be used when cultural and biological controls fail to suppress pest populations below economic damage levels. Floral and faunal diversities represent the cornerstone in the strategy of managing pests and diseases under organic production system. Crop protection program in organic farming needs to be documented to allow inspectors to file their reports, which are essential for the certification process. The documents needed are a well-written plan, copies of scouting records and protocols used in monitoring of different pests, and provision of pest management guidelines, according to the organic standards, if available. For optimizing pest management tactics in organic farming, future research priorities and recommendations would include:

i. Long-term ecological studies on ecosystem biodiversity to elucidate its potential role in pest management

ii. Testing more plant protection products including plant extracts and microbial preparations for use in pest population suppression

iii. Exploitation of inherited resistance in different crops against plant herbivores

iv. Strengthening participatory research approach with organic farmers and encouraging citizen science to optimize existing practices and develop new techniques

Author details

Hamadttu Abdel Farag El-Shafie
Date Palm Research Center of Excellence, King Faisal University, Saudi Arabia

*Address all correspondence to: elshafie62@yahoo.com

References

[1] IFOAM (International Movement of Organic Agriculture Movements), 2018. Definition of Organic Agriculture. Available from: https://www.ifoam. bio/en/organic-landmarks/definition-organic-agriculture [Accessed: 20 October 2018]

[2] Hole DG, Perkins AJ, Wilson JD, Alexander IH, Grice PV, Evan AD. Does organic farming benefit biodiversity? Biological Conservation. 2005;**122**:113-130

[3] Gomiero T, Pimentel D, Paoletti MG. Environmental impact of different agricultural management practices: Conventional vs. organic agriculture. Critical Reviews in Plant Sciences. 2011;**30**:95-124

[4] Henning J, Baker L, Thomassin PJ. Economics issues in organic agriculture. Canadian Journal of Agricultural Economics. 1991;**39**:877-889

[5] Honek A. The effect of plant cover and weather on the activity density of ground surface arthropods in fallow field. Entomological Research in Organic Agriculture. 1997;**15**:203-210

[6] Edwards-Jones G, Howells O. The origin and hazard of inputs to crop protection in organic farming systems: Are they sustainable? Agricultural Systems. 2001;**67**:31-47

[7] Drinkwater LE, Letourneau DK, Workneh F, van Bruggen AHC, Shennan C. Fundamental differences between conventional and organic tomato agroecosystem in California. Ecological Applications. 1995;**5**:1098-1112

[8] Westerman PS, Wes JS, Kropff MJ, van der Werf W. Annual losses of weed seeds due to predation in organic cereal fields. Journal of Applied Ecology. 2003;**40**:824-836

[9] Hajjar R, Jarvis DI, Gemmill-Herren B. The utility of crop genetic diversity in maintaining ecosystem services. Agriculture, Ecosystems & Environment. 2008;**123**:261-270

[10] Letourneau DK, Goldstein B. Pest damage and arthropod community structure in organic vs. conventional tomato production in California. Journal of Applied Ecology. 2001;**38**:557-570

[11] Van Bruggen AHC. Plant disease severity in high-input compared to reduced input and organic farming systems. Plant Disease. 1995;**79**:976-984

[12] Theunissen J. Application of intercropping in organic agriculture. Entomological Research in Organic Agriculture. 1997;**12**:251-259

[13] Matyjaszczyk E. Prevention methods for pest control and their use in Poland. Pest Management Science. 2015;**71**:485-491

[14] Nicholas JM. Biological and cultural controls. Non-pesticide alternatives can suppress crop pests. California Agriculture. 2005;**59**:23-28

[15] Latifian M, Rad B. Efficacy of cultural control for date palm borer management. Indian Journal of Plant Protection. 2017;**45**(1):7-11

[16] Jervis MA, Kidd NAC. Host-feeding strategies in hymenopteran parasitoids. Biological Reviews. 1986;**61**:395-434

[17] Hill DS. Pests of Crops in Warmer Climates and their Control. The Netherlands: Springer Science + Business Media, B. V; 2008. p. 704

[18] Reilly D, Reilly A. Cultivation, chronicle of a new date palm grower. Development of a date industry in Australia. Emirates Journal of Food and Agriculture. 2014;**26**(11):1000-1013

[19] Stern VM, Smith RF, van den Bosch R, Hagen KS. The integrated control concept. Hilgardia. 1959;**29**:81-101

[20] Olson S. An analysis of the biopesticides market now and where it is going. Outlooks on Pest Management. 2015;**26**(5):203-206

[21] Matyjaszczyk E. Plant protection means used in organic farming throughout the European Union. Pest Management Science. 2018;**74**:505-510

[22] Bozsik A. Studies on aphicidal efficiency of different stinging nettle extracts. Anzeiger für Schädlingskunde Pflanzenschutz Umweltschutz. 1996;**69**:21-22

[23] Marchand PA. Basic substances: An opportunity for approval of low-concern substances under EU pesticide regulation. Pest Management Science. 2015;**71**:1197-1200

[24] Commission Regulation (EC) No 889/2008 of 5 September 2008 laying down detailed rules for the implementation of Council Regulation (EC) No 834/2007 on organic production and labelling of organic products with regard to organic production, labelling and control Official Journal of European Union L 250; 2008. 51. 1-84

[25] FAO (Food and Agriculture Organization). Crop production and trade data, food and agriculture organization of the United Nations; 2015. Available from: http://faostat3.fao.org [Accessed: 21 November 2018]

[26] El-Shafie HAF, Abdel-Banat BMA, Al-Hajhoj MR. Arthropod pests of date palm and their management. CAB Reviews. 2017;**12**:049. DOI: 10.10.1079/PAVSNNR201712049

[27] El-Shafie HAF, Abdel-Banat BMA. Non-arthropod pests of date palm and their management. CAB Reviews. 2018:13, 020

[28] Sedra MYH. Management of diseases of date palm. In: El-Bouhssini M, Faleiro JR, editors. Date Palm Pests and Diseases Integrated Management Guide. Beirut, Lebanon: International Center for Agricultural Research in the Dry Areas (ICARDA); 2018. pp. 105-159

[29] Faleiro JR. A review of the issues and management of the red palm weevil *Rhynchophorus ferrugineus* (Coleoptera: Rhynchophoridae) in coconut and date palm during the last one hundred years. International Journal of Tropical Insect Science. 2006;**26**:135-150

[30] El-Shafie HAF, Faleiro JR. Semiochemicals and their potential use in pest management. In: VDC S, editor. Biological Control of Pest and Vector Insects. Rijeka, Croatia: InTech Publisher; 2017. pp. 3-22

[31] El-Shafie HAF, Faleiro JR. Optimizing components of Pheromone-baited trap for the management of Red palm weevil, *Rhynchophorus ferrugineus* (Coleoptera: Curculionidae) in date palm agro-ecosystem. Journal of Plant Diseases and Protection. 2017;**124**(3):279-287

[32] Sallam AA, Elshafie HAF, Al-Abdan S. Influence of farming practices on infestation by red palm weevil *Rhynchophorus ferrugineus* (Olivier) in date palm: A case study. International Research Journal of Agricultural Science and Soil Science. 2012;**2**(8):370-376

[33] Ali AAS, Hama NN. Integrated management for major date palm pests in Iraq. Emirates Journal of Food and Agriculture. 2016;**28**(1):24-33

[34] Nixon RW, Wedding RT. Age of date leaves in relation to efficiency of photosynthesis. Proceedings of the American Society for Horticultural Science. 1956;**67**:265-269

[35] Al Mashhadani ASS. The calendar of the date palm care in Abu Dhabi Emirate. In: Fifth International Date Palm Conference. Abu Dhabi, United Arab Emirates; 2014, 2014. pp. 1-14

[36] Ben SM. Field observations in date palm and their importance for reducing pest infestation. In: El-Bouhssini M, Faleiro JR, editors. Date Palm Pests and Diseases Integrated Management Guide. Beirut, Lebanon: International Center for Agricultural Research in the Dry Areas (ICARDA); 2018. pp. 161-171

[37] Latifian M. The effects of cultural management on the lesser date moth (*Batrachedra amydraula* Myer) infestation. Emirates Journal of Food and Agriculture. 2012;**24**:224-229

[38] Kehat M, Blumberg D, Greenberg S. Experiments on the control of the raisin moth, *Cadra figulilella* Gregs. (Phycitidae, Pyralidae), on dates in Israel. Israel Journal of Agricultural Research. 1969;**19**:121-128

[39] Latifian M, Rahnama AA, Amani M. The effects of cultural management on the Date spider mite (*Oligonychus afrasiaticus* McG) infestation. International Journal of Farming and Allied Sciences. 2014;**3**:1009-1014

[40] Blumberg D. Review: Date palm arthropod pests and their management in Israel. Phytoparasitica. 2008;**36**:411-448

[41] El-Shafie HAF. Alternatives to methyl bromide for disinfesting date moth, *Cadra cautella*, in stored dates. Outlooks on Pest Management. 2017;**28**(1):17-20

[42] El-Shafie HAF, Abdel-Banat BMA. The frugivorous white-eared bulbul bird, *Pycnonotus leucotis* depredating date fruits: Biology, feeding ecology and management. Outlooks on Pest Management. 2018;**29**(4):153-157

[43] Al-Khatri SAH. IPM of dubas bug. In: El-Bouhssini M, Faleiro JR, editors. Date Palm Pests and Diseases Integrated Management Guide. Beirut, Lebanon: International Center for Agricultural Research in the Dry Areas (ICARDA); 2018. pp. 68-75

[44] Peterson G, Allen CR, Holling CS. Ecological resilience, biodiversity, and scale. Ecosystems. 1998;**1**:6-18

4

Accounting for the Impact of Sustainable Agriculture: The Role of Community Based Organization and Local Governance Structures in Promoting Sustainable Agriculture

Terrence Thomas, Cihat Gunden and Befikadu Legesse

Abstract

This chapter explores the role of community-based organizations (CBOs) in promoting and sustaining an organic/sustainable food production system. The chapter argues that CBOs offer a unique platform for this purpose considering their potential to promote collective impact and overcome our ancestral tendencies that tend to discourage sustainable behavior. The chapter discusses the role of local governance in creating the institutional support that drives collective systemic impacts. The chapter uses data collected via a telephone survey of a census of Community Action Agencies (CAAs) to assess the level of support for sustainable agriculture, organic methods of production, and the responsibility of citizens in supporting locally produced food. The results show that CBOs believe in organic/sustainable and community-based food production system. CAAs support for a sustainable food production system speaks to their potential to serve as a linchpin in their communities for promoting sustainable agricultural production systems and ensuring collective impact.

Keywords: collective impact, community-based agriculture, participative governance, organic food production, community-based organizations

1. Introduction

Clarifying exactly what a concept represents provides the information needed for identifying its constituent elements and distinguishing it from other concepts. Description of an object or thing provides insight into the nature of what that thing is and what it can do. Since what a thing can do depends on what it is, insights into its nature enables us to hypothesize about potential courses of action regarding that thing. Or, to be more specific, insights developed from clarifying the definition of organic agriculture or, more broadly, a sustainable agricultural production system, enables us to design courses of action that lead to a more enduring and fruitful relationship with our food system. Organic agriculture is inextricably linked to a sustainable food production system; simply because the ultimate motive for adopting organic production practices in agriculture is to achieve a more sustainable food production system.

In earlier work [1] synthesized the work of [2, 3] in proposing this definition of sustainable agriculture as: *the practice of agriculture to produce food and fiber that meets the needs of the current population without compromising the capacity of the ecological capital, on which it depends, to support the needs of future populations.*

And organic agriculture following Codex Alimentarius Commission as *"Organic agriculture is a holistic production management system which promotes and enhances agro-ecosystems health including biodiversity, biological cycles, and soil biological activity. It emphasizes the use of management practices in preference to the use of off-farm inputs, taking into account that regional conditions require locally adapted systems. This is accomplished by using, where possible, cultural, biological, and mechanical methods, as opposed to using synthetic materials, to fulfill any specific function within the system."* (Quoted in [4], p. 6)

Given the definitions and the arguments presented above, it can be deduced that organic agriculture is the instrument through which people working within the limits of the overarching ecological system can achieve a sustainable food production system. It is generally agreed that the current conventional methods of food production are unsustainable at current levels of resource consumption [2, 5]. The need for a sustainable system of food production becomes even more urgent if the aspira-tions of the millions of people in developing countries for a first world lifestyle is taken into account. Many scholars believe that a first world lifestyle for everyone is not possible given our current endowment of resources, for example [5]. Therefore, our survival depends on more than just innovation in markets and science. If all this is true, our relationship with each other and the environment is central to progress toward sustainable food production system and ultimately our survival.

One indispensable aspect of progress in advancing a sustainable production system is the development and implementation of standards of production and marketing of organic food and fiber. The development of these essential standards is a political process as revealed by the interaction of multiple stakeholders in the USDA's design and promulgation of standard in the organic food and fiber production system [6]. As [6] argued, markets can work to weaken or strengthen standards. Whether this happens in a positive or negative direction will depend on the relative political strength of the committed stakeholders and the non-committed stakeholders. For example, [6] points out that stricter standards can act as a barrier to uncommitted farmers, in which case, well-established committed organic farmers might have an incentive to push for stricter standards in order to safeguard their market share. On the other hand, if farm businesses not committed to the organic philosophy and practice are able to exert influence in the market and related regulating agencies, there will be pressure to weaken the standards that may encourage a large number of uncommitted businesses to enter the marketplace. Then the possibility exists that we could end up with a quasi-organic/sustainable food production system, which would put in peril any hope of attaining levels of sustainability that would ensure future food security. Another crucial aspect is overcoming the usual delay and resistance associated with changing deeply held values and or the adoption of innovations [5]. Societies often cling stubbornly to the values that have served them well in the past, even when these same values are demonstrably inappropriate for the present [5, 7].

Corporate industrial agriculture with deep vested interest in conventional food and fiber production models, and their lobbying power buttressed by their deep seated belief in the power of the free market may hinder or slow the rate of change toward the adoption of a organic/sustainable food and fiber production system. Thus, in a capitalist system, the tendency to focus on markets and profits can derail or impede progress toward the ideal by confusing organic production for its own sake with achieving the ultimate goal of a sustainable food system. This is where CBOs can play a really critical and pivotal role in exerting bottom up pressure to maintain standards and promote organic/sustainable values. CBOs are pivotal for creating and sustaining a collective impact because they provide a platform through which strategies can be applied to overcome ancestral tendencies that tend to dis-courage sustainable behavior [8]. Research suggests that strong community ties and group identity tend to promote sustainable behavior [9] and small interdependent social groups will foster pro-environmental behavior [10].

The following statistics on a few key indicators provide a glimpse of the global impact of organic agriculture: As of 2016, the number of hectares (in millions) under organic production worldwide: Oceana 27.3, Europe 13.5, Latin America 7.1, Asia 4.9, North America 3.1, Africa 1.8. In 2016, 57.8 million hectares were under organic production compared to 11 million hectares in 1999, even though the 57.8 million hectares represent just 1.2% of total crop land worldwide. The number of producers engaged in organic production was 2.7 million in 2016 up from 200,000 in 1999, and total sales amounted to 89.7 billion US dollars up from 17.9 billion in 2000. A reasonable inference to be drawn is that 57.8 million hectares under production represent this many acres of increased soil fertility, farm and field diversity. Even more significant, in terms of impact, is the increase in the number of farms producing higher added value and the increased income that flows from this value [11]. These statistics indicate movement in the right direction, even though slow, if progress is judged by just the percentage of crop land under organic cultivation. To achieve faster and sustained progress requires applying an approach that leads to collective impact instead of isolated impact.

It is difficult to achieve progress in the wide adoption of organic/sustainable food production system without achieving collective impact. More generally, impact for the purpose of this chapter is defined as a change in condition that supports a desirable change in behavior. And where appropriate supporting values become infused in the value system of the individual or group from which it operates to reinforce the desired behavior. For the purpose of this chapter we distinguish between isolated and collective impact. In isolated impact, a single organization is assumed to have the solution to the problem, and works in isolation to address one aspect of the problem. Further, it is often assumed that the solution can be scaled up and applied to address the problem in disparate contexts. On the other hand, collective impact assumes that the problem is complex and cannot be solved by any single organization working in isolation. Here, multiple organizations learn, share and act together from the same perspective and agenda [12] to produce system wide and emergent solutions instead of localized solutions with narrow impacts.

CBOs facilitate the development of the type of social community infrastructure that supports collective impact, hence their potential for promoting organic/sustainable food production system. Additionally, because of their cohesiveness and strict commitment to core values, community-based organizations can check the tendency of the market to water down standards, and serve as a source of steadfast support and champions of the organic/sustainable food production philosophy. [13–16] have argued convincingly in support of the capacity of alternative agricultural models (referred to as, community supported agriculture (CSA), community-based agriculture, civic agriculture and farmers markets) to support holistic community development and food production in an environmentally sustainable manner. CBOs have played a pivotal role in promoting and supporting these production models rooted in local communities.

2. Rationale for local action in designing sustainable agricultural production systems

Forces operating in the current global economic and political environments tend to favor and catalyze local community action in solving problems affecting local communities—especially those having their genesis in the local political, social, economic and natural environments. For example, market failure, devolution due to the pressures of globalization [17], a shift to open macro economies and the inabil-ity of the state to offer protection from powerful global market forces [18] have combined to create opportunities for local action by civil society, for example CBOs. Under these conditions, governments at all levels are too preoccupied with trade issues, security issues, global financial markets, competing for investments to create jobs and balancing budgets to shepherd local communities. In such situations, many local communities are left to fend for themselves [19].

Additionally, a reflective and proactive citizenry living in the same information-rich environment as their leaders realize that neither the church nor the state nor other bulwarks of authority are omnipotent, and that leaders are more or less ordinary people. Consequently, citizens living in a global world assign less significance to the guidance of their leaders and institutions and have opted to become more self-regulating [20]. Proactive and reflective citizens in a complex globalizing political and social environments are more concerned with economic and political questions and issues about which they feel politicians neglected to address. In this situation, people turn to groups which promise to offer what conventional politics is unable to deliver. Thus, people are increasingly becoming more involved in single issue groups, which play a pivotal role in raising important issues and problems that may otherwise go unnoticed in conventional political circles until it is too late [21]. Since sustainable agricultural practices are specific to local conditions, and since there are many diverse stakeholders with a vested interest in sustainable agricultural systems, decisions regarding the design and development of sustainable agricultural production systems are best made utilizing participative approaches. Therefore, to advance organic agriculture/sustainable food system in the current context, it is not likely that top down solutions would be prescribed, and, it if they were prescribed, it is unlikely that they would produce desirable outcomes. Thus, under the current socio-political context, locally based participative action seems more likely to succeed in advancing the development of a sustainable agricultural production system. Two other reasons for local bottom up action make clear the indispensable role of CBOs in promoting and sustaining the practice of organic/sustainable food system. First, from an evolutionary perspective, [8] explains that adaptive instincts tend to dictate human behavior. That is, humans tend to exhibit patterns of behavior adapted to our past environment but which behavior is inappropriate in our present one. For example, humans prioritize their self-interest at the expense of cooperating with the group. Thus, our adaptive instincts would dictate that we cooperate with and imitate behavior that is practiced by those with whom we share a kindred spirit or kinship relationship [22]. Accordingly, because of the kindred spirit that exist and is fostered in a CBO, a network of local CBOs would be more effective in promoting organic/sustainable production practices than top down prescriptions that originate from organizations and agencies to which people do not feel a special connection. Modern communities with a densely connected and codependent social networks resembling those that would exist in ancestral communities are found to be more effective in preserving communal resources, and by implication promoting sustainability [9, 23]—Note italics our emphasis. We believe that CBOs provide the substrate for nurturing such networks.

[24, 25] community framework offers a rationale for the role of community-based organizations in supporting a sustainable food production system. It posits that communities are embodied rather than abstract, i.e., residents are attached to place by several institutions and community-based organizations [26]. Businesses are linked via a web of institutional and community-based organizational networks [27, 28], which form an integral part of the community's problem solving capability [24, 25]. From our perspective, an embodied community provides valuable institutional support for sustainable food production system via a community network of CBOs—their mode of governance facilitates collective or participatory decision-making. CBOs serve as a repository of community experience, values and indigenous knowledge because they are well connected to the community. Such a repository provides a readily available reference which establishes a context for interpreting and evaluating information and action with respect to addressing current or future problem situations. As community agents, CBOs provide a critical mass for action by serving as the hub that brings community members together to deal with problems.

Designing and supporting a sustainable food system is a complex process. It involves synthesizing knowledge from different sources, working within a complex and dynamic natural and social environment, and dealing with several stakeholders with competing interests. Complex phenomena are not amenable to rule of thumb or recipe-like solutions. Because a sustainable agricultural production system is linked to local conditions-closely tied to social, economic and political infrastruc-ture—developing a sustainable food production system means working with a pro-cess that accommodates many different perspectives and value systems, and attracts information, expertise and other resources from diverse sources. Additionally, the process must be participative, which means honoring openness, trust, transparency [28, 29], and must include a willingness to accommodate conflicting value positions [30]. Finally, to negate our evolutionary tendencies toward prioritizing and pursu-ing selfish ends, participants must feel a sense of belonging or kinship with the group or strongly identify with the group. Strong group identity will likely encour-age greater self-sacrifice for the communal good by activating a psychological sense of kinship [9].

In summary, CBOs are ideally suited for mobilizing support for developing organic/sustainable agricultural food system throughout communities for these reasons: (1) CBOs have unique knowledge of and connections with the community; (2) CBOs provide a platform on which strategies that support sustainable behavior can be developed to overcome ancestral tendencies that discourage sustainable behavior; (3) the character of a sustainable food production system is determined by local conditions, which requires significant local resources—CBOs with their unique knowledge of community have the potential to mobilize local support and resource; (4) CBOs adaptive governance structure equips them to deal with a dynamic and uncertain environment; ad (5) a network of CBOs informed by the same perspective, working from a collective agenda, learning together and sharing ensures collective impact.

3. Relevance of governance structures

Governance may be viewed from several perspectives: (1) the creation or adoption of means and processes for guiding planning, decision-making, implementing decisions, and ensuring accountability and responsibility for actions taken [31]; (2) as the capacity of an organization to stay on course in a turbulent and changing world [28], (3) self organizing inter-organizational networks [32]; and

(4) multi-level governance, which represents the dispersion of authority to supranational, e.g. the UN, sub national authorities (states, counties, local municipalities, development districts) and informal networks (non-governmental organizations and community-based organizations) [33].

Implicit in all the above views of governance is the authority, official or otherwise bestowed, to make decisions with regards to the allocation of resources for the production of goods or services for a particular constituency. We regard the definition by [31] as generic process views of governance, since it identifies the core concepts of governing without suggesting how these activities should be prosecuted. Conceptually, one can locate the way governance as a process activity is practiced on a continuum. At one end would be a highly bureaucratic regulatory approach to enacting governance, at the other would be a participative open approach to governance. [29] concept of governance is a process view of governance, which specifies a participative approach for discharging governance activities. In the participative model, decision-making is decentralized; freedom, autonomy, trust, transparency, continual learning and creativity are nurtured. In contrast, in the bureaucratic regulatory process model, decision making is centralized; freedom, autonomy, trust, transparency and learning are constrained. The two perspectives by [32, 33] offer a structural political view of governance, i.e., the relationship among the entities involved in discharging the process of governance. Any of these structural forms could conceivably discharge their governance function in either a bureaucratic regulatory or participative mode. Considering the above background and the work of [34, 37]; we use the term governance in this chapter to mean action taken by groups or communities to address problems in the public sphere that cannot be handled either by individuals acting alone or by markets and government.

In our field studies across the southeastern states we have observed many community-based organizations (CBOs) and neighborhood groups that provide valuable and indispensable social services for community residents. CBOs such as North Carolina Coalition of Rural Farm and Families, Alabama Watch, Rural Georgia Development Collaborative, Friends of Children of Mississippi exemplify governance as self-organizing inter-organizational networks. They are self-organizing because they are autonomous and self-governing [32]; implying that they are not controlled by any superior power and were not brought into being by official edict. These networks operate to fill a void in the provisioning of goods and services resulting from the failure of the state and market to provide similar goods and services. In other words, they are self-organizing inter-organizational networks which practice governance as per [31] process definition. They operate in the public sphere without the designated formal authority of government. The network status of these groups derives from the fact that they interact with each other and with government agencies and private entities in exchanging information and garnering resources; they depend on this interaction to survive. An example would be the Georgia rural development collaborative comprising eight independent CBOs agreeing to work in a partnership. The collaborative also interacts with government agencies and foundations as they practice governance in rural Georgia. They receive technical assistance and funding from foundations and government agencies and share information among themselves, government agencies and foundations. Another example would be The North Carolina Coalition of Rural Farms and Families, a grouping of six small CBOs. In their case, they interact with cooperative extension, USDA agencies, with each other and with foundations as they work to provide services to small farmers in Eastern North Carolina. Many scholars of political science believe that this form of governance—the interactive, social, political network mode of governing—is evolving to be the dominant form of governance, eclipsing governing by a super-ordinate authority as in conventional

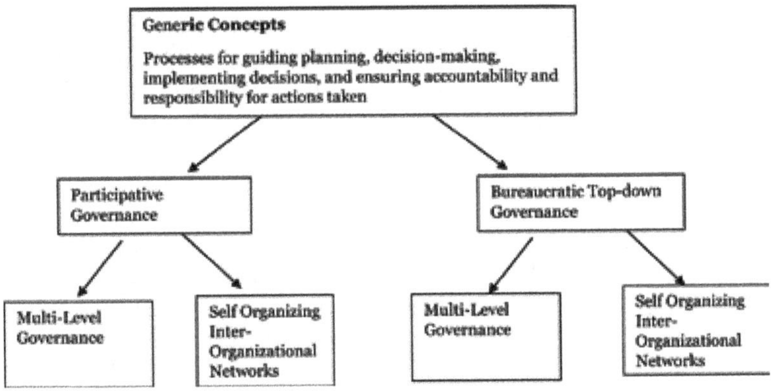

Figure 1.
Schematic representation of the relationship among concepts, process and structure of governance.

government [32, 33]. Within these networks, it is conceivable that groups may adopt a bureaucratic or a participative form of governance.

Multilevel governance may be viewed as being similar to the networks described above with one minor difference; some members of the network are not self-organiz-ing since they were establish under the auspices of federal, state, or local government or some other influential entity, e.g., a foundation. Nonetheless, they interact in a network fashion and discharge the role of governance described above. So then, multilevel governance can include self-organizing networks as well as quasi govern-mental groups (groups established under the auspices of government or government agencies) or groups established under the auspice of large foundations. Community action agencies would be good examples of community-based organizations estab-lished under the auspices of government, these agencies were established under the economic opportunity act of 1964. The relationship among the various concept of governance may be depicted as shown in **Figure 1**. Generic concepts describe the activities that are the focus of governance in general. Entities may operationalize these functions of governance either as a bureaucratic top-down or participative pro-cess. The actual operations of governance may be carried out by entities organized in a multi-level governance or self-organizing inter-organizational network structure.

4. What difference does it make which governance model a community or organization adopts to coordinate the production of goods and services?

What difference does it make which governance model a community or organization adopts to coordinate the production of goods and services? Many scholars of organizational theory believe that the particular form of governance process adopted by a particular entity affects their ability to adapt to changes in the environment in which they operate—their task environment. Organizational theorists believe that organizations that adopt the perspective of governance as coordination and control-the bureaucratic regulatory model—are inflexible and unresponsive to their task environment. Bureaucratic organizations lack the participative culture that nurtures freedom, transparency, commitment, creativity and continuous learning among members. These latter qualities are indispensable for promoting participation, innovation and responsiveness in addressing complex diverse issues and the varied perspectives of an increasingly heterogeneous stakeholder—these are key capabilities for survival in a continually changing, complex and turbulent environment. An environment that requires increasing integration with the ecological sustenance base in order to achieve sustainability. Consequently, these qualities determine the capacity of the organization or the community to adapt to its ever changing environment.

Under a bureaucratic type of governance, decisions are always made by manage-ment at a level above where the work is actually done. Thinking and doing are seen as separate tasks carried out by different individuals. Decision making is centralized and autonomy and freedom to be creative are curtailed. Here, the task of gover-nance is discharged by a privileged few on behalf of the organization. Such organi-zations are tightly integrated and emphasize control to maintain order and protect the organization from external threats that would disrupt established structures and ways of doing things. The stability of these organizations depends on the extent to which they can be insulated from disrupting forces. These organizations operate as closed systems which react to change by attempting to manage or transform the environment in an adversarial or competitive manner rather than responding to the environment in a proactive manner [29].

Globalization, the proliferation of communication possibilities, continual technological change, the easy movement of technology and capital across countries, and the need to reorder our relationship with the sustenance base (the ecological realignment of our industrial, economic and social institutions) create unprecedented complexity and dynamism that require organizations to continually adjust and adopt. To increase the odds of survival, an organization must become more participative, i.e., decision-making is decentralized; freedom, autonomy, trust, transparency and continual learning and creativity are nurtured. In this context, participative organization processes-freedom, autonomy, openness, learning and innovation-create the flexibility the organization needs to become an adaptive and open system as opposed to being closed and rigid. As an open system, participative organizations develop a symbiotic relationship with the environment—influencing the environment and being influenced by it. The interface of the participative organization and the environment becomes the "focal point" of activity where the purpose and mission of the organization achieve meaning [29]. In this sense then, an organization achieves meaning when it responds adaptively to the needs existing in the environment, which includes not only the need for products but also for quality service, non-inva-sive, unsustainable use of resources, collaborative partnering with stakeholders and the participation of stakeholders in charting the course of the organization as wells as steering the organization on course. The central role of governance is to define purpose and chart a course for achieving the defined purpose. However, achieving purpose in a turbulent and dynamic environment is a function of adaptive capacity, and since adaptive capacity is a function of participative processes, then, the role of governance in a dynamic environment is to create conditions under which participa-tive processes can take root and flourish. Trust, freedom, autonomy, creativity, and openness are the touchstones of participative governance. These principles enable organizations such as CBOs and communities to deal effectively with complexity and change because each member is vested in the participatory process, which nurtures commitment to working to develop solutions to problems.

As discussed in [9, 10], our ancestral tendencies can thwart our ability to advance sustainable practices. Moreover, in many cases the strategies we pursue are ones that do not match these tendencies—in that they do not take into account our natural tendency to promote and prioritize our own self-interests above that of the group or common good. They recommend strategies that take advantage of these natural tendencies (which resemble our ancestral mode of behavior) such as creating small dense interdependent social networks and foster group identities. CBOs provide a platform to facilitate the implementation of these strategies that promote sustainable behavior.

5. Material and methods

5.1 Population and sample

In this section, our objective is to answer the question: What is the level of support among CBOs in the southeastern black belt states (BBS) for sustainable agricultural practices? To answer this question, we conducted a telephone survey of community action agencies (CAAs) located in eleven Black Belt States in the south-eastern U.S. Namely, Alabama, Arkansas, Florida, Georgia, Louisiana, Mississippi, North Carolina, South Carolina, Tennessee, Texas, and Virginia. BBS was defined as a state with an African American population that is equal to or greater than 12% of the population of the state. Following [34], we defined community-based organiza-tions as nonprofit civic entities that are locally controlled; and whose mission is to serve a particular constituency that is tied to a defined locality. These entities com-prise groups of people who interact directly, frequently and in multi-faceted ways to deliver service to their constituency [34]. We chose to use CAAs as our population of CBOs because they have a long operational history as a group, and they were spe-cifically established by federal mandate to address poverty, by engaging the com-munity in the problem solving process. Additionally, an easily accessible data base was available, and they fit neatly the profile of CBOs as defined in this chapter. The sample frame for CAAs was obtained from U.S. Department of Health and Human Services Web site. We retrieved lists of CAAs for the BBS listed above and combined them into a single master roster containing 315 CAAs. Since the population size is a relatively small one, and all the CAAs on our list were accessible, we decided to do a census instead of drawing a random sample from this small sample frame.

5.2 Instrumentation

Respondents to our interview were CAA leaders. Based on our discussion above on the practice of sustainable agricultural production system at the community level, we asked the following questions:

1. In your opinion, how important is it for more farmers to use organic methods for producing food? As we have explained previously, sustainable agricultural production is practiced at the community level via CSAs or community-based agriculture. Drawing on the literature, [35] defined a CSA as community-based organizations of consumers and producers. This collective of producers and consumers focus on using organic and sustainable methods to produce their products.

2. We also asked respondents to use a five-point scale anchored with "strongly agree" through "strongly disagree" to indicate the extent of their agreement with the following statements: Part of the duty of a good citizen is to buy locally grown farm produce. Vibrant community-based farming is more likely to keep family farmers on the land than large corporate farming (factory farming, large plantation). Community-based farming is more likely to do a better job of preserving the quality of the land than large corporate farming. Small farms are better for the environment than large corporate farms. These questions are based on the rationale presented above, in addition to the follow-ing considerations: A persistent critique of corporate commodity agriculture is that it has depleted the natural resource base and degraded the environ-ment from which it draws its support [2, 14]. The advent and growth of civic agriculture or community-based agriculture systems is seen as a response to

the socioeconomic and ecological concerns associated with corporate commodity agriculture or, more broadly, conventional food production systems [2, 14, 36, 37]. Community-based agriculture with its emphasis on holistic and locally based agriculture systems has catalyzed regionally based economic activity, the focus of which is to reinvigorate rural communities and economies and improve farmer income [13]. The growing interest and belief in the potential of civic agriculture systems to bolster rural communities and their economies rests on the findings of [2, 24, 38], which indicate that several small locally owned and operated businesses (farms) are positively correlated with economically vibrant communities and superior income equity. These findings speak to the thesis that a sustainable agricultural production system must meet economic and social criteria in addition to addressing ecological concerns (see **Figure 1**).

3. CAA leaders were asked to indicate how they felt about farmers cooperatives. Their responses were measured on a five point Likert-type scale, anchored with very unfavorable to very favorable.
4. Leaders were also asked to use a "yes" or "no" response to indicate whether their organization supports community-based food production, the local farmers' market, and whether they encouraged clients to participate in urban agriculture.

The questionnaire was reviewed by faculty of the Applied Survey Laboratory at North Carolina A&T State University and two leaders of CAAs. The response rate for the survey was approximately 39%. We acknowledge that the results are probably biased because of the relatively low response rate. However, because CAAs are probably subjected to similar socializing influences with regard to the variables of the study, we believe that the low response rate is not a very serious problem. [39] suggested that discrepancies and bias due to non-response are a greater threat for variables denoting characteristics of an entity than for those variables that represent opinions, attitudes or processes. Nonetheless, the results should be interpreted with this weakness in mind.

6. Results

6.1 Support for community-based agriculture

Figure 2 shows that 48% of CAAs provide program support for community based food production, 53% support local farmers market and 36% support urban agriculture.

The data in **Figure 3** show that approximately 85% of CAAs rated the importance of farmers using organic methods to produce food greater than a 5 on a 7 point scale. We see this as indicating that CAAs believe that it is important for farmers to use organic methods, which is in contrast to the relatively small number (48%) of CAAs that offer program support for community-based food production.

Response is based on a 7-point scale Where (1) = not important and (7) = extremely important.

The response pattern in **Figure 3** also shows that over 80% of CAAs believe that it is important for farmers to use organic methods of production. The response pattern shown in **Table 1** below indicates that CAAs overwhelmingly agree that community-based farming is more likely to do a better job of preserving the land:

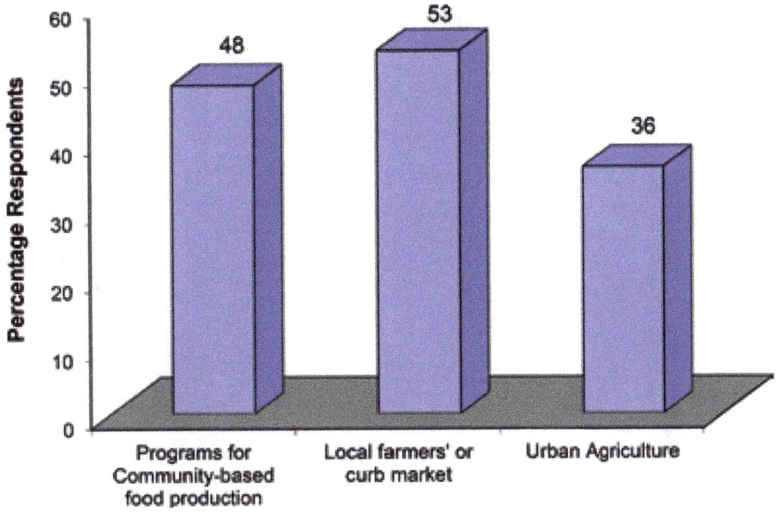

Figure 2.
Support for community-based agriculture (N = 124).

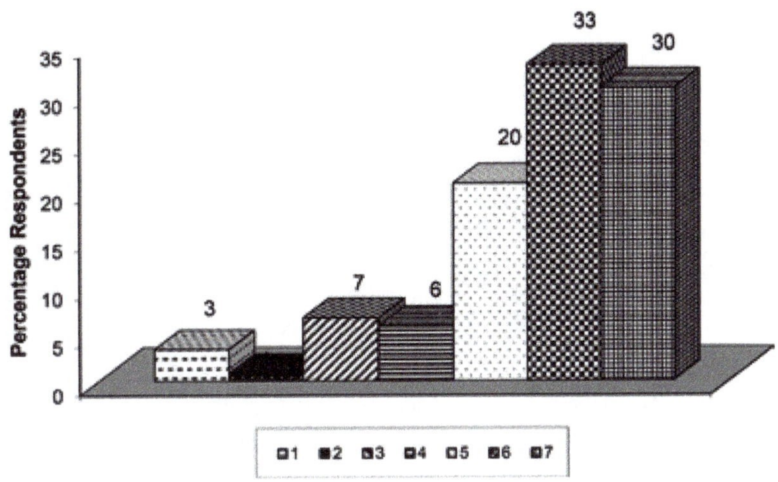

Figure 3.
How important is it for more farmers to use organic methods for producing food? (N = 122).

Items	Ratings (%)				
	1	2	3	4	5
Land	1	2	7	71	20
Environment	2	6	8	67	18
Cooperative*	3	2	11	59	25
Promote	2	1	5	69	23

Response is based on a 5-point scale, where (1) = strongly disagree and (5) = very strongly disagree.(1) = very unfavorable and (5) = very favorable.*

Table 1.
Summary of descriptives for selected variables.

	Model 1		Model 2		Model 3		Model 4	
	b	e^b	b	e^b	b	e^b	b	e^b
Program	2.380***	10.806	2.411***	11.150	2.420***	11.251	2.414***	11.173
Urban	1.208**	3.347	1.187**	3.278	1.159**	3.187	1.146**	3.146
Land	0.801**	2.228	0.641	1.898	0.563	1.755	0.510	1.666
Environment			0.213	1.237	0.208	1.232	0.203	1.225
Cooperatives					0.160	1.173	0.155	1.167
Promote							0.099	1.104
Constant	−8.620***		−8.818		−1.353		−1.191	
−2 log likelihood	117.071		116.698		116.425		116.369	
Chi-square	45.289***		45.662***		45.936***		45.991***	
R square	0.319		0.321		0.322		0.323	

*Note: b = unstandardized coefficient estimate; e^b = exponential of b (the odds ratio);*** $p < 0.001$.*
** $p < 0.05$.
Program = CAAs with programs that support community-based food production; urban = CAAs which encourage its clients to participate in urban agriculture (grow fresh fruits and vegetables for home use); land = opinion on the statement that community-based farming preserves the quality of land better than large corporate farms; environment = opinion on the statement that community-based farming preserves the environment better than large corporate farms; cooperatives = extent of favorable perception of farmers' cooperatives; promote = opinion on the duty of citizens to buy locally grown farm produce.

Table 2.
Logistic regressions results: Factors that influence the likelihood of supporting local farmers market.

71% agree and 20% strongly agree. The pattern also shows that there is very strong agreement among CAAs' leadership that small farms are better for the environment than large cooperate farms. 67% of CAAs' leadership agrees with the statement, and 18% strongly agree. Only 15% of CAAs' leadership can be collectively categorized as strongly disagree, disagree or are indifferent to the statement that small farms are better for the environment. A similar pattern is also evident among CAA leaders with respect to their opinion concerning the duty of citizens to purchase (promote) locally grown produce. Here, 69% agree and 23% strongly agree that it is the duty of good citizens to purchase (promote) locally grown farm products. CAA leaders' pattern of response to the statement that vibrant community-based farming is more likely to keep family farmers on the land than large corporate farming is similar to the overall pattern response shown in **Table 1**. Generally, the data in **Figures 2** and **3** and **Table 1** show that CAAs believe in organic/sustainable and community-based food production system. CAAs support for a sustainable food production system speaks to their potential to serve as a linchpin in their communities for promoting sustainable agricultural production systems and ensuring collective impact.

6.2 Predicting support for sustainable agricultural production systems

Table 2 shows the results of logit models using the stepwise procedure (backward deletion) in SPSS. The use of stepwise procedures when the object of the analysis is prediction and there is no formal theory to guide the selection of variables to enter the model. The overall goal of the procedure is to maximize R^2 while minimizing the number of predictors. In our case, we employed common sense logic. We used the arguments in the instrumentation section to explain the relevance of questions to the study objective and the size of the correlation with the dependent variable. Based on this, we selected the initial set of six variables shown in **Table 2**. In all models, all variables have the anticipated sign in the right direction, i.e., all the variables should have a positive effect on the likelihood of CAAs supporting farmer's market. The best model is Model 1, with three variables predicting CAAs support for local farmers' market. Those CAAs with programs supporting community based food production, those that encourage clients to participate in urban agriculture and those that believe that community-based farming is better than corporate farming for preserving the quality of the land are more likely to support the local farmers market. These three variables have coefficients of 3.380, 1.208 and 0.801 respectively. The model Chi-square 45.289 was significant at the 0.001 level. These variables produced R2 of 0.319, odds ratio of 10.806, 3.347 and 2.228 respectively. The size of the odds ratios indicates that there would be substantial improvement in support for farmers markets with a unit change in the independent variables.

For example, CAAs with programs supporting community-based agriculture are almost 11 times more likely to support farmers markets. We believe that support for farmers markets is the most meaningful measure of CAAs overall support for community-based agriculture, since this form of support translates into income for farmers and the community in general through the multiplier effect.

7. Discussion

The data show that there is moderate support for community based agriculture (see **Figure 3**). 53% of CAAs report that they support farmers markets and 48 and 36% report program support for community-based and urban agriculture respectively. These results are encouraging, given that community-based agriculture is not seen as a program priority for CAAs, considering the demand on their resources for other programs to address persistent poverty in the BBS [40]. The overwhelm-ingly strong positive opinion among CAAs concerning the use of organic methods, the role of citizens in supporting farmer's markets, the value of community-based farming in preserving the land, environment and family farms and their favorable view of cooperatives indicate that CAAs have the potential for providing strong institutional support for the development and promotion of sustainable agricul-tural production systems at the community level. In conducting 40 listening session with CBOs across 9 states in the Black Belt region, we discovered that advocacy is a core component of their programs. Thus, they possess the requisite experience and skills to advance sustainable agricultural production systems. CBOs represent a form of social capital and their networks foster coordination and cooperation for the common good and the promotion of sustainable behavior [9, 10, 24, 41]. Social capital is able to reduce transaction cost associated with collective action directed at solving complex problems. Increased social capital is linked with movement toward sustainable agriculture. Collective action facilitated by community organizations such as CBOs can make a difference in achieving goals because the farmer and community are more proactive in solving their own problems and are no longer dependent on the whims of government or other outside entities [20, 42]. In the context of developing and promoting sustainable agricultural systems, CBOs and their networks provide the institutional support that empowers communities to be more self-regulating and to act independently, collectively and proactively.

Promoting and developing organic/sustainable agriculture is unlike solving a technical problem, although the tendency is to treat it like a purely technical problem. A technical problem by definition is straightforward because the solution is known and protocols for implementing solutions are well defined and results are predictable and in many cases a single organization has the capacity to solve it, for example producing a crop of corn or building a bridge. On the other hand, developing and promoting organic/sustainable food production system is akin to solving an adaptive problem. An adaptive problem is complex. Its solution is not known or well understood and even when solutions are known, it requires several organizations working in unison to solve it. Developing a sustainable food production system is a collective impact initiative that seeks to find a solution to an adaptive problem. Such an initiative requires many stakeholders—a network of organizations—from different sectors learning and working together to systematically address the system of variables that will define a solution to the problem. In addition, all involved stakeholders must be committed to changing their own behavior in order to adapt to the change they seek to bring about [12]. CBOs, as we have discussed, are indispens-able members of this network.

Author details

Terrence Thomas[1*], Cihat Gunden[2] and Befikadu Legesse[3]

1 North Carolina A&T State University, Greensboro, NC, USA

2 Ege University, Izmir, Turkey

3 Smart-Eco Consulting, Silver Spring, MD, USA

*Address all correspondence to: twthomas@ncat.edu

References

[1] Thomas TW, Gunden C. Organic agriculture, consumers, community and the environment. In: Pilipavicius V, editor. Organic Agriculture. Rijeka, Croatia: Intech Publications; 2014. ISBN 980-953-307-1113-6

[2] Ikerd J. Small farms: The foundation for long-run security. In: Paper Presented at "A Time to ACT: Providing Educators with Resources to Address Small Farm Issues". Effingham and Peoria, IL: University of Illinois, Agroecology/Sustainable Agriculture Program; 2002. Available from: http://www.ssu.missouri.edu/faculty/jikerd/papers/I11Small.html

[3] World Commission on Environment and Development. From One Earth to One World: An Overview. Oxford: Oxford University Press; 1987

[4] Borron S. Building Resilience for an Unpredictable Future: How Organic Agriculture Can Help Farmers Adapt to Climate Change. Rome: Food and Agriculture Organization of the United Nations; 2006

[5] Diamond J. Collapse: How Societies Choose to Fail or Succeed. New York: Viking; 2005

[6] Allen P, Kovach M. Capitalist composition of organic: The potential of markets in fulfilling the promise of organic agriculture. Agriculture and Human Values. 2000;**17**:221-232

[7] Schein EH. Organizational Culture and Leadership. California: Jossey-Bass; 2004

[8] Griskevicius V, Cantu SM, van Vugt M. Evolutionary bases for sustainable behavior: Implications for marketing, policy and social entrepreneurship. Journal of Public Policy & Marketing. **31**(11):115-128

[9] Van Vugt M. Averting the tragedy of the commons: Using social psychological science to protect the environment. Current Directions in Psychological Science. 2009;**18**(3):169-173

[10] Gardner GT, Stern PC. Environmental Problems and Human Behavior. Pearson; 2002

[11] Willer H, Julia L, editors. The World of Organic Agriculture: Statistics and Emerging Trends. Frick/Bonn: Research Institute of Organic Agriculture (FiBL)/IFOAM-Organics International; 2018

[12] Kania J, Kramer M. Embracing emergence: How collective impact addresses complexity. Stanford Social Innovation Review. 2 January 2013:1-7

[13] Delind LB. Place, work, and civic agriculture: Fields for cultivation. Agriculture and Human Values. 2002;**19**:217-224

[14] Lapping MB. Big places, big plans. In: Furuseth O, editor. Perspectives on Rural Policy and Planning. Hampshire: Ashgate Publishing Limited; 2004

[15] Lyson TA, Guptill A. Commodity agriculture, civic agriculture and the future of U.S. farming. Rural Sociology. 2004;**69**:370-385

[16] Flora CB. Sustainability of agriculture and rural communities. In: Francis CA, Flora CB, King LD, editors. Sustainable Agriculture in Temperate Zones. NY: John Wiley & Sons; 1990

[17] Kamarck EC. Globalization and public administration reform. In: Nye JS Jr, Donahue JD, editors. Governance in a Globalizing World. Washington, DC: Brookings Institution Press; 2000. pp. 229-252

[18] Brown LD, Khagram S, Moore MH, Frumkin P. Globalization, NGOs, and

multisectoral relations. In: Nye JS Jr, Donahue JD, editors. Governance in a Globalizing World. Washington, DC: Brookings Institution Press; 2000. pp. 271-298

[19] Flora CB, Flora JL. Entrepreneurial social infrastructure: A necessary ingredient. The Annals of the American Academy of Political and Social Science. 1993;**529**:48-58

[20] Snyder DP. Five meta-trends changing the world. In: Jackson RM, editor. Annual Editions: Global Issues 06/07. Dubuque, IA: McGraw-Hill; 2**007**,

[21] Giddens A. Runaway World. New York: Routledge; 2003

[22] Harbaugh W, Ulrich M, Burghart D. Neural responses to taxation and voluntary giving reveal motives for charitable donations. Science. 2007;**162**:1243-1249

[23] Ostrum E. Governing the Commons: The Evolution of Institution for Collective Action. Cambridge, UK: Touchstone Books; 1990

[24] Lyson TA, Torres RJ, Welsh R. Scale of agricultural production, civic engagement, and community welfare. Social Forces. 2001;**80**:311-327

[25] Irwin M, Tolbert C, Lyson T. How to Build Strong Towns. American Demographics. 1997;**19**:42-47

[26] Bagnasco A, Sabel CF. Small and Medium Size Enterprises. London: Pinter; 1995

[27] Piore MJ, Sabel CF. The Second Industrial Divide. New York: Basic Books; 1984

[28] Cludts S. Organization theory and ethics of participation. Journal of Business Ethics. 1999;**21**:157-171

[29] Collier J, Esteban R. Governance in the participative organization: Freedom, creativity and ethics. Journal of Business Ethics. 1999;**21**:173-188

[30] Habernas J. Justification and Application: Remarks of Discourse Ethics, Translated by Ciaran Cromin. Cambridge: MIT Press; 1993

[31] Chaskin RJ, Garg S. The issue of governance in neighborhood-based inititiatives. Urban Affairs Review. 1997;**32**(5):631-651

[32] Rhodes RAW. The new governance: Governing without government. In: Political Studies Association. Blackwell Publishers. 1996;**44**(4):652-667

[33] Hooghe L, Marks G. Unraveling the central state, but how? Types of multi-level governance. The American Political Science Review. 2003;**97**:233-243

[34] Bowles S, Gintis H. Social Capital and Community Governance. Amherst, MA: University of Massachusetts; 2000

[35] Cone CA, Myhre A. Community supported agriculture a sustainable alternative to industrial agriculture. Human Organization. 2000;**59**(2):187-196

[36] Goodman D, Redclift M. Refashioning Nature: Food Ecology and Culture. London: Rouledge; 1991

[37] Ikerd J. The need for a systems approach to sustainable agriculture. Agriculture, Ecosystems & Environment. 1993;**46**:147-160

[38] Goldschmidt WR. As You Sow. Montclair, NJ: Allanheld, Osmun; 1978

[39] Babbie E. The Practice of Social Research. Belmont, CA: Thompson Wadsworth; 2006

[40] Carl Vinson Institute of Government. Dismantling Persistent Poverty in the South. Athens, GA: University of Georgia; 2002

[41] Putnam R. Making Democracy Work: Civic Traditions in Modern Italy. Princeton: Princeton University Press; 1993

[42] Flora CB. Social capital and sustainability: Agriculture and communities in the great and corn belt. Research in Rural Sociology and Development. 1995;**6**:227-246

5

Pedometric Tools Applied to Zoning Management of Areas in Brazilian Semiarid Region

Helena Saraiva Koenow Pinheiro,
Pedro Armentano Mudado Xavier,
Lúcia Helena Cunha dos Anjos, Cesar da Silva Chagas
and Waldir de Carvalho Júnior

Abstract

Brazilian semi-arid region is a recent frontier in the country for agribusiness. The objective of this study is to apply pedometric tools to zoning areas with distinct potential and limitations to agricultural purposes. The research was set in three main steps: (i) to compile a database with all complete profile data collection; (ii) to analyze the vertical variability of soil properties and select a set of soil key properties useful to define the land potential and limitations; and (iii) to classify the area according to potential for agriculture, considering a medium technological level of the farmers The quantitative methods applied are supported by geographic information systems (GIS) and spatial statistics. The soil data compilation was based on legacy data, with corresponding topographical data and information from remote sensing images of the area. Tree-based and geostatistical algorithms were applied to predict the spatial variability of the soil key properties. The definition of management zones was based on Iso Cluster and Maximum Likelihood Classification tools. The results pointed three different management zones according to risks of salinization and requirements for irrigation control. The approach showed to be a simple and useful way to select and recommend primarily potential areas for agriculture based on soil properties.

Keywords: AQP package, pedometrics, soil-depth functions, soil key properties, zoning management areas

1. Introduction

Soils are essential to human survival, and they provide a wide diversity of ecosystem services. Among them, soils are at the base of food production, carbon storage and cycles to atmosphere, water availability, turning biomass into valuable nutrients and degrading toxic elements, and supporting biodiversity [1]. Soil surveys are a relevant source of soil information and environmental data for many uses in agriculture and, otherwise, are fundamental for evaluating land capability and preventing degradation [2]. Ramalho Filho and Pereira [3] highlight the importance of assessing regional conditions and soil properties to recommend adequate land-use management.

However, in Brazil, even though this is well-known, the soil class information is not commonly taken into account in planning or land-use zoning. One of the problems is the small scale of available maps from systematic soil surveys (1:250,000–1:1,000,000) in the country; this small scale is not suitable for recommendations regarding soil management at the farm level [4]. This limits the implementation of land-use zoning and soil sustainable management practices and policies.

The northeast region of Brazil presents complex and heterogeneous natural features in relation to rainfall, soil, and vegetation. It represents a huge challenge for the use and management of soil and water in sustainable agricultural systems. The soil degradation observed in some areas of this region is due to irregular rainfall, soil fertility conditions, and population pressures in a typically fragile environment [5].

Despite of the environmental variability, the northeast region of Brazil has a unique biome with a particular vegetation type (named "caatinga"). The majority of soils present shallow superficial horizons and profiles and very low organic carbon contents [6–8]. The rainfall deficit together with an inappropriate soil management for agricultural use can lead to desertification processes and to increase soil salinity in many areas, further degrading the land. Also common is the presence of rocks on the soil mass and/or surface layers and soils with high clay activity, both implying strong limitations for tillage. Thus, knowing the distribution of soil classes and properties is a requirement for land-use zoning.

The manipulation and analysis of a large soil profile collection, in the existing datasets, can be difficult due to the wide variability of horizon depths and thickness, as well as the variability among and between soil classes and orders. Addressing this issue, studies about the variability of soil properties along with depth were developed by [9–15]. Beaudette et al. [16] introduced the algorithm for quantitative pedology (AQP) package, which gathers tools to produce standard profile sketches highlighting differences in soil properties between horizons. It allows to standardize profile sketches, according to Munsell color chart, and dataset harmonization by applying the slice-wise algorithm, among other useful tools to express variations inside a given soil profile collection.

Quantitative studies have been developed in soil science through numerical modeling relationships between environmental variables and soil, which are applied to a geographic dataset to create a preliminary or predictive map [17, 18]. All these techniques are known as pedometrics [17], which represents an interface between statics and soil science, supporting predictive analysis related with soil classes and attributes [19, 20]. Based in pedometric concepts, the digital soil mapping (DSM) allows to study the relationships between soil and forming factors represented by spectral bands from remote sensing data, surface numerical models, and thematic maps [17]. Remote sensing data deserves an important place in digital soil mapping, particularly in flat landscapes, where morphometric covariates have lesser influence in soil formation [21].

Landsat spectral bands, particularly in the short-wave infrared range (SWIR), are commonly used to represent the environmental covariates of parent material and/or soil. Different mineral assemblages will have different spectral reflectances, which may be separable by analyzing bands 1 5 and 7 [22, 23]. The normalized difference vegetation index (NDVI) values range between −1 and +1. High positive values usually indicate the occurrence of dense green vegetation, pointing to an appropriate supply of water and nutrients. Low values express limited photosynthetic activity, and negative ones correspond to sparse lacking ground coverage [24]. The soil enhancement ratios of Landsat spectral band ratios 3:2, 3:7, and 5:7

have been interpreted to accentuate carbonate radicals, ferrous iron, and hydroxyl radicals, respectively, in exposed soil and geologic materials [25].

The identification of homogeneous zones of management consists in grouping areas whose environmental characteristics, as relief and soil, are similar. The design of homogeneous zones must be done according to the soil management, considering as well the risk of degradation, and it contributes to the sustainable land-use planning. Sánchez and Silva [26] pointed out that management areas present similar response when applied similar agronomic practices and they are subject to the same risks and limitations of agricultural use.

The cluster analysis deals with segmentation of a set of *N* objects into clusters (groups) in a fashion that the same type of datasets falls in a cluster that is different from those with dissimilar datasets [27]. The results of cluster analysis reveal internal data structure and improve understanding of data. There are many cluster algorithms that are available for data partitioning into. This analysis was used to delineate management zone methods [28], and the authors prefer to apply fuzzy cluster algorithms, and they are commonly used to define management zones for precision agriculture. To develop management zones [29], the cluster analysis was conducted by Management Zone Analyst 1.0.1 (Agricultural Research Services, University of Missouri, Columbia), with a combination of variables (elevation-Electrical Conductivity-EC, ECah-%Na; elevation-ECah-%Na, elevation-pH 1:1-%Na).

The hypothesis of this study is that the use of soil legacy data allied with modern tools that are able to analyze large datasets can assist the definition of soil potential and limitations to agricultural practices and thus assist the land-use planning and zoning. The work was developed from a legacy dataset of soil surveys made by the *Companhia de Desenvolvimento do Vale do São Francisco* (CODEVASF) [30] in the Brazilian semiarid region.

The specific goals of this study are (i) to identify soil key properties for definition of different management zones based on area databased and literature; (ii) to harmonize the dataset according predefined properties; (iii) to produce soil key property maps through digital soil mapping; and (iv) to create a map zoning the area according the potential to agricultural uses. This approach was used to establish the main trends from the profile collection by means of soil-depth functions and to select soil key properties adequate for the area (flatland landscape and semiarid climate), in this way, separating the areas according to their agriculture potentials.

2. Materials and methods

2.1 Study area and dataset

The study was conducted in an area located between 9°53′0″ and 9°36′30″ S and 40°34′30″ and 40°23′30″W (with 34,437.82 hectares), in the municipality of Juazeiro in Bahia State, northeast region of Brazil (**Figure 1**).

The biome in the region studied was the caatinga [31], an ecoregion characterized by xeromorphic vegetation with broadleaf thorny shrubs and trees that shed their leaves seasonally, cactus plants, and sparse arid-adapted grasses [32]. The climate is characterized as semiarid with dry winter and rainy summer, and the coldest month mean temperature remains above 18°C (BSwh' in the Köppen classification). The average annual rainfall is 400 mm, with the rainy season extending from November to April (highest precipitation in March); and average annual temperature is around 26°C. The xerothermic indexes are between 200 and 150, comprising of 7 to 8 dry months.

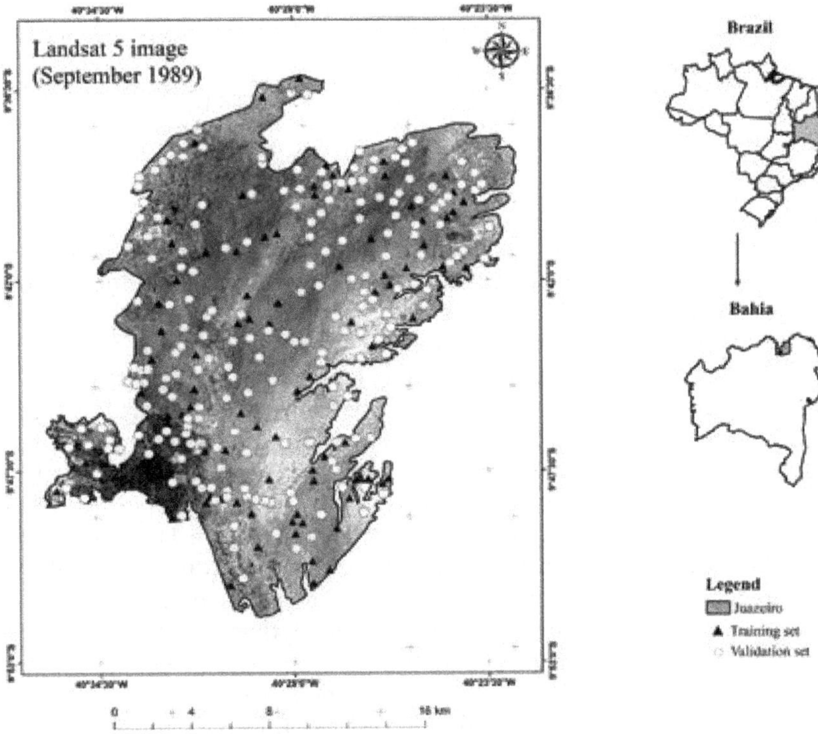

Figure 1.
Study area and location of soil profiles, over Landsat image (band 3).

At the study site, a portion of the original vegetation has been removed, and signs of land degradation such as bare soils are common. The landscape is mainly compounded by flat surfaces, with maximum slopes of 8%, and plateaus are common in the region. Geology comprises of limestone rocks from the caatinga formation, gneiss-granite rocks of the Caraiba-Paramirim complex, and recent colluvium/alluvium sediments [33].

The soils types in the northeast region of Brazil show large variation, according to parent material and relief, from shallow and with high content of basic cations (Ca, Mg, K, and Na) to deep and leached profiles [34]. Dominant soil classes according to the World Reference Base for Soil Resources [35] are Vertisols, Cambisols, and Planosols, with smaller extension of Regosols, Acrisols, and Luvisols.

2.2 Soil input data and covariates

The soil properties used in this study were sand, clay, and cation exchange capacity (CEC), determined according to [36]. The selection of soil properties was based on two main conditions: (i) presenting variability according to depth and ii) having importance to agricultural management (e.g., clay or sodium (Na) content).

From the original dataset, with 523 profiles, the ones with complete morphological description and analytical data were selected, performed by *Companhia de Desenvolvimento do Vale do São Francisco* (CODEVASF) in 1989 [29]. So, the input dataset comprised of the topsoil (0–20 cm) and subsoil (20–60 cm) layers for 290 soil profiles from a soil survey (legacy data). The sampling was performed according to the requirements of national soil survey service, correspondent to the detailed soil survey level [37, 38].

As covariates, Landsat 5 TM spectral bands were used in this study (**Table 1**), from September 1989, available in http://www.dgi.inpe.br/CDSR/. The NDVI values range between −1 and + 1. In the study area, most of the NDVI values were below 0.1, indicating little vegetation cover in the study area [39].

2.3 Modeling procedures

Quantitative soil properties, such as clay and sand content, can be estimated along the depth of soil profile through slice-wise algorithm [16] implemented in R software through algorithm for quantitative pedology (AQP) package [46]. The functions calculate an estimative of the main trend of values, represented by median and the variation interval by quartiles (25th and 75th of data distribution). The values shown in a column corresponds to percentage of soil profiles contributing with the function at each depth.

Initially, Pearson's linear correlation analysis was used to measure the linear association between variables, to determine among the soil properties defined by AQP, which correlated with environmental covariates (**Table 1**). This analysis was implemented in R [46], through the function cor.test, according to [42, 44, 47]. In Pearson's correlation the p-value defines whether or not two variables are statistically correlated, and for this study, it was defined that values lower than 0.02 indicate that the correlation is significant.

The modeling procedure to execute the random forest (RF) prediction was performed in R software, through randomForest (RF). RF is a nonparametric technique developed by Breiman [48] as an extension of classification and regression tree (CART) systems, to improve the performance of the predictors. To implement the RF models, three parameters are necessary: number of trees in the forest (ntree); minimum amount of data in each terminal node (nodesize); and number of covariates used in each tree (mtry) [49]. The ntree value was set to system default (500), although more stable results can be achieved with a larger number [50]. The

Covariates	Spatial resolution (m)	Spectral range (μm)*	References
Band 1	30	0.450–0.515	[23, 39–42]
Band 2	30	0.525–0.605	
Band 3	30	0.630–0.690	
Band 4	30	0.755–0.900	
Band 5	30	1.550–1.750	
Band 7	30	2.090–2.350	
NDVI (Band 4 - Band 3)/(Band 4 + Band 3)	30	—	[11, 23, 41–44]
Band 3/Band 2	30	—	
Band 3/Band 7	30	—	
Band 5/Band 7	30	—	
GSI (Band 3 - Band 1)/(Band 3 + Band 2 + Band 1)	30	—	[45]

Table 1.
Covariates used in spatial prediction of soil properties.

**Landsat 5 satellite*

nodesize value was set to 5 for each terminal node, as usually selected in regression studies. The mtry value chosen in this study was according to Liaw and Wiener [49], which proposes an amount corresponding to one third of the total number of predictor variables for regression problems.

Although Na showed a significant correlation at the two depths, according to the Pearson correlation analysis, the preliminary results using the random forest (RF) model were very unsatisfactory. Thus, exceptionally for this property, the RF model has been replaced by ordinary kriging (OK). Semivariograms were used to analyze the spatial structure of the Na, and to generate predictive maps, in both depths. The OK was performed in R software, through krige function [46]. OK model is the most familiar type of kriging and provides an accurate estimate for an area around a measure sample [51].

The model's performance was evaluated based on independent validation set, which was not used in the training procedure. Thereby, the 290 soil samples were randomly divided into 2 independent datasets in the R software; one of these was used in the training process (200 soil samples) and another for the validation process (90 soil samples). The analysis of the model's performance was based on the correlation between the measured values (validation samples) and estimated values, calculated by the coefficient of determination (R^2), the root mean square error (RMSE), and mean error (ME), presented as Eqs. (1) and (2):

$$RMSE = \sqrt{\frac{1}{n}\sum_{i=1}^{n} d_i^2} \tag{1}$$

$$ME = \frac{1}{n}\sum_{i=1}^{n} d_i \tag{2}$$

where "d" is the difference between the observed and estimated values and "n" is the number of samples used in the validation process.

The RMSE is a measure of the overall error of the estimation and commonly is used to estimate the error or uncertainty in places where the error was not measured directly; thereby, the higher the values of RMSE, the greater the differences between the datasets [52]. The ME gives the bias and allows evaluation of overestimation (positive values) or underestimation (negative values); values close to zero are preferable.

2.4 Definition of management zones

Management zones were defined in this study according to potential for agriculture, considering variability of soil key properties along profile depth, importance of soil properties for the land management, and the performance of the models to predict the spatial variation of the properties. Based on the maps for the selected soil key properties, an unsupervised classification was performed by using a series of input raster bands (Na, CEC, clay, and sand) using the Iso Cluster and Maximum Likelihood Classification tools from ArcGIS Desktop 10.3.

The Iso Cluster tool uses a modified iterative optimization clustering procedure, also known as the migrating means technique. The algorithm separates all cells into the user-specified number of distinct unimodal groups in the multidimensional space of the input bands; the iso prefix of the isodata clustering algorithm is an abbreviation for the iterative self-organizing way of performing clustering. In the clustering process, during each iteration, all samples are assigned to existing cluster centers, and new means are recalculated for every class. The optimal number of classes to specify is usually unknown. Therefore, it is advised to enter a conservatively high number, analyze the resulting clusters, and rerun the function with a reduced number of classes [53].

3. Results and discussion

3.1 Algorithm for quantitative pedology (AQP)

The AQP package allows to gather a set of functions to work and to analyze large soil profile collections. Depth functions of soil key properties used to distinguish the profiles are presented in **Figure 2** and **Figure 3**. The percentage values plotted along the profile shows the relative quantity of profiles used in the soil-depth function to calculate the statistics (median and quartiles) at each depth.

The study area presents shallow soils, sometimes with high base saturation; but the main limitation for agricultural use is the soil texture and high clay activity, which will influence the soil moisture to adequately manage the soils due to their high plasticity.

The pH and the calcium (Ca) content presented a linear trend of median values along depth (**Figure 2**), even though a smaller number of soil profiles (less than 25%) were used in the estimative for deeper than 150 cm depth. However, different patterns were shown for potassium (K), sodium (Na), electrical conductivity (EC), and cation exchange capacity (CEC). The soil key properties Ca, K, and CEC showed more variability around the median value, thus presenting a better predictive potential to distinguishing soils with different parent materials and high clay activity, respectively.

It is pertinent to select which soil key properties could indicate differences in soil behavior along depth and are relevant for land management. From the analyses of **Figure 3**, it is possible to conclude that clay and sand contents have an opposite and large variability among the profiles of the collection; thus, they have great potential to aid in the definition of soil management zones. Since both properties are

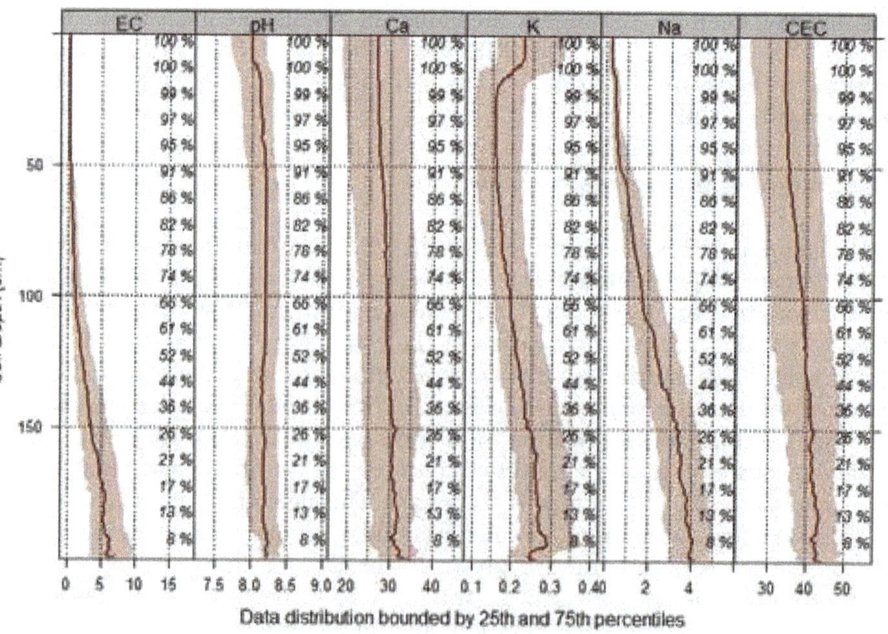

Figure 2.
Depth functions for soil key chemical properties. EC, electrical conductivity; CEC, cation exchange capacity.

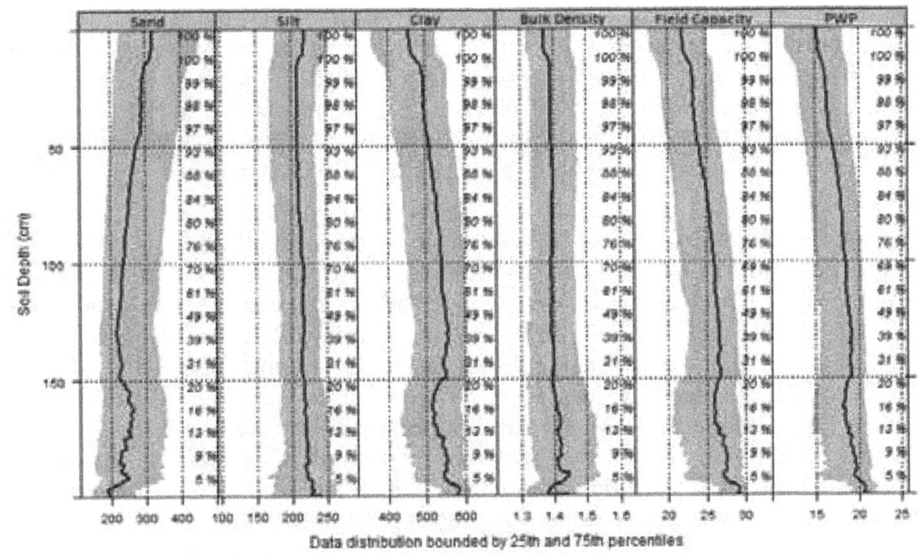

Figure 3.
Depth functions for soil key physical properties. PWP, permanent wilting point.

important to irrigation and mechanization practices, they were selected as criteria for agricultural zoning.

3.2 Predictive models

3.2.1 Pearson's correlation

The Pearson correlation coefficients (**Table 2**) showed that in general, the environmental covariates were significantly correlated with all soil properties analyzed ($p < 0.05$). The sand content was significantly correlated with covariates b5, b7,

Variables	Soil properties							
	Sand		Clay		CEC		Na	
Depth (cm)	0–20	20–60	0–20	20–60	0–20	20–60	0–20	20–60
b1	0.303	0.212	0.236	0.599	0.289	0.667	0.001	0.000
b2	0.198	0.321	0.431	0.177	0.848	0.723	0.001	0.000
b3	0.281	0.396	0.631	0.274	0.360	0.687	0.001	0.000
b4	0.037	0.040	0.009	0.072	0.002	0.016	0.019	0.000
b5	0.000	0.001	0.000	0.002	0.000	0.001	0.099	0.001
b7	0.000	0.002	0.000	0.003	0.000	0.000	0.152	0.001
NDVI	0.000	0.000	0.000	0.000	0.000	0.000	0.000	0.000
b3/b2	0.267	0.218	0.712	0.291	0.028	0.051	0.108	0.041
b3/b7	0.000	0.000	0.000	0.000	0.000	0.000	0.000	0.000
b5/b7	0.702	0.990	0.274	0.718	0.000	0.001	0.617	0.082
GSI	0.000	0.000	0.000	0.000	0.649	0.493	0.010	0.000

Table 2.
Pearson's correlation (p-value) between soil properties (sand, clay, CEC, and Na) and environmental covariates (b1, b2, b3, b4, b5, b7, NDVI, b3/2, b3/b7, b5/b7, and GSI).

NDVI, b3/b7 ratio, and GSI (**Table 2**). These results do not agree with [44], who found no correlation between sand content and the Landsat 5 TM image data. Only the covariate b3/b7 was strongly correlated (r = −0.58 and −0.49, topsoil and subsoil, respectively) with the sand content [54], whereas the NDVI was moderately correlated (r = 0.39 and 0.36, topsoil and subsoil, respectively) and the other covariates weakly correlated, with r values below 0.27 (positive or negative) (**Table 3**).

Inverse relationships were observed between clay content and the environmental covariates, but magnitudes were the same as those observed for sand (**Table 2**), except for the covariate b4 (topsoil) which had no correlation with sand; similar results were reported by [42, 55]. The most relevant covariates were b3/b7 (r = −0.56 and −0.51, topsoil and subsoil, respectively) and NDVI (r = −0.36 and −0.37, topsoil and subsoil, respectively). These results vary among authors in the literature. Significant correlations were obtained between clay, the NDVI index, and b3/b2 and b5/b7 band ratios by [44], while there was no correlation between clay and the b3/b7 ratio band. On the other hand, Ahmed and Iqbal [56] found significant correlations between clay and bands 4 and 6 using Landsat 5 TM.

The CEC was significantly correlated with the covariates b4, b5, b7, NDVI, and b3/b7 and b5/b7 ratios (**Table 2**). Only the covariate b3/b7 was strongly correlated (r = −0.45 and −0.44, topsoil and subsoil, respectively) with the CEC [54], whereas the other covariates were weakly correlated, with r values below 0.27 (positive or negative) (**Table 3**). A strong correlation was observed by [55] between the CEC and covariates measured by ASTER spectral bands (1–8). The Na content was significantly correlated with most of the covariates, except with b7 (topsoil) and b3/b2 and b5/b7 ratios (**Table 2**). However, none of the covariates had a strong or moderate correlation with Na content (**Table 3**), which may explain the very low performance of the random forest model in the prediction of this soil property.

The study performed by Demattê et al. [57] highlighted that the correlation with a particular spectral band is directly related with soil characteristics in specific regions, explaining the differences between study cases.

Variables	Soil properties							
	Sand		Clay		CEC		Na	
Depth (cm)	0–20	20–60	0–20	20–60	0–20	20–60	0–20	20–60
b1	0.06	0.07	−0.07	−0.03	−0.06	−0.03	0.19	0.32
b2	−0.08	−0.06	0.05	0.08	−0.01	0.02	0.20	0.34
b3	−0.06	−0.05	0.03	0.06	−0.05	−0.02	0.19	0.32
b4	0.12	0.12	−0.15	−0.11	−0.18	−0.14	0.14	0.27
b5	0.23	0.20	−0.25	−0.19	−0.24	−0.20	0.10	0.20
b7	0.21	0.18	−0.24	−0.18	−0.27	−0.23	0.09	0.19
NDVI	0.39	0.36	−0.36	−0.37	−0.22	−0.23	−0.21	−0.27
B3/b2	0.07	0.07	−0.02	−0.06	0.13	0.11	−0.10	−0.12
b3/b7	−0.58	−0.49	0.56	0.51	0.45	0.44	0.24	0.29
b5/b7	−0.02	0.00	0.06	0.02	0.21	0.20	−0.03	−0.10
GSI	−0.27	−0.26	0.21	0.24	0.03	0.04	0.15	0.23

Table 3.
Pearson's correlation coefficient (r) between soil properties (sand, clay, CEC, and Na) and environmental covariates (b1, b2, b3, b4, b5, b7, NDVI, b3/2, b3/b7, b5/b7, and GSI).

3.2.2 Predictive models

In the training process, the RF model only maintains the covariates that had moderate or strong correlation with the soil properties (**Table 3**). The results obtained by the predictive models (RF and OK) using an independent dataset for validation (90 samples) are illustrated in **Figure 4**. In this study, the results obtained by RF models in the topsoil layer were higher than those obtained by

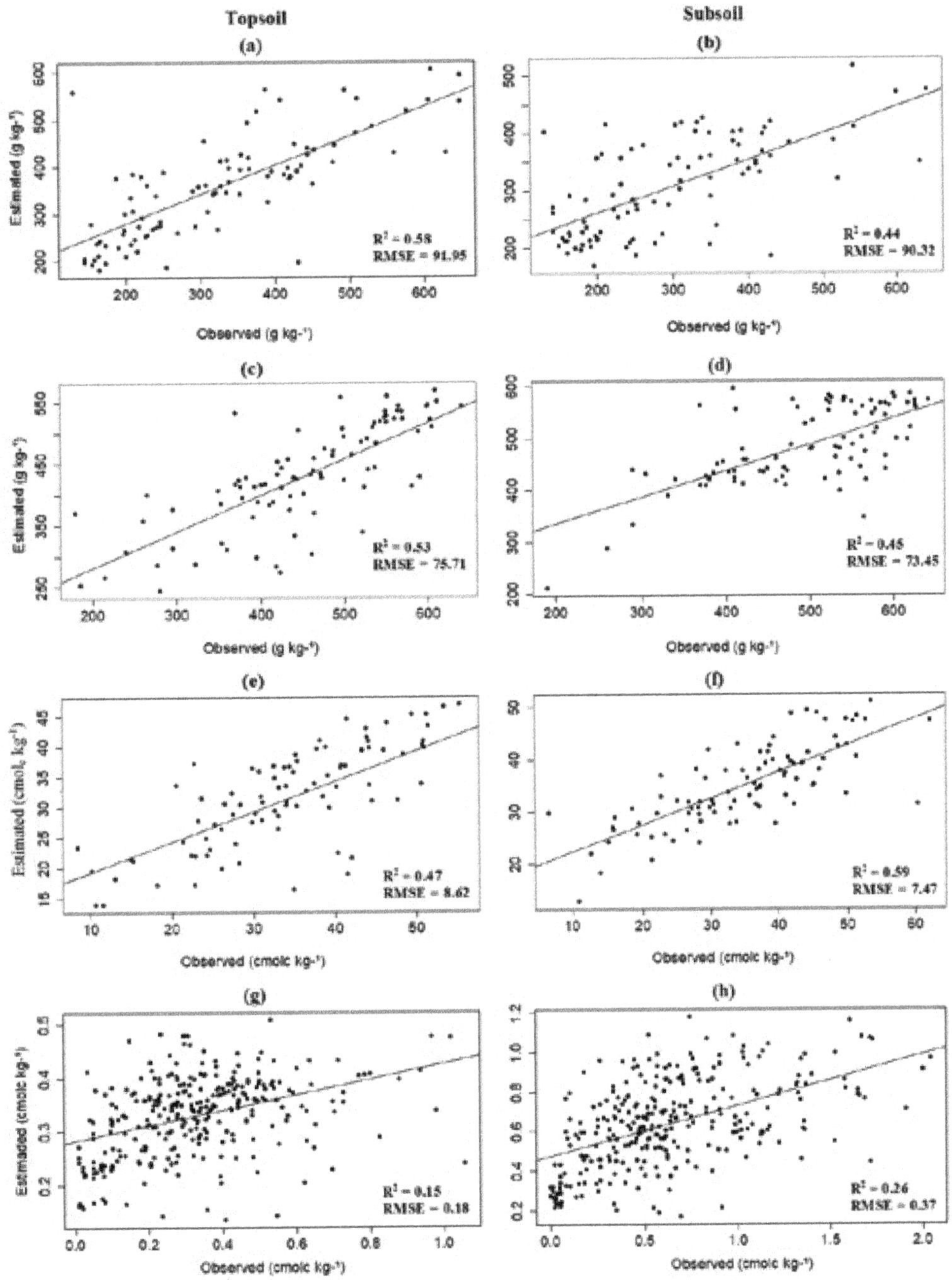

Figure 4.
Results obtained by prediction models using independent validation samples. (a) Topsoil sand—RF; (b) subsoil sand—RF; (c) topsoil clay—RF; (d) subsoil clay—RF; (e) topsoil CEC—RF; (f) subsoil CEC—RF; (g) topsoil Na—OK; (h) subsoil sand—OK.

subsoil for sand (R^2 = 0.58 and 0.44 and RMSE = 91.95 and 90.32 g kg^{-1}, topsoil and subsoil, respectively) and clay (R^2 = 0.53 and 0.45 and RMSE = 75.71 and 73.45 g kg^{-1}, topsoil and subsoil, respectively). The results were higher in subsoil than topsoil for CEC (R^2 = 0.44 and 0.59 and RMSE = 8.62 and 7.47 $cmol_c$ kg^{-1}, topsoil and subsoil, respectively) and Na content (R^2 = 0.15 and 0.26 and RMSE = 0.18 and 0.37 $cmol_c$ kg^{-1}, topsoil and subsoil, respectively) (**Figure 4**).

The results achieved by RF models were superior (topsoil) or similar (subsoil) to [58] that used terrain properties derived from a digital elevation model, for sand (variance explained = 30%) and clay (variance explained = 43%). In a study in Nigeria [59], percentages of variance explained (Var_{ex}) for RF models were of 48–49% for sand and 53–56% for clay in the top soil layer (0–15 cm). These values are inferior or similar to those obtained in this study for sand, similar for clay in the topsoil and superior in the subsoil. Therefore, the RMSE results for sand (19.26–19.67 g kg^{-1}) and clay (13.11–13.59 g kg^{-1}) were lower than those in the present study for the topsoil and subsoil layers (**Figure 4**). Lower Var_{ex} values were reported by [60] for sand (33–35%) and clay (31–35%).

The Var_{ex} obtained for CEC using the validation samples was 47% (R2 = 0.47) for topsoil layer and 59% (R2 = 0.59) for subsoil layer. The goodness of fit estimated by RMSE was 8.62 and 7.47 $cmol_c$ kg^{-1} (topsoil and subsoil, respectively). Regard-ing the model's performance, the results can be considered satisfactory; besides that, few studies in literature have used RF to predict soil CEC, and no one has used only remote sensing data as main covariates. The present results showed worst performance from statistical indexes when compared with those obtained by Lagacherie et al. [61], which reached 79% for Var_{ex} for the layer between 15 and 30 cm and 3.4 $cmol_c$ kg^{-1} for RMSE, using as input data terrain attributes and hyperspectral data in the visible and near infrared (AISA-Dual) with 5 m of spatial

Statistics	Dataset				Predictive model			
					Random forest			Ordinary kriging
	Sand	Clay	CEC	Na	Sand	Clay	CEC	Na
	g kg^{-1}		cmolc kg^{-1}		g kg^{-1}			cmolc kg^{-1}
	Topsoil							
Min	75	133	5.71	0.01	175	235	12.30	0.14
Max	802	640	61.41	3.59	626	594	54.33	0.57
Mean	334	449	33.29	0.37	338	449	34.27	0.34
SD	144	113	11.91	0.37	103	78	8.52	0.07
CV	43	25	36	100	30	17	25	206
	Subsoil							
Min	100	160	6.59	0.01	165	204	10.69	0.17
Max	672	673	62.15	7.99	576	623	51.28	1.26
Mean	308	484	33.99	0.71	308	483	35.24	0.65
SD	125	102	11.54	0.69	85	76	7.86	0.21
CV	41	21	34	97	28	16	22	32

Min, minimum; Max, maximum; SD, standard deviation; CV, coefficient of variation

Table 4.
Descriptive statistics from soil samples and predictive models for topsoil and subsoil.

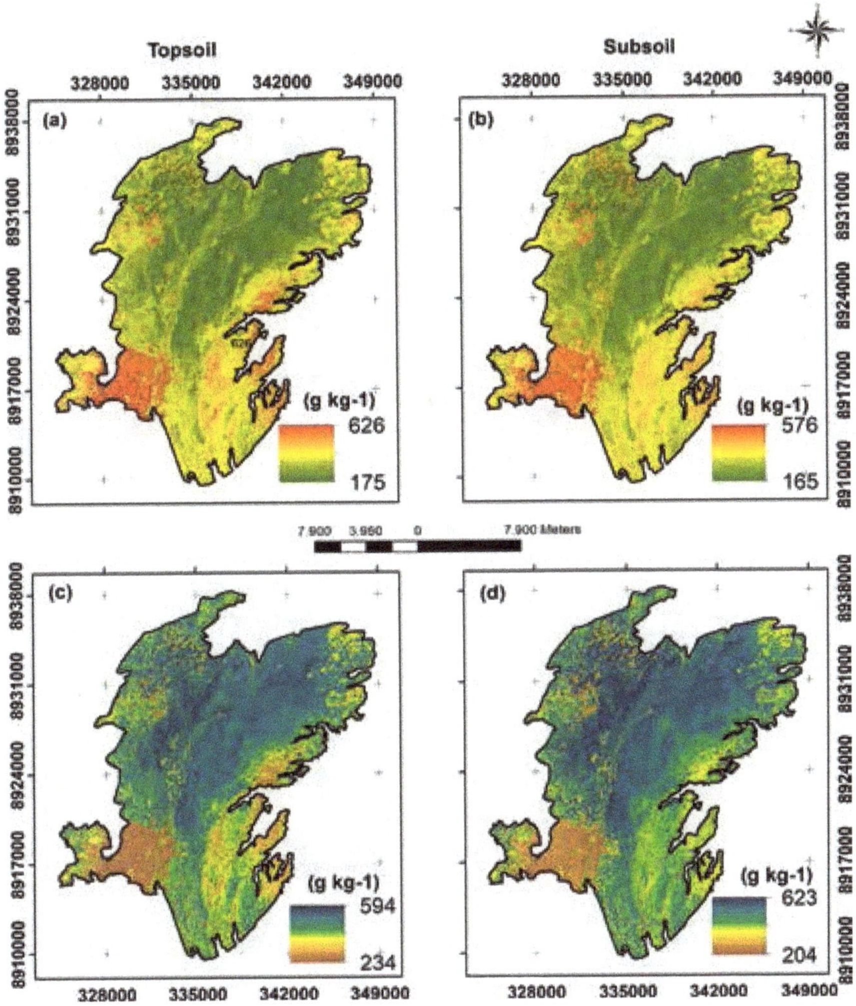

Figure 5.
Spatial distribution of soil physical properties estimated by the RK models. (a) Topsoil sand; (b) subsoil sand; (c) topsoil clay; (d) subsoil clay.

resolution. The difference from this study may be related to the coarser spatial resolution of images from Landsat 5 (30 m), in comparison with Lagacherie et al. [61] who used hyperspectral data (5 m). The influence of spatial resolution in the prediction of soil properties is reported by other studies [62, 63].

However, in this study the performance of metrics for CEC prediction was better when compared to the values (9.49 $cmol_c\ kg^{-1}$) achieved by [60]. The poor performance was explained by the authors as due to the small-scale variation of parental material and erosion/deposition rates, which were not captured by the spatial resolution of the covariates (100 m). The authors also highlighted importance of input dataset to improve the models' performance.

The semivariogram obtained for Na content (**Figure 4**) provides a description of the spatial dependence and indicates processes related with the spatial distribution [65]. For both depths, the best semivariogram was the exponential model. The semivariograms for topsoil and for subsoil (**Figure 4**) used to estimate the Na content present a R2 of 0.26 and RMSE of 0.37 $cmol_c\ kg^{-1}$ and R2 of 0.15 and RMSE of 0.18 $cmol_c\ kg^{-1}$, respectively. The descriptive statistics of soil properties prediction for the predictive models are presented in **Table 4**.

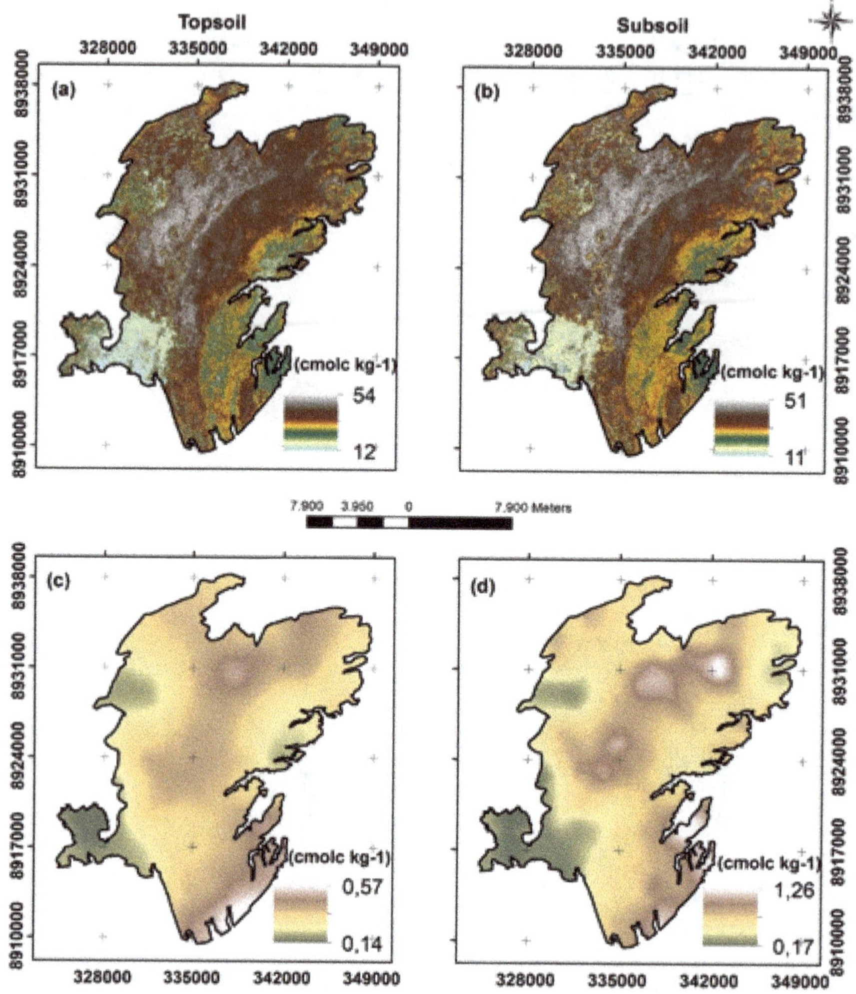

Figure 6.
Spatial distribution of soil chemical properties estimated by the RF and OK models. (a) Topsoil CEC; (b) subsoil CEC; (c) topsoil Na content; (d) subsoil Na content.

The RF model produced predicted values for sand, clay, and CEC within the range of the original values (**Table 4**), with a smaller standard deviation and coefficient of variation, as expected for this model [64]. The same was true for Na content in the OK model. However, the map produced presented a smaller range of values than the original dataset, which, according to [65], means that this model had low accuracy to describe the spatial variation of Na content, corroborating results found when predicting this property at both depths. The mean values for sand, clay, CEC, and Na content in both depths (**Table 4**) are closer to the values of the original dataset. The CV for both models was smaller than the CV from the original dataset, except for Na content in the topsoil.

These results could be explained by the moderate correlation of soil properties with the covariates and predominance of short-scale variations that could not be modeled from the set of profiles used. The results were considered satisfactory, except for Na content, and they can be ascribed to the physical interference of soil properties in the incident and reflected energy. However, the quantification of soil properties using an orbital sensor is not an easy task due to complexity of soil dynamics and formation [66]. The spatial distribution of soil properties according to the predictive models is shown in **Figures 5** and **6**.

3.3 Definition of the management zones

Based on the maps created for each soil key property, the zoning procedure was applied by using Iso Cluster and Maximum Likelihood Classification tools. The descriptive statistics and the map produced are presented on **Table 5** and **Figure 7**, respectively.

Management zones	MIN	MAX	Range	Mean	STD
Na (0–20 cm)					
1	0.14	0.57	0.43	0.35	0.05
2	0.14	0.57	0.43	0.33	0.08
3	0.14	0.57	0.43	0.29	0.09
Na (20–60 cm)					
1	0.17	1.21	1.05	0.72	0.16
2	0.17	1.25	1.09	0.62	0.19
3	0.17	1.26	1.10	0.52	0.29
CEC (0–20 cm)					
1	26.90	54.33	27.43	41.88	3.87
2	15.17	47.03	31.86	31.55	4.40
3	12.30	35.81	23.50	20.85	3.63
CEC (20–60 cm)					
1	24.32	51.28	26.96	41.70	4.29
2	14.39	48.22	33.83	33.18	4.46
3	10.69	37.09	26.41	23.15	4.48
Clay content (0–20 cm)					
1	401.4	593.7	192.3	520.1	25.9
2	321.2	536.7	215.5	427.2	27.7
3	234.2	501.3	267.1	313.0	38.3
Clay content (20–60 cm)					
1	457.3	623.0	165.7	552.4	26.0
2	325.8	574.9	249.0	457.2	30.9
3	204.2	488.3	284.1	362.4	55.3
Sand content (0–20 cm)					
1	175.0	370.9	195.9	242.7	31.6
2	221.0	538.1	317.1	369.1	37.2
3	230.7	626.1	395.5	512.3	51.8
Sand content (20–60 cm)					
1	164.5	321.4	156.9	228.8	24.3
2	201.5	518.3	316.9	335.7	37.9
3	209.9	576.3	366.4	447.0	47.5

MIN, minimum; MAX, maximum; STD, standard deviation

Table 5.
Descriptive statistics of the management zones separated by layers 0–20 and 20–60 cm.

The conditions to define the agricultural zones have to be adjusted according to available data and heterogeneity of soil properties in the study area, in this case located in a semiarid region.

Zone 1 presents the greater amount of Na in both layers, and it is associated with greater values of CEC as indicated by the mean values, than Zones 2 and 3. The same condition is verified for clay content, where Zone 1 has the greatest mean value in both layers (0 – 20 and 20 – 60 cm). As expected, sand content shows the

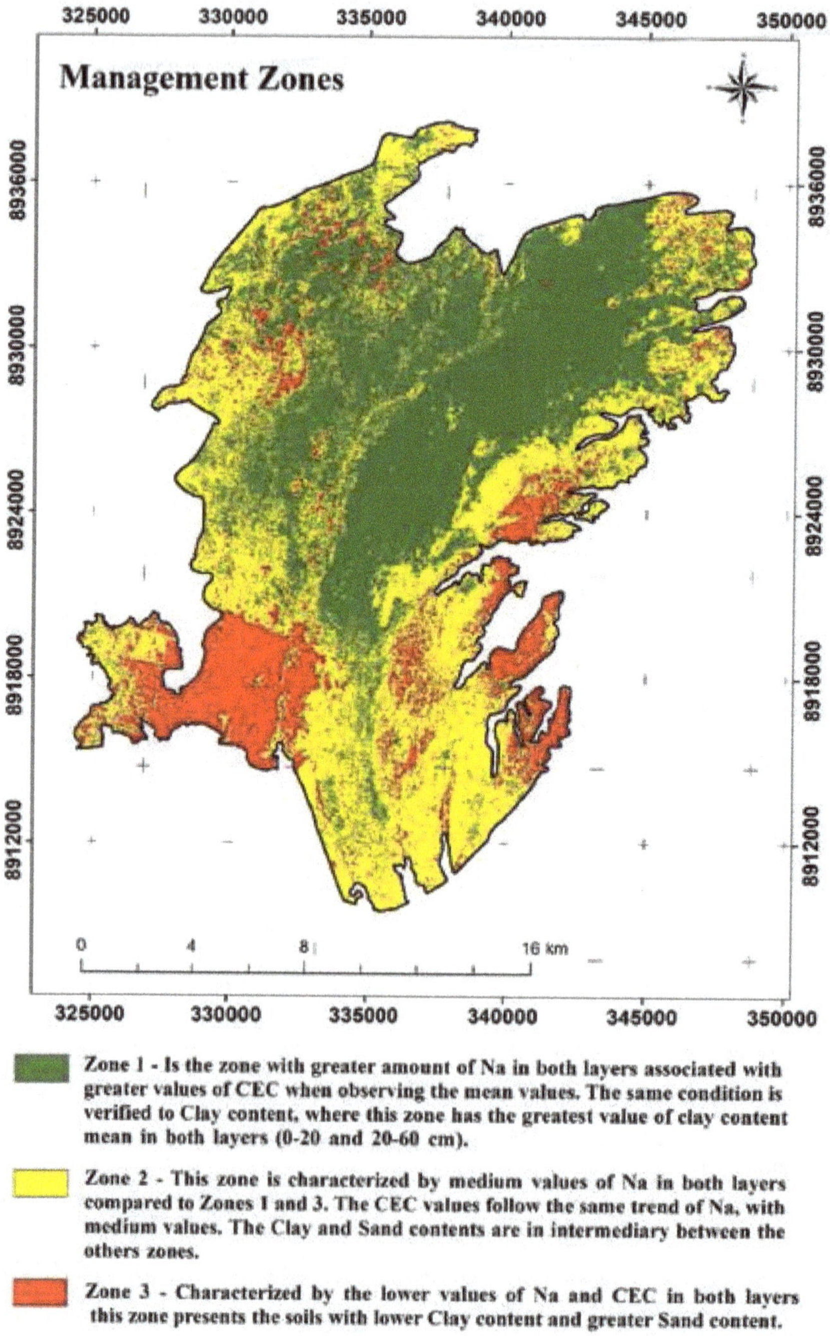

Figure 7.
Zoning of the study area according to selected soil key properties.

opposite pattern. Zone 1 occupies 42.6% of the area (1467 ha), and it has greater risk of salinization due to the higher mean values of Na. Even though Zone 1 has the greater mean values of CEC, it is associated to the high Na content and high content of clay, and possibly high clay activity (2:1). These properties limit the soil potential and irrigation management, further increasing salinization risks.

Zone 2 is characterized by medium values of Na in both layers, when compared to Zones 1 and 3. The CEC values follow the same trend of Na, with medium values, and clay and sand contents are intermediary. This zone occupies 41.7% of the area with 1436 ha. Zone 2 presents medium effort and limitation to soil management, considering risks of salinization and requirements for irrigation control.

Zone 3 is characterized by the lowest values of Na and CEC in both layers; it has the soils with lowest clay contents and proportionally greatest sand contents. This zone occupies 15.7% of the area with 539 ha. Zone 3 has the lowest mean values of CEC and Na, which decrease risks of salinization; plus, the lower clay content is less limiting to soil mechanization.

4. Conclusions

The approach based on the combination of orbital remote sensing data, with random forest (RF) and OK models to predict properties of soils in the Brazilian semiarid region, was considered satisfactory for sand, clay and CEC contents, and unsatisfactory for Na content.

The use of remote sensing data to map soil properties in flat relief areas, where common topographic covariates may not result in adequate maps by digital soil technics, is an advance. However, more research is needed to improve the quality of the covariates used as an input data.

The zoning procedures through multivariate analysis allowed to identify areas with higher salinization risks and where the management of irrigation is more complex. In the same way, the approach allowed to recognize zone of soils with lower sodium content and clay activity, which are preferred to grow annual crops.

Author details

Helena Saraiva Koenow Pinheiro[1*], Pedro Armentano Mudado Xavier[1], Lúcia Helena Cunha dos Anjos[1], Cesar da Silva Chagas[2] and Waldir de Carvalho Júnior[2]

1 Soil's Department, Agronomy Institute, Federal Rural University of Rio de Janeiro (UFRRJ), Seropédica, RJ, Brazil

2 Embrapa Solos, Rio de Janeiro, RJ, Brazil

*Address all correspondence to: lenask@gmail.com

References

[1] FAO, ITPS. Status of the World's Soil Resources (SWSR)—Technical Summary. Rome, Italy: Food and Agriculture Organization of the United Nations and Intergovernmental Technical Panel on Soils; 2015

[2] Beek KJ. Land Evaluation for Agricultural Development: Some Explorations of Land-Use Systems Analysis with Particular Reference to Latin America. Wageningen: International Institute for Land Reclamation and Improvement; 1978. 333p. (ILRI Publication, 23)

[3] Ramalho Filho A, Pereira LC. Aptidão agrícola das terras do Brasil: potencial de terras e análise dos principais métodos de avaliação. Rio de Janeiro: Embrapa Solos; 1999. 36p. (Documentos, 1). (in Portuguese)

[4] Streck EV, Kämpf N, Dalmolin RSD, Klamt E, Nascimento PC, Schneider P, et al. Solos do Rio Grande do Sul. 2.ed ed. rev. e ampl. Porto Alegre: Emater/RS; 2008. 222p. (in Portuguese)

[5] Araújo Filho JÁ. Manejo Sustentável da Caatinga. Recife: Projeto Dom Helder Camara; 2013. 200 p. (in Portuguese)

[6] Lepsch IF. Formação e Conservação dos Solos. São Paulo: Oficina de Textos; 2002. 178p. (in Portuguese)

[7] Boletim informativo do Núcleo Regional Nordeste da Sociedade Brasileira de Ciência do Solo. Cruz das Almas, BA: NRNE/SBCS; 2017:1(1) (in Portuguese)

[8] de Castro SS, Hernani LC. Solos frágeis: caracterização, manejo e sustentabilidade. Brasília, DF: Embrapa; 2015. 367 p. (in Portuguese)

[9] Bishop TFA, McBratney AB, Laslett GM. Modelling soil attribute depth functions with equal-area quadratic smoothing splines. Geoderma. 1999;**91**:27-45. DOI: 10.1016/S0016-7061(99)00003-8

[10] Mishra U, Lal R, Slater B, Calhoun F, Liu D, Meirvenne MV. Predicting soil organic carbon stock using profile depth distribution functions and ordinary kriging. Soil Science Society of America Journal. 2007;**73**:614-621

[11] Malone BP, Minasny B, McBratney AB. Mapping continuous soil depth functions in the Edgeroi District, NSW, Australia, using terrain attributes and other environmental factors. In: Purves R, Gruber S, Straumann R, editors. Proceedings of Geomorphometry. Zurich: University of Zurich. 2009. pp. 90-97

[12] Malone BP, McBratney AB, Minasny B. Empirical estimates of uncertainty for mapping continuous depth functions of soil attributes. Geoderma. 2011;**160**:614-626

[13] Lacoste M, Minasny B, McBratney AB, Micho D, Viaud V, Walter C. High-resolution 3D mapping of soil organic carbon in a heterogeneous agricultural landscape. Geoderma. 2014;**213**:296-311

[14] Wiese L, Ros I, Rozanov A, Boshoff A, de Clercq W, Seifert T. An approach to soil carbon accounting and mapping using vertical distribution functions for known soil types. Geoderma. 2015;**263**:264-273

[15] Liu F, Rossiter DG, Song XD, Zhang GL, Yang RM, Zhao YG, et al. A similarity-based method for three-dimensional prediction of soil organic matter concentration. Geoderma. 2016; **263**:254-263

[16] Beaudette DE, Roudier P, O'Geen AT. Algorithms for quantitative

pedology: A toolkit for soil scientists. Computers and Geosciences. 2013;**52**: 258-268

[17] McBratney AB, Odeh IOA, Bishop TFA, Dunbar MS, Shatar TM. An overview of pedometric techniques for use in soil survey. Geoderma. 2000;**97**: 293-327. DOI: 10.1016/S0016-7061(00) 00043-4

[18] Ten Caten A. Mapeamento digital de solos: metodologias para atender a demanda por informação espacial em solos [thesis]. Universidade Federal de Santa Maria; 2011. 108p. (in Portuguese)

[19] Silveira CT, Oka-Fiori C, Santos LJC, Sirtoli AV, Silva CR. Pedometria apoiada em atributos topográficos com operações de tabulação cruzada por álgebra de mapas. Revista Brasileira de Geomorfologia. 2012;**13**(2):125-137. (in Portuguese)

[20] Coelho FF, Giasson E. Métodos para mapeamento digital de solos com utilização de sistema de informação geográfica. Revista Ciência Rural, Santa Maria. 2010;**40**(10):2099-2106. (in Portuguese)

[21] Chagas CS, Carvalho Júnior W, Pinheiro HSK, Xavier PAM, Bhering SB, Pereira NR, et al. Mapping soil cation exchange capacity in a semiarid region through predictive models and covariates from remote sensing data. Revista Brasileira de Ciência do Solo. 2018;**42**:1-12, e0170183. DOI: 10.1590/18069657rbcs20170183

[22] Sirtoli AE. Atributos do terreno e índices espectrais integrados por redes neurais artificiais [thesis]. Curitiba, PR: UFPR; 2008. 114p. (in Portuguese)

[23] Boettinger JL, Ramsey RD, Bodily JM, Cole NJ, Kienast-Brown S, Nield SJ, et al. Landsat spectral data for digital soil mapping. In: Hartemink AE, McBratney AB, Mendonça-Santos ML, editors. Digital Soil Mapping with Limited Data. New York: Springer-Verlag; 2008. pp. 192-202

[24] Sumfleth K, Duttmann R. Prediction of soil property distribution in paddy soil landscapes using terrain data and satellite information as indicators. Ecological Indicators. 2008;**8**:485-501

[25] Amen A, Blaszczynski J. Integrated Landscape Analysis. Denver, CO: U.S. Department of the Interior, Bureau of Land Management, National Science and Technology Center; 2001. pp. 2-20

[26] Sánchez OR, SILVA TC. Zoneamento ambiental: uma estratégia de ordenamento da paisagem. Cad. Geociências, Rio de Janeiro. 1995;**14**: 47-53. (in Portuguese)

[27] Verma RR, Manjunath BL, Singh NP, Kumar A, Asolkar T, Chavan V, et al. Soil mapping and delineation of management zones in the Western Ghats of coastal India. Land Degradation and Development. 2018; **29**(12):4313-4322

[28] Reyes J, Wendroth O, Matocha C, Zhu JF. Delineating site-specific management zones and evaluating soil water temporal dynamics in a Farmer's field in Kentucky. Vadose Zone Journal. 2019;**18**(1):1-19

[29] He YB, DeSutter T, Norland J, Chatterjee A, Casey F, Clay D. The measurement, prediction, and development of soil management zones in low-relief sodic soils. Precision Agriculture. 2018;**19**(5):858-875

[30] Companhia de Desenvolvimento dos Vales do São Francisco - Codevasf. Projeto Salitre: levantamento detalhado de solo e classificação de terras para irrigação. Codevasf; 1989. (Projetos técnicos, 15). (in Portuguese)

[31] Veloso HP, Filho ALRR, Lima JCA. Classificação da vegetação brasileira,

adaptada a um sistema universal. Rio de Janeiro: IBGE; 1991. (in Portuguese)

[32] Xue Y, Sellers PJ, Kinter JL, Shukla J. A simplified biosphere model for global climate studies. Journal of Climate. 1991;**4**:345-364

[33] Souza DJ, Kosin M, Melo RC, Santos RA, Teixeira LR, Sampaio AR, et al. Mapa geológico do Estado da Bahia —escala 1:1.000.000. Salvador: CPRM; 2003. Versão 1.1. Programas Carta Geológica do Brasil a milionésimo e Levantamentos geológicos básicos do Brasil (in Portuguese)

[34] Vieira RMSP, Cunha APMA, Alvalá RCS, Carvalho VC, Ferraz Neto S, Sestini MF. Land use and land cover map of a semiarid region of Brazil for meteorological and climatic models. Revista Brasileira de Meteorologia. 2013; **28**:129-138

[35] IUSS Working Group WRB. World Reference Base for Soil Resources 2014. World Soil Resources Reports 106. Rome: Food and Agriculture Organization of the United Nations; 2014

[36] EMBRAPA. Centro Nacional de Pesquisa de Solos. Manual de métodos de análise de solo/Centro Nacional de Pesquisa de Solos. 2nd ed. rev. atual.— Rio de Janeiro; 1997. 212 p.: il. (EMBRAPA-CNPS. Documentos; 1) (in Portuguese)

[37] EMBRAPA. Centro Nacional de Pesquisa de Solos. Procedimentos normativos de levantamentos pedológicos. Brasília: EMBRAPA-SPI; 1995. 101p. (in Portuguese)

[38] EMBRAPA. Serviço Nacional de Levantamento e Conservação de Solos. Critérios para distinção de classes de solos e de fases de unidades de mapeamento: normas em uso pelo SNLCS. Rio de Janeiro: EMBRAPA-SNLCS; 1988. (in Portuguese)

[39] Liao K, Xu S, Wu J, Zhu Q. Spatial estimation of surface soil texture using remote sensing data. Soil Science and Plant Nutrition. 2013;**59**:488-500

[40] Bishop TFA, McBratney AB. A comparison of prediction methods for the creation of field-extent soil property maps. Geoderma. 2001;**103**:149-160

[41] Wu C, Wu J, Luo Y, Zhang L, De Gloria SD. Spatial prediction of soil organic matter content using cokriging with remotely sensed data. Soil Science Society of America Journal. 2009;**73**: 1202-1208

[42] Chagas CS, Carvalho Junior W, Bhering SB, Calderano Filho B. Spatial prediction of soil surface texture in a semiarid region using random forest and multiple linear regressions. Catena. 2016;**139**:232-240

[43] Saunders AM, Boettinger JL. Incorporating classification trees into a pedogenic understanding raster classification methodology, Green River Basin, Wyoming, USA. In: Lagacherie P, McBratney AB, Voltz M, editors. Digital Soil Mapping: An Introductory Perspective. Developments in Soil Science. Vol. 31. Amsterdam: Elsevier; 2007. pp. 389-399

[44] Carvalho Junior W, Lagacherie P, Chagas CS, Calderano Filho B, Bhering SB. A regional-scale assessment of digital mapping of soil attributes in a tropical hillslope environment. Geoderma. 2014;**232**:479-486

[45] Xiao J, Shen Y, Tateishi R, Bayaer W. Development of topsoil grain size index for monitoring desertification in arid land using remote sensing. International Journal of Remote Sensing. 2006;**12**:2411-2422

[46] R Development Core Team, R A Language and Environment for Statistical Computing. R Foundation for Statistical Computing, Vienna, Austria.

2013. Available from: http://www.R-project.org/isbn 3-900051-07-0 [Accessed: 08 May 2015]

[47] Ciampalini R, Lagacherie P, Hamrouni H. Documenting Globa lSoilMap.net grid cells from legacy measured soil profile and global available covariates in Northern Tunisia. In: Minasny B, Malone BP, McBratney AB, editors. Digital Soil Assessments and Beyond. London: CRC Press/Balkema; 2012. pp. 439-444

[48] Breiman L. Random Forests. Technical Report for Version 3. 2001. Available from: http//oz.berkeley.edu/users/breiman/randomforest2001.pdf [Accessed: 28 December 2014]

[49] Liaw A, Wiener M. Classification and regression by random forest. R News. 2002;**2**(3):18-22

[50] Grimm R, Behrens T, Märker M, Elsenbeer H. Soil organic carbon concentrations and stocks on Barro Colorado Island—Digital soil mapping using Random forests analysis. Geoderma. 2008;**146**:102-113

[51] Pang S, Li TX, Zhang XF, Wang YD, Yu HY. Spatial variability of cropland leadand its influencing factors: A case study in Shuangliu county, Sichuan province,China. Geoderma. 2011;**162**: 223-230

[52] Holmes KW, Chadwick OA, Kyriakidis PC. Error in a USGS 30-meter digital elevation model and its impact on terrain modeling. Journal of Hydrology. 2000;**233**:154-173

[53] ArcGIS 10 Help. Available from: http://desktop.arcgis.com/en/arcmap/10.3/tools/spatial-analyst-toolbox/iso-cluster-unsupervised-classification.htm [Accessed: 29 June 2019]

[54] Cohen J. Statistical Power Analysis for the Behavioral Sciences. Hillsdale, NJ: Erlbaum; 1988

[55] Souza Junior JG, Demattê JA, Araújo SR. Modelos espectrais terrestres e orbitais na determinação de teores de atributos dos solos: potencial e custos. Bragantia. 2011;**70**:610-621. (in Portuguese)

[56] Ahmed Z, Iqbal J. Evaluation of Landsat TM5 multispectral data for automated mapping of surface soil texture and organic matter in GIS. European Journal of Remote Sensing. 2014;**47**:557-573

[57] Demattê JAM, Fiorio PR, Ben-Dor E. Estimation of soil properties by orbital and laboratory reflectance means and its relation with soil classification. Open Remote Sensing Journal. 2009;**2**:12-23

[58] Ließ M, Glaser B, Huwe B. Uncertainty in the spatial prediction of soil texture: Comparison of regression tree and random forest models. Geoderma. 2012;**170**:70-79

[59] Akpa SIC, Odeh IOA, Bishop TFA, Hartemink AE. Digital mapping of soil particle-size fractions for Nigeria. Soil Science Society of America Journal. 2014;**78**:1953-1966

[60] Vaysse K, Lagacherie P. Evaluating digital soil mapping approaches for mapping GlobalSoilMap soil properties from legacy data in Languedoc-Roussillon (France). Geoderma Regional. 2015;**4**:20-30

[61] Lagacherie P, Sneep AR, Gomez C, Bacha S, Coulouma G, Hamrouni MH, et al. Combining VisNIR hyperspectral imagery and legacy measured soil profiles to map subsurface soil properties in a Mediterranean area (cap-bon, Tunisia). Geoderma. 2013;**209-210**: 168-176

[62] Smith MP, Zhu AX, Burt JE, Stiles C. The effects of DEM resolution and neighborhood size on digital soil survey. Geoderma. 2006;**137**:58-69

[63] Ruiz-Navarro A, Barberá GG, García-Haro J, Albaladejo J. Effect of the spatial resolution on landscape control of soil fertility in a semiarid area. Journal of Soils and Sediments. 2012;**12**:471-485

[64] Hengl T, Heuvelink GBM, Kempen B, Leenaars JGB, Walsh MG, Shepherd KD. Mapping soil properties of Africa at 250 m resolution: Random forests significantly improve current predictions. PLoS One. 2015;**10**(6):1-26, e0125814. DOI: 10.1371/journal.pone.0125814

[65] Liao KH, Xu SH, Wu JC, Ji SH, Qing LIN. Cokriging of soil cation exchange capacity using the first principal component derived from soil physico-chemical properties. Agricultural Sciences in China. 2011;**10**: 1246-1253

[66] Demattê JAM, Galdos MV, Guimarães RV, Genú AM, Nanni MR, Zullo J Jr. Quantification of tropical soil attributes from ETM+/LANDSAT-7 data. International Journal of Remote Sensing. 2007;**28**:3813-3829

6

Organic Farming: A Good Production Decision for Slovenian Small Size Farms and Farms in the Areas with Restrictions/Limitations or Natural Obstacles for Agriculture?

Anton Perpar and Andrej Udovč

Abstract

The negative effects of intensive agriculture on the environment, human health, natural resources, etc. require the use of more sustainable farming methods. The decision for organic farming is based on various factors: the personal levels of development and consciousness of the operator of the farm and members of his family, the socio-economic and size structure of the farm, the type of area in which the farm is located, the availability of the market and the demand for organic products, etc. Organic farming in Slovenia appears to be particularly suitable for smaller farms, since it is more labor intensive, and payments from agricultural policy contribute to the economic efficiency of the farm, despite lower yields. Especially in protected areas, water protected areas, or less favored areas for agriculture, organic farming can be a more convenient farming method. The traditional farming practices that exist are already extensive and the available measures can contribute to farm income and compensate farmers for different services that they provide for the society. The multifunctionality of organic farming (and agriculture in general), i.e., ensuring also the environmental and social roles, is only possible if farming is economical at the same time.

Keywords: organic farming, sustainability, conversion, organic products, ANCs, Slovenia

1. Introduction

The world today faces many challenges globally. It is characterized by extremely rapid technological progress, changes in political and economic relations, increasing income disparities, increasing global climate change, and, consequently, environmental burdens and changes in the natural environment. All this has implications for agriculture and the countryside, the society's attitude to them, and views on food production, the methods and technologies used, and the safety and quality of produced food. Adequate habitat for humans, plants, and animals, the quality of natural resources (soil, water, air, and ecosystems), and safe food are the basic conditions for living on the planet. Global trends are causing a decline in the quality and availability of natural resources that are being consumed by today's civilization. Trends and projections for population growth (10 billion people are expected to live on Earth by 2050 [1]) will also require increasing food needs. The current attitudes to agricultural production resources, social and climate changes, and increasing pollution of basic natural resources (soil, water, and air) require the strategic and economical management of these resources in order to enable the present and future generations to survive.

Agriculture faces many challenges as well as requirements related to natural resources, production technologies and methods, food safety, and quality. The society also has high expectations for rural areas. Agriculture is expected to be productive, competitive and economically attractive, resilient, and environmentally sustainable. The countryside, which is largely characterized by agriculture, is expected to be attractive to nonagricultural populations and to various economic activities. Food and the environment are becoming increasingly important areas of interest in modern society. Most countries are also increasing their emphasis, including in the light of their current experience, to ensure an adequate level of self-sufficiency in food products and to ensure food security. It should increasingly be based on locally produced food, with known and controlled origins, short retail chains, which also leaves a smaller environmental footprint while providing employment and adequate income for local growers.

Modern consumers are giving more attention to nutrition and changing eating habits. It is important that the food is healthy and safe, the supply is undisturbed, and the way in which the food is produced or processed is supervised, so the ethical aspects of food production are also becoming important. Developed countries are also becoming more aware of the problem of large quantities of discarded food, which requires a different attitude toward food, reducing quantities of discarded food, better management of food surpluses, and greater awareness among consumers, traders, and producers.

Agriculture can also have negative impacts on the environment and climate, such as greenhouse gas emissions, negative impacts on water quality and water resources, and on natural ecosystems and biodiversity.

Organic agriculture, on a global, European, and Slovenian scale, is gaining in importance. The number of organic farms, the share of organically cultivated agricultural land, the amount of organically produced food, and thus the market share of such food are increasing. In Slovenia, organic farming is one of the possibilities of producing safe and quality food, while fulfilling many other roles that agriculture plays beside the production of food.

Organic farming largely meets the expectations and needs of the society in terms of protecting the environment in food production, the production of quality, healthy, and safe foods with high nutritional value, and the sustainable management of nonrenewable natural resources and the livestock of adequate breeding. Usually, organic farms have mixed production that combines crop production and animal husbandry. This allows for efficient resource management, nutrient cycling, biodiversity, environmental protection, and animal-friendly breeding. In both crop and livestock production, prevention of problems rather than their treatment is emphasized. The use of chemically synthesized fertilizers and pesticides, growth regulators, and hormones and the use of genetically modified organisms are not permitted in organic farming. Organic farming is based on rational use of natural resources, crop rotation, production of intermediate crops allowing the binding of nitrogen from the air, animal fertilizers, green fertilization and compost, biotic control of pests, and varieties more resistant to diseases and pests.

Organic agriculture has a clear philosophical approach to farming, and the holistic approach focuses on working together with nature instead of against it [2].

Organic farming contributes significantly to the provision of public goods, in particular to maintaining a sharp improvement in biodiversity, preserving drinking water resources, creating jobs due to the increased need for labor and increased unit labor value, preserving the agricultural cultural landscape, and protecting the environment in general.

Organic and agro-ecological farming methods are based on four key principles: health, ecology, fairness, and care, enunciated by the International Federation of Organic Agriculture Movements (IFOAM) [3] and they guarantee healthy food production, assure environmental protection, and emphasize local resources and food systems [4]. Following these principles (included in various sets of standards, legislations, and production guidelines in different countries or for example on EU level) in practice enhances soil fertility and biodiversity, minimizing land degradation and erosion, chemical pollution, and other negative effects of industrialized agricultural activities.

2. Various aspects of organic farming

Agriculture to be sustainable should be environmentally and socially sensitive, but also economically viable. We can talk about economic, environmental, and socio-cultural aspects of organic agriculture.

2.1 Economic aspects

From an economic point of view, organic farming, as a more extension production method, has mostly higher production costs and lower yield that have to be covered with higher prices for consumers, price premiums, or subsidies [5, 6]. Organic farmers also sell their products through short distribution chains, mostly directly on farm, or on local street markets where they can achieve higher prices instead of supermarkets. They are able to achieve higher margins from the added-value if they processed their organic products at home or in cooperatives. Short food supply not only reduces negative environmental effects but, because of better efficiency, also brings down final prices for consumers and highlights the value of farmers' work.

2.2 Environmental aspects

Environmental aspects are one of the most important characteristics and benefits of organic farming with very positive effects on soil, water, biodiversity, and climate change [5, 8, 9]. The main objectives of organic farming in relation to soil are the maintenance and enhancement of soil life and natural soil fertility, soil stability, and soil biodiversity, prevention of soil compaction and erosion, and the nourishment of plants with natural nutrient circuit. High organic matter in organic soil can be maintained and improved through crop rotation, with crops that fix nitrogen from the air, with green cover, or with manure from livestock production. Strict rules do not allow using external nonorganic inputs (synthetic fertilizers) [5, 7–9].

Another problem of intensive conventional agriculture is water pollution. Organic farming, with strong restrictions in the use of chemically synthesized pesticides and mineral nitrogen fertilizers and lower animal stocking rates on farm, causes much less water pollution and helps to reduce leaching rates With timing tillage properly, farmers can achieve further benefits in order to reduce nutrient leaking.

Organic farming contributes to high degree of biodiversity in terms of domesticated species and floral and faunal diversity on the surface and in the soil. Traditional varieties and breeds, adopted on local conditions, and crop rotation make the whole system even more resilient, especially regarding pest and diseases. The maintenance of natural elements in the landscape, such as hedgerows, strips, and field margins can help to establish the ecological balance and make conditions for the predators of crop pests.

Farming, organic and conventional, has some positive and negative climate effects. The positive effect is that agriculture has a positive impact on carbon dioxide (CO_2), acting as carbon sink. Plants use it for photosynthesis and store it in the soil's organic matter. Due to the high organic matter content in the soil, organic farming as also conventional can contribute to CO_2 reduction. Methane emissions from livestock production have negative effects. Due to the low stocking density, organic farming produces less methane on a hectare unit as conventional farming. There are still potentials to reduce methane emissions with improving manure management, ruminants' diet, or increased productivity that can improve methane emissions on yield basis and reduce production costs [5, 7–9].

2.3 Social aspects

Organic farming has positive effects also on human and animal health and animal welfare. Currently, the questions of food choice has become also an issue from the perspective of public health and motivation of the policy makers in developed economies to improve dietary patterns of the population [10]. Therefore, healthiness of the products in comparison to conventional food options is among the main reasons for organic food purchase. Food safety and well-balanced diet are important prerequisites for good health and well-being [5]. General rules on animal welfare are the same for organic and conventional farming. However, organic farming rules demand still higher level of animal welfare and an assurance of species-specific needs. On organic animal husbandry, animals must have free access to open air that can also interact among them. The number of animals must be adjusted to the avail-able land and sticking density inside buildings and outside to ensure comfort and well-being of the different species. The use of antibiotics and hormones is strictly prohibited and good animal health can be achieved by the selection of appropriate breeds and building of immunological defenses of animals.

Organic farming has the potential to assure also some social aspects, as a kind of positive externalities of its operating, like job opportunities and other contribution to local and regional economies.

One of the characteristics of organic farming is also higher demand for labor in comparison with conventional agriculture [11], due to the need for more manual and mechanical work instead of chemical inputs. Some additional time is needed also to prepare products for market sale or for their distribution to consumers directly or for selling personally on local markets, for their processing at home, etc. On the other hand, this means a contribution to employment in rural areas and makes possible also to keep and operate small farms, which otherwise would not be competitive enough. Other social dimensions of organic farming, beside the influence on labor demand, are also connected with human health, democratic participation, resiliency, biological and cultural diversity, quality of life and human well-being, equity and ethics, and institutions (e.g., farmers' cooperatives and associations) [12]. Benefits for farmers and other local people result from diversification of activities and income on farms (e.g., processing of products and tourism on farms) and through payments for ecosystem services that can have income and quality of life influence. Local residents gain human health benefits through access to fresh, traditional foods and access to landscape for leisure activities and have opportunities for direct relationships with producers. They can not only directly buy products but also help farmers with work as a kind of social gathering. Other wider social benefits can be in the field of democracy, gender perspective, resiliency, quality of rural life, and cultural conservation. Through the active engagement of producers, local residents, and visitors, democratic participa-tion has improved. There are much more opportunities and need for cooperation between different stakeholders to build social capital, to preserve agricultural culture, knowledge, and traditions, to preserve some traditional production and processing techniques, etc.

3. Factors influencing the conversion to organic farming

Several studies in different countries tried to find out the factors that determine or have influence on farmer's decision to convert to organic farming [13–20]. In general, the determinants can be divided into economic and noneconomic factors [17], and most studies that analyzed the adoption or conversion to organic farming confirmed the relevance of both types of factors.

For the decision about farming system, farmers' objectives are important. Objectives can be summarized as economic, environmental, or sociocultural [17].

Factors that influence the decision to convert from conventional to organic farming in the reviewed studies are as follows:

- Farmers' characteristics (age, gender, education, experiences, entrepreneurial spirit, etc.)

- Farm structure (farm size, location, soil type, favorable or not favorable condi-tions for agriculture, available machinery, etc.)

- Farm management (input use, crop rotation, crop diversification, specializa-tion, etc.)

- Exogenous factors (market prices, market size, available subsidies, information access, availability of advisors, support policies, etc.)

- Attitudes, opinions, and preferences (about the environment, life style, health, risk, acceptance within the rural community, etc.)

More authors [14, 15, 21] concluded that the availability of information sources is an important factor in conversion process. Conversion is always connected with some risk. De Cock [22] in his study confirmed that conventional farmers are more risk averse than organic farmers; similar are the findings of Sera et al. [23] and Gardebroek [24] that organic farmers are less risk averse as conventional farmers.

Kallas et al. [17] confirmed an expectation that for the adoption of organic farming location of farm can also play an important role. The location of farms in an unfavorable area for agriculture motivates adoption, and more likely to convert are also farmers who have second economic activity apart from agriculture and small family farms. Farm specialization also can have an influence; for example, farmers whose total farm income comes only from viticulture are less prone to convert. Conversion is more probable if the farm has diversified production or activities.

Organic farms in general usually diversify their activities to reduce all kinds of risks (yield loss, decreasing prices, natural catastrophes, etc.).

Older farmers are less willing to convert [15, 17, 20, 25], are mostly less educated, and are not so ambitious anymore. It is understandable that they are not so ready anymore for adjustments and changes, new learning, investments, new risks, etc. Their decisions are mainly based on economic variables, while the importance of the environmental over the economic considerations is a basic factor in the decision to convert to organic farming.

The results of the Norwegian study [18] did not differ much. Organic farmers in Norway, compared with conventional farmers, mostly have larger farms, are mostly oriented to crop production, are more educated, and are located closer to urban centers. Farmers have different goals for their farms that can differ between conventional farmers, organic farmers, and farmers that are planning to convert. In Norway, main goals of conventional farmers were economic: to "achieve stable and reliable income," to "maximize profit," and to "improve the farm for next generation." Organic farmers rank "sustainable and environmental-friendly farming" in the first place, while conventional farmers put this goal after economic goals. After environmental goals, organic farmers ranked "producing high quality food" and "reliable and stable income." Suitable income was one of the three most important goals by all three groups of investigated farmers (conventional, organic, and farmers that are planning to convert), and it is also seen as a strategy for risk avoidance. All three groups of Norwegian farmers as least important goals ranked "higher private consumption," "increasing equity," and "social contacts." Among the motives for the decision for organic farming, organic farmers choose as most important to "produce high quality food," "higher soil fertility and less pollution problems" as second, and "professional challenges" as third. As least important motives, they defined "natural conditions" and "more stable income." Farmers that are planning to convert to organic farming ranked motives little different; for them, financial motives ("profitability" and "organic farming payments") are most important, while "production of high quality food" and "ideological and philosophic reasons" were less important for them as compared to certified organic farmers. Main goals and motives for organic farmers in Norway are therefore still traditional environment, food quality, and philosophical concerns. On the other hand, financial considerations ("profitability" and "income stability") are important for conventional farmers as well as important motives for conversion for the potential converters.

Studies in other European countries found higher density of organic farms in regions less favorable to agricultural production (e.g., in Austria, Switzerland, and Germany). Results of a study in Germany [26] show that the number of organic farms and higher percentage of organically managed land are negatively correlated to soil quality and positively influenced by organic grassland payments.

Darnhofer et al. [27] studied the reasons and constraints of farmers for converting to organic farming in Austria. They identified five types of farmers: "committed conventional," the "pragmatic conventional," the "environment-conscious but not organic," the "pragmatic organic," and the "committed organic." For "committed conventional" farmers, conventional approach to agriculture is the only sensible way. Their focus is maximizing outputs and profit per hectare, minimizing production costs, intensive use of external inputs, the introduction of the new technologies that make this possible, specialization of the farm, etc. Organic farming for them is not more environmentally friendly than conventional farming and they do not see organic farming as technically and/or economically feasible. They do not even think to convert to organic farming. "Pragmatic conventional" farmers can be seen as potential converters. They are generally not against organic farming, but a conversion is too risky for them, and without tangible benefit of the conversion, they will not implement it. Their main constraints are connected with the technical challenges and needed changes with farm organization, the uncertainty of price, and available market for their products. Some good examples of organic farmers' conversion in the area and available market for organic products can make them to be more open for conversion. Most farmers see economic viability as a necessary condition for conversion, but not a sufficient one. These farmers are more inclined to solutions that do not require conversion, for example, farm diversification or additional off-farm income. The third type of farmers is "environment-conscious but not organic." They are committed to environmentally friendly farming practices, but they do not receive any agri-environmental payments. Because they are not certified organic farmers, they are more flexibly and not subject to controls and are more independent of the regulations, although some of them follow organic standards very closely. The reasons are also bureaucratic demands and costs connected with certification, record-keeping, etc. Some of them are self-declared organic producers (without organic certificate) that have customers willing to pay premiums without organic certificate, because they trust them, as well as the quality of their products. There is no need for them to convert to organic methods. "Pragmatic organic" farmers were motivated to convert by good prospect for security income, mostly through the payments of agri-environmental programs. Financial motives were more important for their conversion as sustainability, health, or ethical aspects. Although financial motives are important for this group of farmers, income-maximization attitude for them is not necessary in the first place [15].

Compensatory payments enable them to "learn by doing." They like to learn and make experiments with new ventures. Their challenges are the diversity of task in organic production, need for the craftsmanship, and always new skill requirements. Organic farming for them is an alternative to conventional farming but close to the "farming economically" thinking [28].

The "committed organic" farmers follow a fundamental philosophy of organic farming: close nutrient cycles, no use of synthetic fertilizers and pesticides, crop rotation, and care for soil health. Organic farming for them is also a social movement and political statement and not only a production technique. Their first considerations are, beside the already mentioned basic principles, producer and/or customer health, ethical norms, and lifestyle; economic considerations are of secondary importance. These groups of farmers are really organic farming believers and mostly pioneers.

4. Organic farming today on global, EU, and Slovenian level

4.1 Organic farming on the global level

At the end of 2017, nearly 70 million hectares of agricultural land were involved in organic farming on a global scale. Comparing to the end of 2016, there is a growth of 20% [29]. The largest organic agricultural area was recorded in Australia with 35.6 million hectares; the second in Argentina with 3.4 million hectares and the third in China with 3 million hectares. Europe together had the second largest area in 2017 (14.6 million hectares). Organic-managed agricultural area increased globally in all continents and in 2017 presented 1.4% of total global agricultural land. Shares of organic agricultural area compared to total agricultural area differ between the countries and were the highest in 2017 in Liechtenstein (37.9%), Samoa (37.6%), and Austria (24%). The global market for organic food in 2017 based on estimations reached around 97 billion US dollars (approximately 90 billion euros). The leading market with organic products were in the USA (40 billion euros), followed by Germany (10 billion euros), France (7.9 billion euros), and China (7.6 billion euros). Organic markets in 2017 showed a growth, in France for example by 18%. Country with the highest amount spent on organic food in 2017 was Switzerland (288 euros per capita), while Denmark had the highest organic market share (13.3% of the total food market). Countries differ also by numbers of organic producers. The total number of organic producers in 2017 has been estimated at 2.9 million. The highest number had India, followed by Uganda and Mexico. The number of producers between 2016 and 2017 increased by almost 5%. Organic land use structure in 2017 was the following: over two-thirds presented grassland or grazing areas (almost 48.2 million hectares), 17% (over 12 million hectares) arable land, and 7% permanent crops (nearly 4.9 million hectares). Arable land use had high-est share for cereals and rice production (almost 38% of all), production of green fodder on arable land (over 23%), oilseed production (10%), and vegetables and dry pulses on the rest. Among permanent crops, the largest share belongs to coffee and olives (each with almost 20% of the organic permanent cropland), followed by nuts, grapes, and tropical and subtropical fruits.

4.2 Organic farming in Europe and European Union

At the end of 2017, organic farming in the whole Europe was carried out on 14.6 million hectares of agricultural land (2.9% of the agricultural area). Compared to 2016, organic farmland has increased by over 1 million hectares. The largest organic agricultural areas were in Spain, Italy, and France. Retail sales of organic products increased by 10.5% compared with 2016 and reached 37.3 billion euros in 2017. Germany had the largest market for organic products (10 billion euros), followed by France (7.9 billion euros) and Italy (3.1 billion euros) [30, 31].

Organic farming in European Union in 2017, by Eurostat data, was carried out on 12.6 million hectares of agricultural land in use in 28 member states, representing 7% of the total utilized agricultural area. The differences between member states are considerable. The largest shares of agricultural land in use for organic farming in 2017 had Austria (23.4%), Estonia (19.6%), and Sweden (19.6%). Since 2013, the number of fully organic agricultural holdings in the EU-28 increased by 30% and represented around 2% of total EU-28 agricultural holdings, while increase in organic area between 2012 and 2017 was 25% [30, 31]. Some member states increased their total organic area between 2012 and 2017 significantly (i.e., Croatia and Bulgaria for over 100%), and some decreased in smaller amount (Poland for 24.5%, the UK for 15.6%, Greece for 11.3%, etc.). Countries with the highest total[1] organic areas in terms of hectares are Spain, Italy, and France. These countries together with Germany in 2017 presented around 55% of the total EU-28 organic areas (**Table 1**).

Total organic area can be used as arable land (production of cereals, root crops, fresh vegetables, green fodder, industrial crops, etc.) or as permanent grassland (pastures and meadows as food for animals) or as permanent crops production (fruit trees, olives, and vineyards). In 2017, in EU-28, arable land represented 44.5% of the total organic crop area, permanent grassland 44.4%, and permanent crops 11%. There are significant differences between EU member states also in

[1] The total organic area is the sum of the "area under conversion" and the "certified area." Before an area can be certified as "organic," it must undergo a conversion process, which may take 2–3 years depending on the crop. For plants and plant products to be certified organic, the production rules must have been applied on the parcels of land during a conversion period of at least 2 years before sowing, or, in the case of grassland or perennial forage, at least 2 years before its use as feed from organic farming, or, in the case of perennial crops other than forage, at least 3 years before the first harvest of organic products.

	Organic area (ha)		Change 2012–2017 (%)
	2012	2017	
EU-28	10,047,896	12,560,191	25.0
Belgium	59,718	83,508	39.8
Bulgaria	39,138	136,618	249.1
Czech Republic	468,670	496,277	5.9
Denmark	194,706	226,307	16.2
Germany	959,832	1,138,272	18.6
Estonia	142,065	196,441	38.3
Ireland	52,793	74,336	40.8
Greece	462,618	410,140	−11.3
Spain	1,756,548	2,082,173	18.5
France	1,030,881	1,744,420	69.2
Croatia	31,904	96,618	202.8
Italy	1,167,362	1,908,570	63.5
Cyprus	3923	5616	43.2
Latvia	195,658	268,870	37.4
Lithuania	156,539	234,134	49.6
Luxembourg	4130	5444	31.8
Hungary	130,607	199,683	52.9
Malta	37	41	10.8
Netherlands	48,038	56,203	17.0
Austria	533,230	620,656	16.4
Poland	655,499	494,978	−24.5
Portugal	200,833	253,786	26.4
Romania	288,261	258,471	−10.3
Slovenia	35,101	46,222	31.7
Slovakia	164,360	189,148	15.1
Finland	197,751	258,672	30.8
Sweden	477,684	576,845	20.8
The United Kingdom	590,011	497,742	−15.6
Norway	55,260	47,042	−14.9
Switzerland	121,013	150,491	24.4

Data source: Eurostat (online data code: org_cropar [31]).

Table 1.
Total organic area in EU-28 and some other European Countries in 2012 and 2017 [31].

this structure: in 13 member states, permanent grassland covered more than 50% of the organic area (i.e., Ireland (95%), Czech Republic (85.5%), and Slovenia (81.4%)), while arable land represented more than 50% of the organic area in 11 EU member states (highly predominant in Finland (99.2%), Denmark (81%), and Sweden (77.9%)). Permanent crops were less presented in the total organic area, the highest shares in 2017 had Malta (48.8%) and Cyprus (46.7%) with

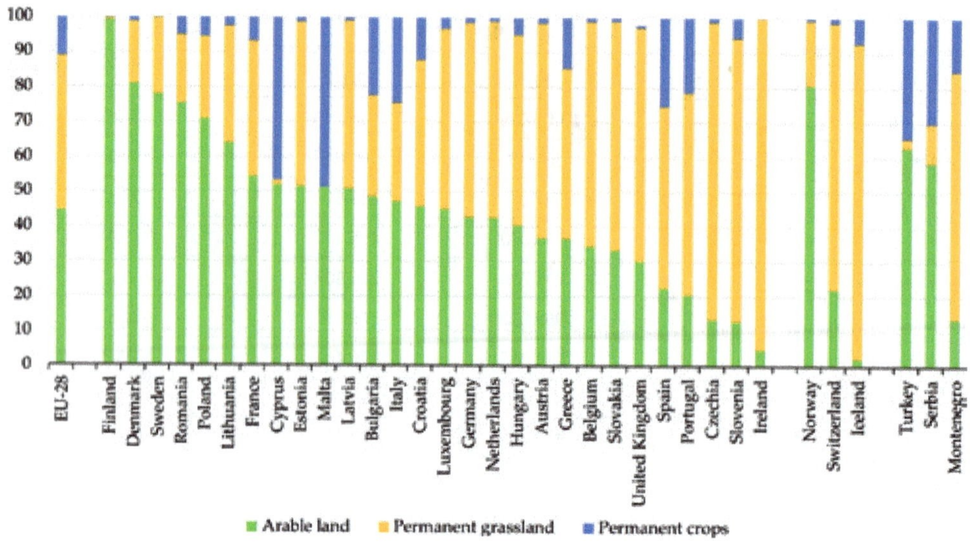

Figure 1.
The structure of organic agricultural land in use by country in 2017 (% of total organic area—fully converted and under conversion). Data source: Eurostat (online data code: org_cropar [31]).

predominated olive trees, and the share was more as 20% in Italy, Spain, Portugal, and Bulgaria and between 10 and 20% in Croatia and Greece. In the rest of the 16 EU member states, permanent crops accounted less than 5% of the total organic areas (**Figure 1**).

The most popular species in organic livestock in EU member states in 2017 were bovines and sheep. The number of organic bovine animals was more than 4 million heads (around 5% of all bovines reported in 2017). The highest shares of organic bovines, dairy cows, and sheep and goats had Latvia (34.1% of total sheep and goat and 23.6% of bovines), Austria (21.7% of total bovines), and Sweden (21.2% of total bovines). The highest share of organic dairy cows in 2017 had Austria (21.2% of all dairy cows), followed by Sweden (16.4%) and Latvia (12.7%). Organically reared pigs had small share of the total reared pigs, and the highest share had Denmark with 3% [31].

4.3 Organic farming in Slovenia

In 2018, 3741 agricultural holdings were included in the organic farming control system (2.9% more than in 2017), of which 3320 were already organic agricultural holdings, with obtained ecological certificate, while the remaining 421 were still in conversion (**Figure 2**) [32]. In 2018, all agricultural holdings in the organic farming control system accounted for 5.4% of all agricultural holdings in Slovenia, and those that already reached the status of organic producers represented 4.8% of all agricul-tural holdings in Slovenia [32, 33].

In 2018, the area of ecological agricultural land in use increased by 1320 ha or 7% compared to 2017 (between 2016 and 2017, increased by 3996 ha or 11%). Organic production in 2018 was carried out on 47,848 ha of utilized agricultural area (on about 10% of all agricultural land in use). Permanent meadows and pastures occupied the largest share in the structure of organic agricultural land in use, over 81% (**Figure 3**). Compared to 2017, in 2018, the area of organic permanent meadows and pastures decreased by 1%, the area of organic vineyards increased the most (by 37% or 124 ha), orchards by 14%, and olive trees by 13%, while the area of organically produced vegetables increased by 11%.

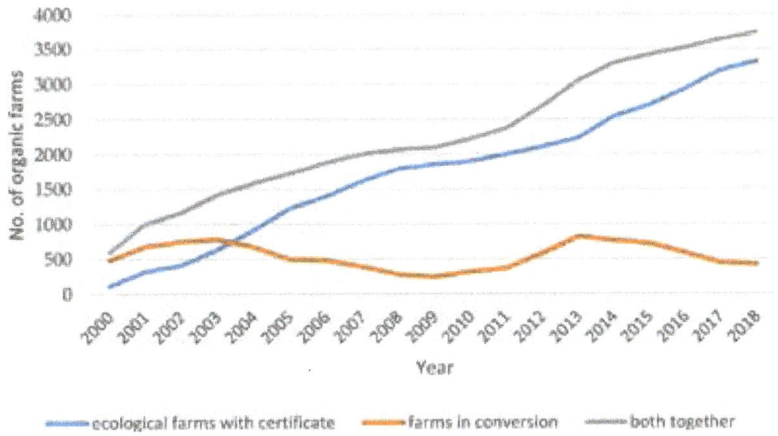

Figure 2.
Dynamic of growth of organic farms in Slovenia between 2000 and 2018. Data source: SiSTAT [33].

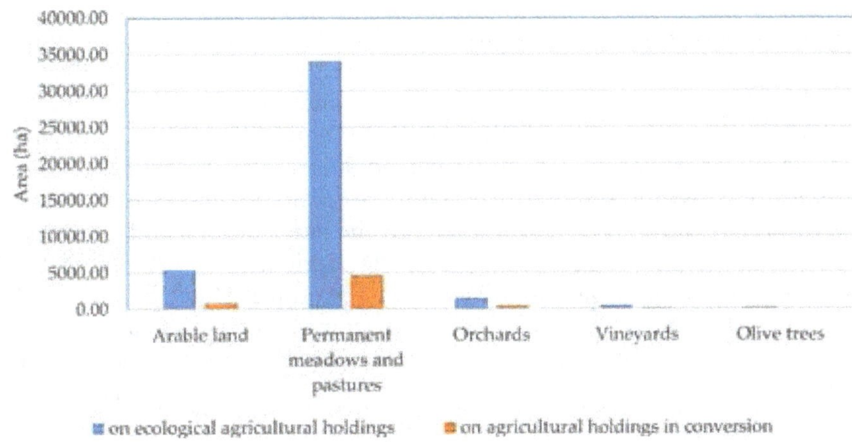

Figure 3.
The structure of organic agricultural land use in Slovenia in 2018. Data source: SiSTAT [33].

In 2018, fruit production was significantly higher than in 2017, almost by 6 times. Due to adverse climatic conditions (frost and hail), fruit production in 2017 was namely very low or even without any products, and also other plant production in 2017 was lower. Better production conditions in 2018 have made it possible to increase the total organic production in arable land and gardens by 27% and in vegetable production by 21%, while the output of organic vineyards' production was 15% higher and in olive groves by 31%.

Due to the large share of grassland, as well as the need for organic fertilizers for organic farming, organic farms usually also keep cattle. The number of individual species kept by organic farms varies, while poultry, cattle, and sheep dominated the number of animals. Between 2016 and 2017, the number of animals on organic farms increased by around 4%. The number of organic honeybees increased by as much as 20%, while the number of rabbits decreased by 11%. In 2018, the number of animals on organic farms decreased by 9%, probably because of worst climatic conditions for the production of animal feed. The only increase compared to 2017 was in the number of beehives by 31% (**Figure 4**). In 2018, the amount of organic products from animal production increased: the total weight of meat by 26% (cattle meat increased by 27%, pig meat by 19%, sheep meat decreased by 41%, and goat

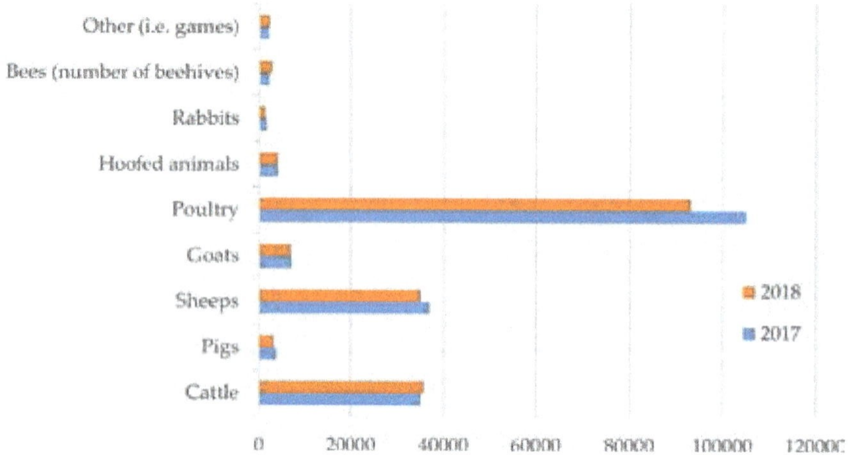

Figure 4.
The number of animals on organic farms in Slovenia in 2017 and 2018. Data source: SiSTAT [33].

meat by 57%). The increase was significant also in the weight of poultry meat by 29% and in the weight of other animals' meat (i.e., games) by 238%. The increase compared with 2017 was evident also in the production of cow's organic milk by 20%, while sheep's and goat's milk decreased (by 2 and 13%). Honey production in 2018 was 41% higher than in 2017 and egg production 26% higher [32].

Despite the growth in number of organic farms in Slovenia, the current development of organic farming is not in line with the expectations and goals set in the Action Plan for the Development of Organic Agriculture in Slovenia by 2015. Problems in the field of organizing and linking producers have been identified as the most problematic [34]. There is still insufficient volume and supply of organic foods, there are not enough qualified organic crop advisers and poor knowledge transfer, in animal products from organic farming it still happens that organic products go into conventional processing, etc.

The Ministry of Agriculture of the Republic of Slovenia will strive to achieve these goals in the future, since organic farming will be an important part of the agricultural development strategy in Slovenia after 2020. "The society expects quality food and smart management of natural resources. Consumers are increasingly putting in front their health and the healthy environment in which they live. Organic farming present also a good entrepreneurial opportunity for farmers, and the ministry intends to continue to raise awareness among consumers of healthy food and the environment, and to encourage the demand for local, organic foods, and, on the other hand, promote the integration and participation in the food production chain" said the Minister of Agriculture of the Republic of Slovenia Dr. Aleksandra Pivec [35].

5. Organic farming and the areas with restrictions/limitations or natural obstacles for agriculture

Organic farming, due to its production characteristics, can be a good choice for farming in areas such as "areas facing natural or other specific constraints" (ANCs), before in the EU called "less favored areas" (LFAs), protected areas (national, regional, or landscape parks), Natura 2000 areas, and water protection areas. Such areas are precisely spatially defined, based on criteria defined by the European or Slovenian legislation.

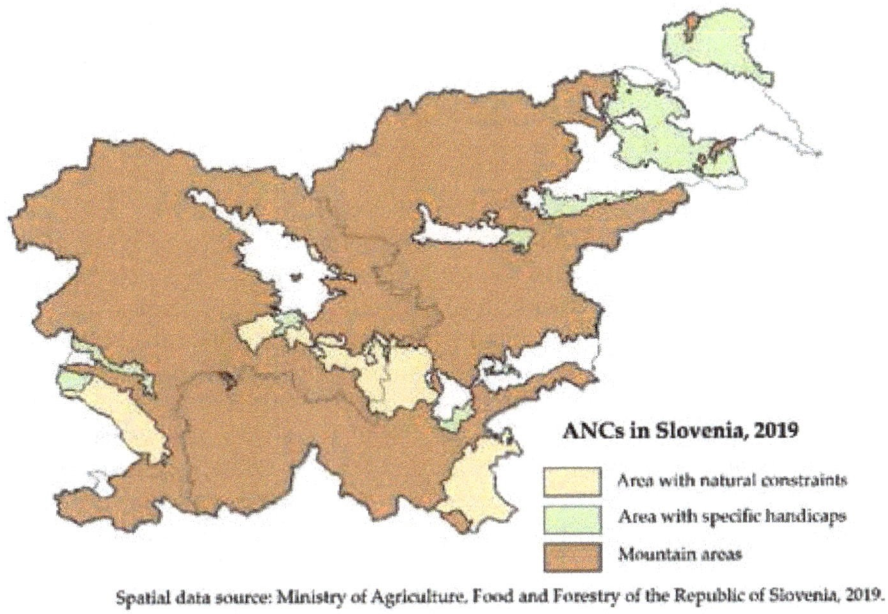

Figure 5.
Areas facing natural or other specific constraints in Slovenia.

5.1 Areas facing natural or other specific constraints (ANCs)

Many farmers in Europe are located in areas that are less favored for agriculture. They are facing natural constraints such as difficult climatic conditions, steep slopes, worst soil quality, and other natural handicaps. Farmers in such areas face many difficulties and have higher production costs but are very important from the viewpoint of sustainability. Most developed countries are paying them to mitigate the risks of land abandonment and thus a possibility of desertification, loss of biodiversity, valuable rural landscape, and other multifunctional roles that farmers play in such areas. ANCs are designated by member states based on EU Regulation 1303/2013 where objective biophysical criteria for the designation are declared. EU Member States have possibility to designate three different categories of ANCs: (1) mountain areas; (2) areas, other than mountain areas, facing natural constraints; and (3) areas affected by specific constraints [36]. In Slovenia, 86.9% of the country area is declared as ANCs or 76.2% of all agricultural land (**Figure 5**) [37]. Farmers with agricultural land in ANCs are eligible to receive compensation payments that are calculated on the basis of differences in costs or income as resulting from natural constraints and in comparison to areas not suffering from those natural or specific constraints.

5.2 Protected areas

In Slovenia, 269,475 ha or 13.3% of the national territory of Slovenia is defined as a protected area characterized by a rich and heterogeneous natural and cultural heritage. Such areas have not only economic but also environmental and social potential; thus, all aspects and prerequisites should ensure sustainable development. Protected areas in Slovenia are represented by one national park (Triglav National Park), three regional parks, 44 landscape parks, and several smaller protected areas such as nature reserves (57) and natural monuments (1164) [38].

In the Nature Conservation Act [39], protected areas are defined as areas of nature with a great abiotic, biotic, and landscape diversity and high density and diversity of natural values.

5.3 Natura 2000 areas

The main objective of the Natura 2000 network in the European Union is to conserve valuable biodiversity for future generations [40]. Slovenia is one of the EU Member States with well-preserved environment and with very rich biodiversity. Great biodiversity is not only the result of different geo-morphological and climatic conditions but also a result of human activities in space. Natura 2000 areas in Slovenia in total encompass 7684 km^2 or approximately 37% of the country territory [41], one of the highest among EU member states. Two thirds of Natura 2000 sites in Slovenia are forested, a substantial part consists of areas with little vegetation above the tree line in the mountains (rock walls), and there are notable grassland areas. A quarter of the total Natura 2000 sites in Slovenia are protected areas (Triglav National Park, regional and landscape parks, nature reserves, or natural monuments).

Implementing of appropriate farming practices in Natura 2000 sites is a prerequisite for the conservation of certain plant and animal species and habitats. The use of agricultural land within Natura sites is regulated by EU regulations and allows farmers to obtain direct payments for farming. There are also certain restrictions, such as prohibiting the plowing of rich grasslands and need for maintaining landscape features (e.g., preserving vegetation zones along watercourses, smaller basins, dry walls, terraces, and tree groups). An instrument for directing the agricultural use of such areas is the agri-environment-climate payments under the CAP (i.e., KOPOP measures in Slovenia), which represent one of the most important ways of directing agricultural activity to a more sustainable way in Slovenia and in the EU. Farmers decide for such measures voluntarily, but they have to implement it on the area covered for at least 5 years. In addition to these measures, some other payments (e.g., ANCs, support for organic farming) that contribute to land cultivation and, in particular, the prevention of grassland overgrowth in Natura 2000 sites can contribute to the achievement of sustainability objectives in Natura 2000 sites. Other CAP Measures such as Cooperation, CLLD, Knowledge Transfer, and Counseling also contribute to the achievement of the conservation objectives of the sites [42].

5.4 Water protection areas

Water protection areas in Slovenia cover about 17% of the country's territory [43]. They are located mainly in the plains, where agriculture is the most intensive. The burden on water and its quality comes not only from agriculture but also from disorderly sewage, wild waste landfills, transport, industry, etc. In Slovenia, groundwater accounts for as much as 98% of all drinking water supply for residents, making efficient protection of groundwater quality of great importance for the health of the population [44]. Most water protection areas in Slovenia are covered by forest (61.1%), followed by grassland (13.6%) and fields (10.9%) [43]. Despite the fact that ecological farming is the most suitable due to the protective objectives on these areas, the share of ecological areas in water protection areas is less than 2%. The reason is likely that these areas are in plains where intensive farming is dominant. In addition, the farm cannot farm organically on water protection areas and conventionally in the remaining ones at the same time. By regulation, only the whole farm can be converted into organic. Due to restrictions on farming, farmers are achieving lower yields on such land. Environmental legislation in the fields

	Utilized agricultural area (UAA) (ha)	Share of utilized agricultural area (%)	Area in organic control (ha)	Share of area in organic control (%)	No. of organic farms	Share of organic farms (%)
ANCs (LFA)	449,000	72.4	28,198	94.5	1922	93.0
Natura 2000 areas	142,067	22.9	6820	22.9	369	17.9
Protected areas	66,293	10.7	1487	5.0	124	6.0
Water protection areas	95,768	15.4	4913	16.5	263	12.7

Data source: [45, 46].

Table 2.
Utilized agricultural area, area in organic control, and no. of organic farms in different types of areas.

of water protection areas and agriculture are consistently implemented and controlled, but on the other hand, the state does not pay farmers yet for a loss of income at the expenses of the restrictions it imposes on the legislation. There is much dissatisfaction among farmers and therefore probably less readiness for organic farming on water protection areas.

The data in **Table 2** show the organic farms by type of area with restrictions/limitations or natural obstacles for agriculture (situation in 2011).

The majority of Slovenian organic farms (93% of all) were in 2011 in areas facing natural or other specific constraints. This is not particularly surprising since such areas make up as much as 85% of the country's territory. Farms in such areas have less favorable conditions for agriculture, so the conversion to organic farming is one of the strategies to survive and improve the economic situation. Due to the high proportion of permanent grassland in ANCs, high share of organic farms in such areas specializes in livestock husbandry [46]. Such a distribution of organic farms indicates that organic farming is more interesting for farms with poor conditions for cultivation and with already extensive farming practices that do not need some special adjustments. Subsidies contribute to their farm efficiency. In 2011, approximately 23% of total area in organic control in Slovenia and 18% of total number of organic farms were located in Natura 2000 areas. This indicates that more incentives to increase a share of organic farming in areas with high biodiversity are needed. Only 13% of organic farms were located in 2011 in water protection areas, and only 16% of the farmlands were organically cultivated.

6. Determinants of organic food consumption

Organic producers depend on the demand for organically produced products. The proportion of consumers purchasing organic food on a regular basis remains low, although consumers have mostly positive attitudes toward organic food as proved in many studies. Decisions about organic food consumption are influenced by different factors. Aertsens et al. [47] made an overview of different studies and exposed the following determinants:

- *Values* such as security (e.g., health is often the strongest motive for purchasing organic food), hedonism (good taste, fun, and enjoyment), stimulation (excitement, novelty, and challenge in life), universalism (understanding, appreciation, tolerance, and protection for the welfare of all people and for nature), benevolence (the desire to do good to others with whom one is often in personal contact), self-direction (choosing, creating, and exploring; e.g., some consume organic food to differ from others and to make a positive self-image and identity), conformity (comply with the expectations of others and violate social expectations or norms), and power (social status, prestige, control, and dominance over people and resources)

- *Attitudes* (e.g., beliefs about health, taste, and environmental consequences)
- *Cognitive and affective components of attitude* (cognitive = thinking, affective = feeling; feelings often tend to predominate in forming personal attitude toward organic food)
- *Emotions* (positive: happiness, love, contentment, and pride; negative: sadness, fear, anger, shame, guilt, empathy etc.)
- *Attitude influencing intention* (attitude toward buying organic food and the intention to buy)
- *Subjective norm or social norms* (subjective norm = social pressure for a person to engage or not to engage in a behavior, social norm = information about what behavior is most appropriate or beneficial)
- *Personal norm and moral norm* (personal norm = an individual's conviction that acting in a certain way is right or wrong; moral norm = morally relevance of a situation)
- *Perceived behavioral control* (people's own perception about the ability to perform a given behavior), perceived barriers (price, availability, lack of trust, and product appearance), and perceived abilities (income impact on the performance of behavior)
- *From intention to behavior* (positive correlation between intensions of buying organic food to behavior)
- *Socio-demographic factors* (gender, age, education, and children in family)
- *Macro-level factors* (the differences between the regions in the world and in the development of their organic markets, the functioning of institutions, the environment people are living in (rural or urban), general knowledge about and trust in organic food, cultural differences, economic and technological factors, etc.)

One of the strongest motives for purchasing organic food is connected with food safety and healthiness (values). Organic consumers believe that organic food tastes better than conventional and also curiosity often stimulates purchase. From universalism point of view, organic farming is mostly perceived by consumers to be more environmental friendly as conventional farming. Some people may consume organic foods that make them different from others and feel some posi-tive self-image and identity. Beliefs about health, taste, and environmental aspects have strong influences on attitude toward buying organic food. Consumer response and behavior are often emotionally conditioned (e.g., fear of contaminated meat determines purchasing behavior of organic meat). All kind of norms (social, personal, and moral) influence purchasing organic products as well and there are some barriers too such as high prices, availability of products, lack of trust in organic certificates, and financial abilities. People with low income have limited or no possibility to buy organic products despite the fact that they want and believe in such products. Socio-demographic characteristics of consumers have less influence on organic food purchases as values [47]. From a gender perspective, more women than men have positive attitudes toward organic food; women generally also concern more about health and healthy food. The age was not recognized as a very significant factor; some studies found the differences between age groups, but more of them did not. The differences are probably influenced by other factors (e.g., macrolevel factors). Families with children are more likely to buy organic food products, while education seems not to play an important role in organic food consumption.

Slovenian consumers purchase mostly in supermarkets or discount stores. One of the latest studies [48, 49] on purchasing habits of Slovenian consumers showed that 28% of interviewed customers often buy organic products, 34% of them occasionally, and 34% very rare or never. There are no significant differences between the groups of customers buying conventional or organic products, except that among customers of organic products share of women is higher. Buyers of organic food are not different by region and type of settlement, but they have higher income. Regular organic food buyers are more likely to buy organic food directly from the farmer, on farm, or on local street market, and to a lesser extent from supermarkets and discount stores. Customers see the supply of organic products in supermarkets as too narrow and mostly not Slovenian origin. One of the strongest motives for purchasing organic food is customers believe that such food is fresh, healthier, of higher quality, and of better taste. Over 40% of organic food buyers believe that such food is completely free of pesticide residues.

The main obstacle for those who do not buy organic food and products is too high price; they also do not trust that organic food is of better quality, healthier, or tastier as conventionally produced food. They even do not trust in their origin and food labels.

The origin of the product is very important for Slovenian buyers in general, so they expect it to be clearly marked on the product, with a clearly defined country of origin (e.g., Slovenia, Austria, etc. and not only as EU origin). Over 80% of customers most trusted in organic products from Slovenia, followed by organic products of Austrian and German origin. Slovenian organic food buyers least trust organic products from Israel and Turkey. They believe that by buying Slovenian food they support Slovenian farmers and contribute to the development of the local food market, to protect the environment, and to the maintenance of cultural landscape and vital countryside.

Buyers of organic food are most attracted to the Slovenian origin, the label "Chosen quality of Slovenia" and the specific indication of the farm where the food comes from. Attractive are also some special offers and lower prices. The highest share of the purchase decision explains the product origin (63%) and the price is slightly more important than the production method. Interesting finding of survey is that Slovenian consumers are ready to pay also 80% higher price for food of Slovenian origin as for comparable foods from abroad. They are even ready to pay more for conventional food with Slovenian origin as for certified organic food from Austria or Italy [48, 49].

7. Conclusions

In terms of the preservation of agricultural production in the entire territory of Slovenia, the preservation of rural settlements and biodiversity, to which the appropriate farming method contributes, areas with limited factors for farming are very important, not just flatlands with the best conditions for agricultural production. In these areas, the promotion of sustainable ways of farming, such as organic farming, is very important and should be supported by appropriate agricultural policy measures and support payments in the future. Data on the volume of organic farming show that the number of farms and the volume of agricultural land in organic production are growing too slowly, especially the share of arable land and permanent crops. The target set by the 2015 organic farming action plan was not achieved. Several farms have also withdrawn from organic farming due to some reductions in payments.

Data on available quantities of Slovenian organic food and consumer needs indicate that demand for this type of food is much higher than supply. Slovenian organic producers can provide only about 20% of the demanded quantities for organic products. The market supply of organic products from Slovenia can, therefore, and should be increased. Direct support to organic producers, support for adaptation and investment for organic farms, farm modernization, greater integra-tion of producers for joint market entry, and more research and knowledge transfer, can contribute to increase the supply of organic products.

The resolution "Our Food, Rural Areas and Natural Resources after 2021" [50], which represents a strategic framework for the development of Slovenian agriculture, food processing, and rural areas, gives organic farming and areas with natural handicaps for agriculture notable attention, which should be followed by agricultural policy measures in the new programming period beyond 2020.

To increase the share of organic farms in Slovenia, as well as in other countries, it is important to know the reasons, objectives, and motives that influence the decision to convert to organic farming. When preparing some effective support policies, it is important to take into account the findings of different studies. General find-ings are that financial motives are very important to attract potential converters. However, many other factors as well influence farmers' decisions. However, if farm-ers are only economically motivated for conversion, short-term benefits through policy payments may go out of the scheme as soon as economic conditions are not so favorable anymore. That has happened in Slovenia as well.

Natural conditions for agriculture, farm and farmer's characteristics, demand for organic products among consumers, yield loss, price premiums, market prices, consumers' purchasing power, and other factors that influence the decision about farming system must be well understood. As conversion is a strategic decision of farmer and his/her family, it is necessary that support policies are long-term oriented and provide the conditions for the success of the farm business. Adoption of organic farming is not only economically motivated but also different noneconomic factors were found as important. The probability of conversion to organic farming is influenced also by the farmers' attitudes to the environment, food quality and health, lifestyle, and other ideals. This should also be taken into account when promoting organic farming and encouraging farmers to convert.

Decision for organic farming in areas with natural handicaps or restrictions seems appropriate, especially on farms where grassland predominates. The readi-ness for conversion increases if the level of payments for organic farming also increases and there are already successful organic farms in the area [51]. Farms can improve the economics of farming through higher direct and rural development payments for organic farming and the implementation of agri-environment-climate measures. Another possibility is linking organic farming with tourism or processing of products on the farm and sold them direct on the farm or in the local environ-ment to local people, visitors of tourist farms, and local institutions (schools, kindergartens, hospitals, etc.). Indeed, ensuring the environmental and social role of organic farming is only possible if farming is economic at the same time.

We can agree with Meemken and Qaim [8] that "organic farming is not paradigm for sustainable agriculture and food security, but smart combinations of organic (especially in less favored, water protection, and protected areas) and improved conventional methods could contribute toward sustainable productivity increases in global agriculture".

Author details

Anton Perpar* and Andrej Udovč
Department of Agronomy, Biotechnical Faculty, University of Ljubljana, Slovenia

*Address all correspondence to: anton.perpar@bf.uni-lj.si

References

[1] Reganold J. Can we feed 10 billion people on organic farming alone? The Guardian. 2016. Available from: https://www.theguardian.com/sustainable-business/2016/aug/14/organic-farming-agriculture-world-hunger.[Accessed: 20 June 2019]

[2] ANEK. Akcijski načrt razvoja ekološkega kmetijstva v Sloveniji do leta 2015. 2005. Available from: http://www. mkgp.gov.si/fileadmin/mkgp.gov.si/pageuploads/ssk/Akcijski_nacrt-ANEK. pdf. [Accessed: 04 May 2011]

[3] IFOAM. Principles of Organic Agriculture Preamble. 2005. Available from: http://www.ifoam.org/organic_facts/principles/pdfs/IFOAM_FS_Principles_forWebsite.pdf. [Accessed: 15 March 2010]

[4] Vaarst M. Organic farming as a development strategy: Who are interested and who are not? Journal of Sustainable Development. 2010;**3**(1): 38-50. DOI: 10.5539/jsd.v3n1p38

[5] The Role of Organic Farming in European Agriculture. Copa-Cogeca. 2009. Available from: https://copa-cogeca.eu [Accessed: 10 May 2019]

[6] Nemes N. Comparative Analysis of Organic and Non-organic Farming Systems: A Critical Assessment of Farm Profitability. Food and Agriculture Organization of the United Nations, Natural Resources Management and Environment; 2009

[7] Tuomisto HL, Hodge ID, Riordan P, Macdonald DW. Does organic farming reduce environmental impacts? A meta-analysis of European research. Journal of Environmental Management. 2012;**112**:309-320. DOI: 10.1016/j. jenvman.2012.08.018

[8] Meemken EM, Qaim M. Organic agriculture, food security, and the environment. Anual Review of Resource Economics. 2018;**10**:39-63. DOI: 10.1146/annurev-resource-100517-023252

[9] Moudrý J Jr, Moudrý J. Environmental aspects of organic farming. In: Pilipavicius V, editor. Organic Agriculture Towards Sustainability. Rijeka: IntechOpen; 2014. pp. 247-274. DOI: 10.5772/58298

[10] Kuhar A, Slabe A, Juvančič L. Determinants of purchasing behaviour for organic and integrated fruits and vegetables: The case of the post socialist economy. In: Reed M, editor. Organic Food and Agriculture. New Trends and Developments in the Social Sciences. Rijeka: IntechOpen; 2012. pp. 19-38. DOI: 10.5772/1526

[11] D'Amario A, Marzoli F, Martino F, Morettini M. Social aspects of organic farming. In: ENAOS 2005—4th ENAOS Summer Meeting, Warsaw and Culavia-Pomerania. Available from: http://www.enoas.org/pol05t/006e.html. [Accessed: 28 May 2019]

[12] Bacon CM, Getz C, Kraus S, Montenegro M, Holland K. The social dimensions of sustainability and change in diversified farming systems. Ecology and Society. 2012;**17**(4):41. DOI: 10.5751/ES-05226-170441

[13] Knowler D, Bradshaw B. Farmers' adoption of conservation agriculture: A review and synthesis of recent research. Food Policy. 2007;**32**(1):25-48

[14] Rigby D, Young T, Burton M. The development of and prospects for organic farming in the UK. Food Policy. 2001;**26**:599-613

[15] Padel S. Conversion to organic farming: A typical example of the diffusion of an innovation? Sociologia Ruralis. 2001;**41**(1):40-61

[16] Padel S, Lampkin N. Conversion to organic farming: An overview. In: Lampkin N, Padel S, editors. The Economics of Organic Farming: An International Perspective. Wallingford: CAB International; 1994. pp. 295-313

[17] Kallas Z, Serra T, Gil JM. Farmers' objectives as determinants of organic farming adoption: The case of Catalonian vineyard production. Agricultural Economics. 2010;**41**:409-423. DOI: 10.1111/j.1574-0862.2010.00454.x

[18] Koesling M, Flaten O, Lien G. Factors influencing the conversion to organic farming in Norway. International Journal of Agricultural Resources, Governance and Ecology. 2008;**7**(1/2):78-95. DOI: 10.1504/IJARGE.2008.016981

[19] Best H. Organic farming as a rational choice. Empirical investigations in Environmental decision making Rationality and Society. 2009;**21**(2): 197-224. DOI: 10.1177/1 043463109103899v

[20] Udovč A, Perpar A. The characteristics of conventional and organic farmers in Podravska region. Acta Agriculturae Slovenica. 2007;**89**(1):81-93

[21] Parra C, Calatrava J. Factors related to the adoption of organic farming in Spanish olive orchards. Spanish Journal of Agricultural Research. 2005;**3**(1): 5-16. Available from: http://www.inia.es/GCONTREC/PUB/005-016-(8905-Factors_rela_1161759971234.pdf. [Accessed: 26 April 2019]

[22] De Cock L. Determinants of organic farming conversion. In: Paper Prepared for Poster Presentation at the XIth International Congress of the EAAE, «The Future of Rural Europe in the Global Agri-Food System», Copenhagen, Denmark, August 24-27, 2005. Available from: https://www.researchgate.net/publication/23509544. [Accessed: 26 April 2019]

[23] Serra T, Zilberman D, Gil JM. Differential uncertainties and risk attitudes between conventional and organic producers. The case of Spanish COP farmers. Agricultural Economics. 2008;**39**(2):219-229. DOI: 10.1111/j.1574-0862.2008.00329.x

[24] Gardebroek C. Comparing risk attitudes of organic and non-organic farmers with a Bayesian random coefficient model. European Review of Agricultural Economics. 2006;**33**(4): 485-510. DOI: 10.1093/erae/jb1029

[25] Anderson JB, Jolly DA, Green RD. Determinants of farmer adoption of organic production methods in the fresh-market produce sector in California: A logistic regression analysis. In: 2005 Annual Meeting, July 6-8, 2005, San Francisco, California 36319, Western Agricultural Economics Association. Available from: https://ideas.repec.org/p/ags/waeasa/36319.html. [Accessed: 28 May 2019]

[26] Bichler B, Lippert C, Häring AM, Dabbert S. Die Bestimmungsgründe der räumlichen Verteilung des ökologischen Landbaus in Deutschland. Berichte über Landwirtschaft. 2005;**83**(1):50-75. Available from: http://orgprints.org/5110/2/5110-bichler-et-al-2005-raeumliche_Verteilung.pdf. [Accessed: 15 April 2019]

[27] Darnhofer I, Schneeberger W, Freyer B. Converting or not converting to organic farming in Austria: Farmer types and their rationale. Agriculture and Human Values. 2005;**22**:39-52. DOI: 10.1007/s10460-004-7229-9

[28] van der Ploeg J. Revitalizing agriculture: Farming economically as starting ground for rural development. Sociologia Ruralis. 2000;**40**(4):497-511

[29] Willer H, Lernoud J, editors. The World of Organic Agriculture. Statistics & Emerging Trends. FiBL&IFOAM-Organics International; 2018. Available from: http://www.organic-world. net/yearbook/yearbook-2018.html [Accessed: 26 April 2019]

[30] Organic farming in the EU. A fast growing sector. European Commission. EU Agricultural Markets Briefs, No. 13, March 2019. Available from: https://ec.europa.eu/info/sites/info/files/food-farming fisheries/farming/documents/market-brief-organic-farming-in-the-eu_mar2019_en.pdf [Accessed: 16 April 2019]

[31] Eurostat database. Available from: https://ec.europa.eu/eurostat/data/database [Accessed: 16 April 2019]

[32] Organic farming, Slovenia. In: 2018, 21% more organic vegetables produced in Slovenia than the year before. 2018. Available from: https://www.stat.si/StatWeb/en/News/Index/8195 [Accessed: 14 March 2019]

[33] SiSTAT database. Available from: https://pxweb.stat.si/SiStatDb/pxweb/sl/30_Okolje/30_Okolje__15_kmetijstvo_ribistvo__11_15619_ekolosko_kmet/?tablelist=true [Accessed: 14 March 2019]

[34] Zaključno poročilo delovne skupine za spremljanje izvajanja Akcijskega načrta razvoja ekološkega kmetijstva do leta 2015 (ANEK). Republika Slovenija, Ministrstvo za kmetijstvo in okolje. 2012. Available from: https://www.gov.si/gone?url=http://mkgp. arhivspletisc.gov.si/fileadmin/mkgp.gov. si/pageuploads/podrocja/Kmetijstvo/Ekolosko_kmetijstvo/ANEK_zakljucno_porocilo.pdf [Accessed: 14 March 2019]

[35] Ministrstvo: Povečati obseg eko-kmetijstva. Dolenjski list, 13. 3. 2019. Available from: https://www.dolenjskilist.si/2019/03/13/215617/novice/dolenjska/Ministrstvo_Povecati_obseg_eko_kmetijstva/ [Accessed: 16 March 2019]

[36] ANCs (Areas facing natural or other specific constraints). Available from: https://ec.europa.eu/agriculture/rural-development/areas-facing-natural-or-other-specific-constraints_en [Accessed: 16 May 2019]

[37] Reforma OMD uspešno zaključena, sledijo spremembe v sistemu točkovanja GERK-ov v OMD. Available from: https://www.kgzs.si/novica/reforma-omd-uspesno-zakljucena-sledijo-spremembe-v-sistemu-tockovanja-gerk-ov-v-omd-2019-03-20 [Accessed: 25 March 2019]

[38] Naravni parki, naravni rezervati in naravni spomeniki. Available from: https://www.gov.si/teme/nov-tema-39/ [Accessed: 25 March 2019]

[39] Zakon o ohranjanju narave (ZON). Uradni list RS, št. 56/99. Available from: http://www.pisrs.si/Pis.web/pregledPredpisa?id=ZAKO1600 [Accessed: 25 March 2019]

[40] Natura 2000. Available from: https://ec.europa.eu/environment/nature/natura2000/index_en.htm [Accessed: 28 May 2019]

[41] Natura 2000 v Sloveniji. Available from: http://www.natura2000.si/o-naturi-2000/natura-2000-v-sloveniji [Accessed: 28 May 2019]

[42] Program upravljanja območij natura 2000 (2015-2020). Vlada Republike Slovenije. 2015. Available from: http://www.natura2000.si/fileadmin/user_upload/LIFE_Upravljanje/PUN__ProgramNatura.pdf [Accessed: 28 August 2019]

[43] Kavčič S, Vidan D. Kazalci okolja v Sloveniji: Raba tal na vodovarstvenih območjih. Projektna naloga. Univerza v Ljubljani, Filozofska fakulteta, Oddelek za geografijo. 2014

[44] Glavan M, Pintar M, Urbanc J. Izzivi kmetovanja na vodovarstvenih območjih Dravskega polja. Strategija Upravljanja z Vodami. 2014;**25**:75-82

[45] Poročilo o stanju kmetijstva, živilstva in gozdarstva v letu 2010. Ljubljana: MAFF (Ministry of the Republic of Slovenia of Agriculture, Forestry and Food). Available from: http://www.mkgp.gov.si/si/splosno/vstopna_stran/aktualne_teme/porocilo_o_stanju_kmetijstva_zivilstva_in_gozdarstva_v_letu_2009_in_ocena_stanja_v_2010/ [Accessed: 12 August 2011]

[46] Lampič B. Organic farming and sustainability: The Slovenian case study. In: Conference Proceedings. Santiago: UGI. 2011. pp. 1-8

[47] Aertsens J, Verbeke W, Mondelaers K, Van Huylenbroek G. Personal determinants of organic food consumption: A review. British Food Journal. 2009;**111**(10):1140-1167. DOI: 10.1108/00070700910992961

[48] Eko ponudba, povpraševanje in potrošnja. Raziskava med potrošniki. Available from: https://www.nasasuperhrana.si/clanek/raziskave-o-eko-potrosnji-in-potencialu-eko-proizvodnje-v-sloveniji/ [Accessed: 19 June 2019]

[49] Eko ponudba, povpraševanje in potrošnja. Raziskava med pridelovalci/predelovalci EKO proizvodov glede potenciala proizvodnje EKO živil ter razsikava javnih zavodov do naročanja EKO živil. Available from: https://www. nasasuperhrana.si/clanek/raziskave-o-eko-potrosnji-in-potencialu-eko-proizvodnje-v-sloveniji/ [Accessed: 19 June 2019]

[50] Resolucija: »Naša hrana, podeželje in naravni viri po 2021« Strateški okvir razvoja slovenskega kmetijstva, predelave hrane in podeželja. Ministrstvo za kmetijstvo, gozdarstvo in prehrano R Slovenije. Available from: https://www.gov.si/assets/ministrstva/MKGP/DOKUMENTI/KMETIJSTVO/ded1a797fe/Resolucija-Nasa-hrana-podezelje-in-naravni-viri-po-2021.pdf [Accessed: 12 July 2019]

[51] Travnikar T, Juvančič L. Prostorski vzorec vključevanja slovenskih kmetijskih gospodarstev v ekološko kmetovanje. Geografski vestnik. 2018;**90**(2):53-70

7

Sugarcane Production Systems in Small Rural Properties

Mauro Wagner de Oliveira,
Luís Cláudio Inácio da Silveira, Alexandre Bosco de Oliveira,
Márcio Henrique Pereira Barbosa, Manoel Gomes Pereira
and Terezinha Bezerra Albino Oliveira

Abstract

Sugarcane grown in small rural properties of the Zona da Mata region, located in the southeast of the state of Minas Gerais (MG), is generally intended for animal feed and the production of *rapadura*, brown sugar, *cachaça,* and ethanol. This chapter focuses on the authors' experience on technologies recommended to small farmers for the implantation and management of sugarcane plantations. The following issues are addressed and discussed: planning and preparation of the sugarcane plantation; soil sampling and soil fertility assessment; application of lime and gypsum; setting up seedling nurseries; green fertilization in the areas of planting and renewal of sugarcane plantation; soil preparation, planting and chemical fertilization of plant-cane; weed and pest control; chemical fertilization of ratoon; assessment of sugarcane nutritional status; organic fertilization with crop residues and agroindustrial residues; mineralization of sugarcane straw; assessment of broth quality and sugar production; and renewal of the sugarcane plantation.

Keywords: sustainability, nutrient cycling, broth quality of the sugarcane, leaching, organic fertilization

1. Introduction

The sugarcane production systems addressed in this chapter were recommended by the authors for small farms of the Zona da Mata region of the state of Minas Gerais, Brazil. These small rural properties are located at geographical coordinates ranging from 20°45′14″ to 21°11′39″ South and 42°52′55″ and 43°01′04″ West. The altitude ranges from 330 to 650 m. The climate of the region is humid subtropical and varies from Aw to Cwa with rainy summers, according to Köppen classification. The average precipitation of the last 30 years is approximately 1200 mm. There is water surplus from November to March, precipitation is below potential evapo-transpiration from April to September (causing water deficit), and precipitation is again higher than evapotranspiration in October. Therefore, the dry and rainy seasons are well defined in the region.

The predominant soils in the region are Distrophic Red-Yellow Latossol, Ultisol Red-Yellow Dystrophic and Latossolic Cambisol [1]. Although soils have low fertility, their physical composition allows for agricultural activity provided that appropriate techniques are used. The main agricultural techniques recommended these small farmers are aimed at improving the physical and chemical properties of the soil by lime and gypsum application, chemical fertilization, green fertilization, using organic compost, planting of sugarcane varieties with greater yield potential, chemical weed control, and biological pest control. The sugarcane produced in these properties is intended for animal feed and the production of *rapadura*, brown sugar, *cachaça,* and ethanol.

2. Planning and implantation cost of sugarcane plantation

Several technologies can be used in setting up and managing sugarcane plantations in small farms. Those selected and recommended by the authors are focused on maximizing the use of inputs, land and human resources to reduce operating costs and increase crop yields, in addition to helping preserve the environment. The main cost items for setting up 1 ha of sugarcane in the region are shown in **Table 1**.

Cost item	Unit*	Unit price	Quantity	Total price	Participation
		(US$)		(US$)	%
A) Consumption					
Limestone	t				
Gypsum	t				
Seeds of *Crotalaria juncea*	kg/ha				
Fertilizer	kg				
Sugarcane seedlings	t				
Chemical insecticide	L or kg				
Herbicide	L or kg				
Formicide	L or kg				
Biological insecticide	L				
Subtotal (A)					
B) Service					
Land rent	ha				
Soil analysis	sample				
Plowing	h/m				
Harrowing	h/m				
Sowing of *Crotalaria juncea*	h/m				
Incorporation of *C. juncea* into soil	h/m				
Furrowing for planting sugarcane	h/m				
Fertilizer application in the planting furrow	h/m or d/H				
Seedling distribution and pruning	d/H				
Insecticide application on the seedlings	h/h or d/H				

Cost item	Unit*	Unit price	Quantity	Total price	Participation
		(US$)		(US$)	%
Covering of seedlings	h/m or d/H				
Herbicide application	h/m or d/H				
Formicide application	d/H				
Insecticide application	d/H				
Subtotal (B)					
C) Total cost for the implantation of a sugarcane plantation (A+B)					

**t: tonne, Kg/ha: kilograms per hectare, L: Liters, Kg: Kilogram, h/m: machine hours, d/H: work days per man, sample.*

Table 1.
Main cost items for the implantation one hectare of sugarcane in small farms of the Zona da Mata region.

Scenarios	Decrease in yield (%)	Cuts 1st	2nd	3rd	4th	5th	6th	7th	Average	Sugarcane plantation (ha)	Area (%)
		t of natural matter per hectare									
1	10	150	135	122	109	98	89	80	112	1.63	100
2	15	150	128	108	92	78	67	57	97	1.88	115
3	10	120	108	97	87	79	71	64	89	2.04	125
4	15	120	102	87	74	63	53	45	78	2.35	144
5	17.5	120	99	82	67	56	46	38	72	2.52	154
6	20	120	96	77	61	49	39	31	68	2.69	165
7	10	110	99	89	80	72	65	58	82	2.23	137
8	15	110	94	79	68	57	49	41	71	2.56	157
9	17.5	110	91	75	62	51	42	35	66	2.75	169
10	20	110	88	70	56	45	36	29	62	2.94	180
11	10	100	90	81	73	66	59	53	75	2.45	150
12	15	100	85	72	61	52	44	38	65	2.82	173
13	17.5	100	83	68	56	46	38	32	60	3.02	185
14	20	100	80	64	51	41	33	26	56	3.23	198

Table 2.
Simulation of a sugarcane plantation area needed to feed 25 dairy cows (20 kg of natural matter per cow/day) for 365 days and plant-cane yield and yield stability in seven succeeding cuts.

In sugarcane, it is common to evaluate the results obtained by quantifying the production of culms, sugars, or total shoot biomass. In analyzing the production costs, one can use the exchange ratio, which is an economic indicator that shows the exchange capacity of a certain product in relation to the inputs used in production (product/input).

The great advantage of exchange ratio analysis over price analysis is that agricultural products represent the weighted average of several inputs and goods used by the farmer. Thus, it is easier to calculate the variation in producer purchasing power or production system efficiency. If possible, this analysis should cover a long period of time, so that the extent of the variation in results can be measured. The use of spreadsheets (e.g., Excel) is helpful in analyzing simulation results, as well as budgeting and managing sugarcane production costs. The authors have guided farmers to adopt practices that recover and maintain soil fertility, recycle nutrients, and reduce compaction and sealing of the topsoil, combined with activities that enable increased yields in plant-cane and small decreases in subsequent cycles. **Table 2** shows a simulation of sugarcane yield in 14 scenarios, which combine high and medium yields in the plant-cane cycle to low, medium, or high decreases in yield over seven cycles (seven cuts).

Table 2 shows that to feed 25 cows for 365 days (20 kg of natural matter per cow/day), an area of 1.63 ha of a sugarcane plantation with high yield in the plant-cane cycle and 10% decrease in subsequent cycles would be necessary (Scenario 1). On the other hand, an area of about 3.3 ha would be needed for a sugarcane plantation with medium yield in the plant-cane cycle and large decreases in subsequent cycles (Scenario 14). In sugarcane plantations with yields of less than 60 tons of natural matter per ha (about 50 tons of industrializable culms), in addition to decreasing the use of land and labor resources, chemical weed control is generally inefficient, as the crop does not completely cover (shade) the soil, allowing the emergence and growth of invasive species (**Figure 1**). Also, in cases where sugarcane is cut by hand, the worker will be more exposed to snakes and scorpions.

Figure 1.
Contrast between a sugarcane plantation with high yields and excellent weed control, and a plantation with low yields and poor weed control.

3. Selecting sugarcane variety

Choosing the right variety is an important and low-cost technology for the producer. Currently, there are several sugarcane cultivars with proper agronomic and zootechnical characteristics, such as high response to improved soil fertility, erect growth, and resistance to falling, which facilitates harvesting, high culm and sucrose yield, regrowth vigor, resistance to pests and diseases, and good dry matter digestibility.

Characteristic	Variety						
	RB835054	RB855536	RB867515	RB928064	RB975201	SP80-1816	SP80-3280
Yield[1]	High	High	High	High	High	High	High
Maturation	Early	Medium	Medium	Late	Late	Medium	Medium
Sucrose content	High	High	High	Medium	Medium	High	High
Harvest [2]	Apr-Nov	Jun-Nov	Jun-Sep	Sep-Nov	Ago-Nov	Jun-Sep	Jun-Sep
Soil[3]	Medium	Medium	Low	High	Medium	Medium	Medium
Regrowth[4]	Good	Excellent	Good	Good	Good	Excellent	Good
Tillering[5]	Average	Excellent	Average	Very good	Average	Very good	Very good
Toppling[6]	Frequent	Little	Little	Little	Frequent	Little	Rare
Detrashing[7]	Easy	Medium	Easy	Medium	Easy	Medium	Easy
Pilosity	No	No	No	Yes	No	Yes	No
Flowering	Absent	Absent	Medium	Absent	Absent	Absent	Present
Wilting	No	No	Medium	No	No	Little	Yes
Sensitivity to Herbicides[8]	High	Medium	Low	Low	Low	Medium	Medium
Coal[9]	Resistant	Intermediate	Resistant	Resistant	Resistant	Intermediate	Intermediate
Rust[9]	Resistant	Resistant	Resistant	Resistant	Resistant	Resistant	Resistant
Red Streak[9]	Resistant	Intermediate	Susceptible	Resistant	Intermediate	Resistant	Resistant
Leaf scalding[9]	Resistant	Intermediate	Resistant	Resistant	Resistant	Resistant	Resistant

[1]*Yield: Yield per ha of industrializable culms and biomass (High, Medium or Low)*

[2]*Harvest: Recommended months for the harvest of sugarcane for brown sugar and rapadura*

[3]*Soils: Soil fertility requirement (High, Medium or Low)*

[4]*Regrowth: sprout vigor of regrowth under conditions of trampling by machines or animal traction vehicles*

[5]*Tillering: growth speed and soil shading, thus minimizing weed competition*

[6]*Toppling: plant growth habit, which varies from erect to decumbent. Even cultivars with upright culms may topple under high-yield conditions*

[7]*Detrashing: removal of dry leaves or attachment of the sheath to the culm*

[8]*Sensitivity to certain herbicides*

[9]*Diseases of the plant shoot.*

Source: Silveira, L. C. I. (unpublished data).

Table 3.
Agroindustrial, morphological, and tolerance characteristics of seven varieties of sugarcane recommended for cultivation in the small farms of the Zona da Mata region.

The authors of this chapter do not recommend planting one variety of sugar-cane in more than 33% of the total area, even if it has a large number of desirable characteristics. This is because sugarcane production will be greatly compromised in cases of possible breakdown of resistance to disease or sudden decline of the cultivar. Thus, in order to obtain a good quality product, ensure vigorous regrowth and consequently increase the longevity of the sugarcane plantation, the authors recommend that farmers use at least four varieties of sugarcane and adopt measures to maintain soil fertility and cut the sugarcane at the most suitable time for each variety. Once the varieties have been selected, it is necessary to check the quality of the seedlings. It is also important to confirm the health of the seedlings in terms of diseases, pests, and mixture of other cultivars. **Table 3** shows characteristics of the varieties currently most planted in small farms of the Zona da Mata region.

RB867515 has been the variety of sugarcane most cultivated by small produc-ers due to its high yield potential in different edaphoclimatic conditions. As mentioned in **Table 3**, RB867515 is a medium maturing variety with high sucrose content in industrializable culms. It has a low requirement for soil fertility, but it is very responsive to fertilization. Its detrashing is easy and it has no pilosity. **Figure 2** shows dry matter accumulation rate in shoots of three sugarcane varieties (RB855536, RB867515, and SP801816). The study was conducted in soil of medium texture in the city of Mercês, state of Minas Gerais (MG) (latitude 21.260232, longitude 43.298827, and altitude 503 m).

Sugarcane was planted in the first half of February. Following the recommenda-tion of Ref. [1], 5.0 t of dolomitic limestone and 1.5 t of gypsum were applied per ha in September of the year prior to the planting of sugarcane. The soil was plowed and harrowed, followed by the sowing of *Crotalaria juncea*. In early February, at the grain filling stage, *Crotalaria juncea* was incorporated into the soil. A week later, the soil was furrowed and sugarcane was planted. The chemical fertilization consisting solely of phosphorus was applied at the bottom of the planting furrow at a dose of 100 kg of phosphorus per ha (equivalent to 229 kg of P_2O_5 per ha). Chemical weed control with pre-emergent herbicide was used. In mid-September, when the rainy season started again, 200 kg of potassium per ha (equivalent to 240 kg of K_2O per ha) was applied between the rows of sugarcane. Assessments of dry matter

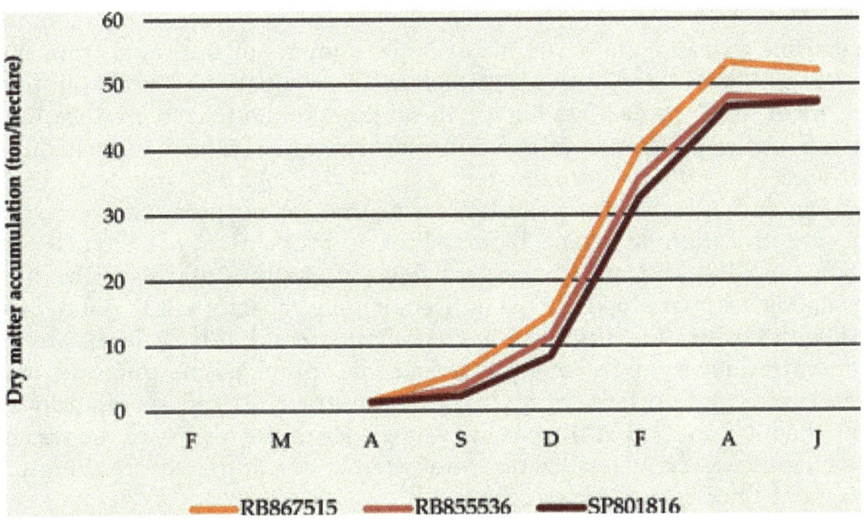

Figure 2.
Dry matter accumulation rate in shoots of three sugarcane varieties planted in February and harvested in July of the following year. Study carried out in the city of Mercês, located in the Zona da Mata region. Source: Oliveira MW (unpublished data).

accumulation were held in April, September, and December of the year of planting of sugarcane and in February, April, and July of the following year. As shown in **Figure 2**, RB867515 produced about 5.0 t more dry matter than the other two variet-ies, which corresponds to approximately 15 t of forage per ha. The average percent-age of culms in forage of these varieties is 85%. Thus, RB867515 produced 12 t of industrializable culms more than the other two varieties.

4. Implantation of sugarcane plantation

Similarly to South Central Brazil, the planting of sugarcane without irrigation in small farms of the Zona da Mata region is essentially done at the beginning of the rainy season (September and October) and at end of the rainy season (February to March). The sugarcane planted at the beginning of the rainy season can be harvested from April to May of the following year (known as "one-year sugarcane"). However, for sugarcane planted from February to March, the harvest will take place about 15–18 months after (known as "one-and-a-half-year sugarcane").

The authors have recommended the planting of "one-year sugarcane" in more fertile soils with smoother and less erosive relief, because there is heavy rainfall during this period. Because sugarcane starts the maximum growth phase in January (when water and thermal availability begin to decrease), nutrient supply should not be a limiting factor to plant development. Therefore, biomass yields exceeding 120 t of natural matter per ha should be reached. However, planting "one-and-a-half-year sugarcane" has been recommended for the more rugged and less fertile soils, since sugarcane will continue to grow in the field for a longer period. Also, the maximum growth phase (**Figure 2**) coincides with the times of greater water and light availability, which results in higher vegetation cover by sugarcane foliage as well as in higher photosynthetic rate and dry matter accumulation. One other great advantage of planting the "one-and-a-half-year sugarcane" is the possibility of growing *Crotalaria juncea* prior to the planting of sugarcane.

5. Evaluation of soil fertility and lime and gypsum application

Sugarcane extracts and accumulates large amounts of nutrients from the soil because it produces large amounts of biomass. In evaluations carried out in small properties of the Zona da Mata region, the authors found that to produce 120 tons of natural matter per ha (about 100 t of industrializable culms), the accumulation of nutrients in shoots is approximately 150, 40, 180, 90, 50, and 40 kg of N, P, K, Ca, Mg, and S, respectively. In the case of iron, manganese, zinc, copper, and boron, accumulation in shoot biomass for a production of 120 t is around 8.0, 3.0, 0.6, 0.4, and, 0.3 kg, respectively [1]. Because of this high nutrient removal, it is necessary to know the nutrient supply capacity of the soil to complement it with fertilization if necessary. On the other hand, if toxic levels are found, the concentration of these elements is reduced by applying limestone and gypsum. The availability and presence of toxic levels of nutrients in soil are typically evaluated by chemical analysis of the topsoil. Knowledge of the history of the area is also of great value, especially fertilization and whether or not there were symptoms of deficiency or toxicity in previous crops.

Soil samples are typically collected at depths of 0–20 and 20–40 cm. The results of the analysis at 0–20 cm have been used to calculate the need for fertilization and liming, while those at 20–40 cm to calculate the need for gypsum. Because these are small areas, the authors have advised producers to collect soil samples using a hole digger and straight shovel, as the use of a straight shovel decreases the variability of soil fertility indexes. Further details on sampling procedures, sample variability, sample drying, and comparison between chemical extractors can be found in Ref. [1]. As previously mentioned, most of the soils of the region are naturally acidic and present low saturation by basic cations such as calcium, magnesium, and potassium. Deficiency of these basic cations combined with high levels of aluminum, iron, and manganese has been detrimental to the growth of the root system and the entire plant. For these reasons, limestone and gypsum applications are recommended by the authors. Several materials have been used as soil acidity correctors, of which the most commonly used is dolomitic limestone. However, calcitic and magnesium lime-stone, as well as calcium and magnesium silicates (referred to as steel slags) are also used. Magnesium oxide content is around 8% in steel slag, while it is less than 5% in calcitic limestone, between 6 and 12% in magnesium limestone, and above 12% in dolomitic limestone. The efficiency of these products in correcting soil acidity depends on particle size, uniform distribution in the field, and soil water availability.

The most used soil analysis method in the region is the one that uses calcium acetate to determine $H^+ + Al^{+3}$. This extractor greatly underestimates the amount of $H^+ + Al^{+3}$, and results in underestimating the cation exchange capacity at pH 7.0 and the limestone dose. For these reasons, the authors have recommended raising the dose of limestone by 1.5–2.0 times. For sugarcane grown in small farms, the recom-mendation is to increase base saturation (V) to 60%. The limestone dose (LD) when using the base saturation method is calculated by the following equation (Eq. (1)):

$$LD\ (t\ ha^{-1}) = [(60-V) \times T] \div RTNP \qquad (1)$$

where V is the current base saturation of the soil, T is the cation exchange capacity at pH 7.0, and RTNP is the relative total neutralizing power of the corrective used.

Dolomitic limestone is recommended when magnesium content at 0–20 cm is less than 0.40 $cmol_c/dm^3$ of soil. On the other hand, if magnesium content at 0–20 cm is greater than 0.40 $cmol_c/dm^3$ of soil, the recommendation is to use the corrective that has the lowest price per ton of RTNP in the crop. Thus, an economic factor is included in the decision making regarding the type of limestone to be used. The use of gypsum has been recommend based on the results of chemical analysis of the 20–40 cm layer. Gypsum has been applied when calcium content is less than 0.40 $cmol_c/dm^3$ of soil or aluminum saturation (m%) is higher than 20%. The usual recommended dose is one-third of the limestone dose (e.g., assuming that the limestone dose is 4.5 t per ha, then gypsum will be 1.5 t per ha). Limestone and gypsum are mixed for subsequent application to the soil. The application of gypsum will lead to the improvement of the root environment of the layers below the topsoil. This effect lasts for several years, which is the reason annual gypsum application is not necessary [1].

In small properties, the application of limestone + gypsum is typically done by hand. A recommended method for these small producers has been to delimit a square or rectangle with the mixture of limestone + gypsum and apply a volume corresponding to the recommended dose in the area. For instance, if the recom-mended dose was 6000 kg (4500 kg of limestone +1500 kg of gypsum per ha) and the density of the limestone + gypsum mixture is 1.25 kg/L, then 4800 L per ha or 0.48 L per m^2 should be applied. Another alternative for small producers to apply limestone + gypsum by hand would be to demarcate areas of 25 m^2 with the mixture itself and apply 12.0 L of limestone + gypsum.

Plowing and harrowing is typically done after the application of limestone + gypsum to incorporate the products into the soil. In most small farms, subsoiling (decompacting soils or breaking compacted layers) has been recommended after plowing and harrowing. This recommendation is based on the land use history of the area, the traffic of machines, implements and animals, the presence of crusts on the surface of the land, and the shallow root system of the natural vegetation. Although it may be an additional burden for the producer, the presence of densified or compacted layers has harmful consequences on water absorption, mineral nutrition, crop development, and longevity of the sugarcane plantation.

6. Green fertilization in "one-and-a-half-year sugarcane"

As previously mentioned in item 4 (implantation of sugarcane plantation), one of the advantages of planting the "one-and-a-half-year sugarcane" is the possibility of a green fertilization prior to the planting of sugarcane. Among the main desirable characteristics of plants used for green fertilization are the following: the possibil-ity of using mechanization from sowing to the harvesting of seeds, the ability to associate with nitrogen-fixing bacteria, rapid growth to control weeds, having mechanisms, or being able to synthesize compounds that help control pests (e.g., nematodes) and diseases, no dormant seeds, and a vigorous and deep root system that assists in the recycling of nutrients from the deepest layers and in soil decompac-tion. Another aspect to be considered is the supply of organic and mineral substrate to soil microorganisms. Thus, green fertilization also contributes to the improvement of the biological quality of the soil [2–4]. Several legumes have these characteristics, but there is generally a preference for *Crotalaria juncea* in South Central Brazil [1].

In the studies conducted by the authors of this chapter in the Zona da Mata region, green fertilization with *Crotalaria juncea* prior to planting the "one-and-a-half-year sugarcane" resulted in increased yield in the plant-cane and first regrowth cycles, which together ranged from 20 to 26 t of culms per ha. In a multiyear analysis, the costs of green fertilization corresponded to 6–12 t of industrializable culms per ha. Thus, the increase in yields covered the costs of growing the legume. Furthermore, there are studies in which increased yields of sugarcane as a result of green fertilization with *Crotalaria juncea* were higher. For instance, in studies conducted over several years in the city of Sales Oliveira, state of São Paulo, Ref. [5] reported increased yields of industrializable culms ranging from 26 to 40 t per ha.

Crotalaria juncea exhibits high growth rates, which result in increased plant height, as shown in **Table 4**. High growth rate associated with increased plant

DAE	Plant height	LAI	DM accumulation	DM accumulation rate
	(cm)	m^2/m^2	(t/ha)	(kg/ha/day)
30	84	2.1	2.2	73
45	178	3.6	4.1	127
60	192	6.5	7.8	247
75	247	8.3	11.9	273
90	313	9.4	14.2	153
105	328	11.3	15.5	87
120	342	8.9	16.2	47

Table 4.
Plant height, leaf area index (LAI), dry matter accumulation (DM accumulation), and dry matter accumulation rate (DM accumulation rate) in shoot biomass of Crotalaria juncea at 30, 45, 60, 75, 90, 105, and 120 days after plant emergence (DAE).

height causes shading of the soil and affects other plants, especially weeds. This is one of the reasons it is used in weed control [1, 4]. Cultural methods are practices that aim to make the crop more competitive than weeds and include reducing planting space, intercropping or rotation with green manure.

Ref. [6] reported excellent results with the use of *Crotalaria juncea*, with weed control percentages greater than 90% in areas with a predominance of grasses, competitive plants, and high nutritional and photosynthetic efficiency. These results were confirmed by Ref. [3], who found that *Crotalaria juncea* was outstanding in terms of soil cover. Plants covered 100% of the soil 50 days after emergence, contributing to the control of erosion and weeds. In addition to the physical effect of shading, *Crotalaria juncea* releases organic compounds from its secondary metabolism (allelopathic compounds), which inhibit weed seed germination or slow down its development [2, 5–7]. Field observations by the authors confirm this allelopathic effect on weeds, verified by the absence of weeds between the planting rows (**Figure 3**). The sowing of *Crotalaria juncea* was carried out in an area adjacent to *Brachiaria* pasture. Therefore, the seed bank of this area should be large.

Crotalaria juncea is extremely sensitive to the length of night (nictoperiod), flowering early under increasing long nights and hence interrupting growth and reducing dry matter accumulation and nutrient cycling, especially of nitrogen [1, 7]. **Table 5** shows the accumulation of dry matter and nitrogen in shoot biomass of *Crotalaria juncea*, as well as plant height was statistically similar for the first three sowing times (beginning of October to beginning of November). For sowing times of mid-November, early and mid-December, there was an average percentage reduction in dry matter accumulation of around 20, 35, and 40% in comparison with the beginning of October.

These reductions were around 6, 13, and 24% for plant height in comparison with that of the first sowing times (**Table 5**). The study was conducted in a Latossolo vermelho amarelo distrófico, which exhibited the following chemi-cal characteristics at 0–20 cm: pH in H_2O = 6.2; 6.0 mg/dm^3 of phosphorus and 59 mg/dm^3 of potassium, (extracted with Mehlich), no exchangeable aluminum and 45% of base saturation. *Crotalaria juncea* is very sensitive to aluminum toxicity and when the soil has exchangeable aluminum, liming should be done prior to sowing [1, 4].

Due to the sensitivity of *Crotalaria juncea* to nictoperiod, the delay in sowing results in early flowering. The authors of this chapter have observed in crops of *Crotalaria juncea* of the Zona da Mata region that plants are able to receive the stimulus for floral induction around 40 days after emergence. Thus, for sowing times

Figure 3.
Crotalaria juncea at the early growth stage and its allelopathic effect on weeds confirmed by the absence of weeds between the planting rows (photo on the right).

Sowing times	DM accumulation (kg/ha)		N accumulation (kg/ha)		Plant height (cm)	
	Year 1	Year 2	Year 1	Year 2	Year 1	Year 2
Early October	14,135 a	14,789 a	273 a	284 a	293 a	305 a
Mid-October	14,768 a	14,845 a	297 a	275 a	311 a	298 a
Early November	14,235 a	13,785 a	268 a	279 a	287a	293 a
Mid-November	11,985 b	11,178 b	220 b	226 b	267 b	256 b
Early December	9,123 c	9,545 c	198 bc	203 c	247 c	236 c
Mid-December	8,523 d	8,037 d	174 c	168 d	217 d	208 d

Means followed by the same letter in the column do not differ statistically from one another other by the Tukey test at 5%.

Table 5.
Accumulation of dry matter (DM accumulation) and nitrogen (N accumulation) in shoot biomass of Crotalaria juncea, and plant height at the grain formation stage according to the sowing time in a study conducted during two agricultural years on a dystrophic red-yellow Latosol (Oxisol).

starting in mid-November, the nights will be increasing in length and early flowering will occur at around 40 days after emergence. The results of studies on sowing times carried out by the authors of this chapter and those found in literature allow us to conclude that to obtain high biomass production in shoots of *Crotalaria juncea*, the sowing of the legume in South Central Brazil should be done from the beginning from October to early November. The incorporation of *Crotalaria juncea* to the soil should be done when the first pods are in the phenological stage of grain filling, at which time the accumulation of dry matter and nitrogen in shoots is the highest [1, 4, 7].

The accumulation of nitrogen in the shoot biomass of *Crotalaria juncea* has also varied according to sowing time. **Table 5** shows that for sowing times from the beginning of October to the beginning of November, nitrogen accumulation in the shoot biomass of *Crotalaria juncea* oscillates around 300 kg/ha. Of the total nitrogen accumulated in the shoot biomass of *Crotalaria juncea*, about 60% originated from the symbiotic associations of the roots with N_2 fixing bacteria, resulting in the contribution of significant amounts of N to the soil-plant system [4, 8] and greater sustainability of the subsequent crop. For comparative purposes, let us consider ammonium sulfate, which is one of the most commonly used nitrogen fertilizers. In 100 kg of ammonium sulfate, there is 20 kg of N. Therefore, it would be necessary to use 1000 kg of ammonium sulfate to obtain 200 kg of N.

The inoculation of *Crotalaria juncea* seeds with nitrogen-fixing bacteria could be a way to increase N_2 biological fixation and nitrogen supply in the soil-plant system. However, research conducted by the authors of this chapter in small farms located in the Zona da Mata region and at sugarcane mills showed that the inoculation of *Crotalaria juncea* seeds with nitrogen-fixing bacteria did not increase N supply in the soil-plant system. Similar results were obtained at EMBRAPA Agrobiologia by Ref. [9], who also found that the inoculants used were not more efficient than the native strains. There was no difference in dry matter and nitrogen accumulation among the treatments with and without inoculation. One of the possible causes could be the high native population of these bacteria in the soils. However, as mentioned by Ref. [10], the fact that the legumes pres-ent high nodulation with native strains does not mean that those bacteria have maximum efficiency, since many of these strains have a high competitive capacity, making it difficult to introduce other strains through seed inoculation. Thus, the authors believe that until more efficient and competitive strains are obtained, the inoculation of the seeds of *Crotalaria juncea* will not result in increased nitrogen biological fixation and accumulation by the plant.

7. Furrowing, fertilization, and planting of sugarcane

Furrowing the soil for the planting of sugarcane is done after plowing and harrowing the land for the incorporation of limestone and gypsum or, after the incorporation of *Crotalaria juncea* in "one-and-a-half-year sugarcane". Furrowing is typically done with fertilizer furrowers that simultaneously open furrows and fertilize. When this implement is not available, it is possible to use a plow with moldboards or discs, using a single disc (straight furrowing if possible). Furrowing should be carried out as close as possible to the distribution of seedlings and planting rows to conserve soil moisture on dry days or avoid the silting of the furrows on rainy days. The spacing between furrows has varied from 0.90 to 1.40 m depending on the distance between the tires of the implement used in the crop fields, on topography, on soil fertility, and on the type of crop. In more fertile areas, wider spacing is used to prevent the sugarcane from tapering and future toppling by the wind. On the other hand, in soils with lower fertility, less fertilized, and on sloped relief, or when cultivars with lower tillering capacity are used, narrower spacing should be used to allow better spatial distribution of plants, more uniform soil cover and increased yields.

The recommended fertilization of the plant-cane is based on the results of the soil analysis at 0–20 cm and the expected yield of the sugarcane plantation. For plant-cane, only phosphate and potassium fertilization are recommended, because studies conducted by the authors of the chapter showed a lack of response to nitrogen fertilization. This low or absent response of the plant-cane to fertilization is widespread for soils grown with sugarcane in Brazil [4]. For the typically low fertile soils of the Zona da Mata region, the recommendation is 100 kg of phosphorus and 200 kg of potassium per ha (equivalent to 229 of P_2O_5 and 240 kg of K_2O per ha). If gypsum is applied to the soil, triple superphosphate should be used to reduce the planting costs, as it is less expensive. More information on fertilizer doses and soil fertility, as well as losses by leaching of both nitrogen and potassium can be found in Ref. [1].

The lack of response of the plant-cane to nitrogen fertilization is due to the mineralization of soil organic matter and the greater nutritional efficiency of the plant-cane root system, compared to the regrowth [1, 4]. Studies conducted by Ref. [11] in the coastal plains of Pernambuco (Northeast Brazil) provide more information regarding the mineralization of soil organic matter. These authors measured carbon and nitrogen mineralization in a Red Yellow Podzolic during the plant-cane cycle. The soil was sampled at 0–20, 20–40, and 40–60 cm prior to planting and 3, 6, 11, and 16 months after planting. Total carbon contents were 6.7, 4.1, and 3.4 g kg^{-1}, while total N contents were 0.7, 0.4, and 0.3 g kg^{-1} at 0–20, 20–40, and 40–60 cm, respectively. The estimated amounts of potentially mineralizable N were 139 and 132 kg per ha at 0–20 and 20–60 cm, respectively, with a mineralization constant of 0.074 per week. Ref. [11] also report that although the soil is considered of low fertility based on the results, the amounts of mineralized organic N would be enough to satisfy the needs of the plant-cane.

Nitrogen uptake and metabolism are strongly influenced by the endogenous availability of phosphorus [1, 12, 13]. In plants with adequate P supply, there is an increase in nitrate uptake from the soil solution and greater nitrate translocation from roots to shoots, increasing the accumulation of amino acids in leaves and roots [4, 12, 14]. Ref. [4] reported research conducted in the state of Minas Gerais, in which the increase of the dose of phosphate fertilization applied in the planting furrow resulted in higher N accumulation in the biomass of the plant-cane. In this case, for each kg of P applied, there was an increase of about 1 kg of N in biomass. These results are caused by changes in N uptake and metabolism, as reported by Refs. [1, 12, 14].

In relation to the planting of sugarcane, an average bud density of 12–15 per meter of furrow is recommended, which is approximately 12–14 t of seedlings per ha. As mentioned previously, one should select good quality sugarcane seedlings (preferably from healthy nurseries) and of first or second cutting at most. It is also important to confirm seedling health in terms of diseases, pests and mixture of other cultivars. The arrangement of the sugarcane within the furrow should preferably be upright with one culm next to the other. The culms are then cut into billets with two or three side branches, which are subsequently covered with soil layer ranging from 5.0 to 8.0 cm and should not exceed 10 cm in thickness. Then, the herbicide application for weed control typically follows the planting of sugarcane.

8. Weed control

Weeds compete with sugarcane during the growth phase for water, light, nutrients, and may exude phytotoxic compounds (allelopathy) and host pests and diseases [4, 15]. Sugarcane exhibits C_4 metabolism, which makes it relatively more competitive in CO_2 assimilation and nutrient use. However, several weeds are from the same family as is sugarcane and therefore have the same metabolism. At the time of harvest, the presence of weeds continues to cause damage, because when sugarcane is cut by hand and without previous burning, the presence of weeds decreases the workers' efficiency and makes them more vulnerable to snake and scorpion bites. The longevity of the sugarcane plantation also decreases when weed control is not effective. In most cases, keeping the crop free from competition with weeds until the vegetation closes the spacing between the rows ensures the harvest of sugarcane without the presence of weeds.

There are several techniques used for weed management. However, the combined use of cultural, mechanical, and chemical methods is the most common [4, 15]. Cultural methods are practices that aim to make sugarcane crop more competitive in relation to weeds and include reducing planting space, intercropping or crop rotation with soybean, peanut, corn, and green manure, as well as the use of high-tillering varieties for faster shading of the soil [4, 15]. In research conducted by the authors in areas heavily infested with Brachiaria, the sowing of *Crotalaria juncea* at the time of the renewal of the sugarcane plantation had a great suppressive effect (due to shading) and reduced the number of seeds produced by the grass.

Plows and harrows are used in the mechanical control during the renewal of the sugarcane plantation. This method is highly efficient, but depends on soil moisture, solar radiation and the predominant species in the area. For weed control between the rows of sugarcane, animal traction plows can be used. Manual weeding, once used, has now been restricted to experimental areas. Mechanical cultivation has limitations, especially because it does not control the weeds of crop rows. Its efficiency in controlling weeds between crop rows may also be greatly reduced depending on the climatic conditions and species, such as some grasses of the *Brachiaria* genus, which reproduce vegetatively. The chemical method has been the most used by small producers (**Figure 4**), with herbicides being used in preemergence or postemergence of weeds and with those in early or developed stages (**Table 6**).

Below are some considerations about herbicides commonly used in sugarcane cultivation.

Figure 4.
The use of herbicides to maintain the crop free from weed competition until the space between rows is closed generally guarantees sugarcane harvest without the presence of weeds.

Ametryn: recommended in pre- or early postemergence application, alone or mixed with other herbicides (diuron, 2,4-D, Tebuthiuron, Clomazone, Monosodium methyl arsenate (MSMA), and others). This herbicide causes little toxicity to the sugarcane crop. The effective control period or residual effect is about 70–100 days. It is effective in controlling *Digitaria horizontalis*, *Brachiaria plantaginea*, *Eleusine indica*, *Portulaca oleracea*, *Amaranthus spp*, *Bidens pilosa* and *Acanthospermum hispidum*. Doses vary from 3.0 to 5.0 L/ha. The smaller doses are recommended for lighter soils or application in moist soil, while higher doses should be used in heavy soils or application in dry soils.

2,4-D: recommended for post-emergence application, alone or mixed with other herbicides (Diuron, MSMA, Tebuthiuron, Ametryn, and others). The effective control period is 25–40 days. It is effective only in the control of dicotyledons. Doses vary from 0.8 to 1.5 l per ha.

Diuron: Recommended in pre- or early postemergence application, alone or mixed with other herbicides (2,4-D, Tebuthiuron, MSMA, and others). It has a residual effect of 150 to 180 days. It is effective in the control of *Digitaria horizontalis*, *Brachiaria plantaginea*, *Eleusine indica*, *Portulaca oleracea*, *Amaranthus spp*, *Bidens pilosa*, *Bidens pilosa* and *Acanthospermum hispidum*. Doses range from 1.8 L/ha (Diuron 500) to 3.2 L/ha (Karmex GRDA).

Diuron + Hexazinone: sold under the names Advance and Velpar K. Advance is mostly used in plant-cane and Velpar K in regrowth. Both are applied in pre- or early postemergence in moist soil. They may cause toxicity to the crop, especially in sandy soils and in conditions of high moisture content. It has broad spectrum of control and residual effect of 90–120 days. It is efficient in controlling *Panicum maximum* jacq. Cv.Colonião, this is a very present and competitive species in the cultivation of sugar cane. The doses of Advance and Velpar K vary from 2.2 to 2.8 and 1.5 to 2.7 kg/ha, respectively.

Action mechanism	Application time	Herbicide (Brand)
Membrane disrupters		
Oxyfluorfen	Pre- or postemergence	Goal
Sulfentrazone	Preemergence	Boral and Solara
Inhibitors of photosynthesis		
Ametryn	Pre- or postemergence	Ametryn, Gesapax, Herbipax, and Metrimex
Atrazine	Pre- or post-emergence	Atrazinex, Gesaprim, Atrazine, and Nortox
Atrazine + Simazine	Pre- or postemergence	Extrazin, Triamex, Primatop, and Herbimix
Diuron	Pre- or postemergence	Karmex, Diuron Nortox
Ametryn + Diuron	Pre- or postemergence	Ametron
Hexazinone + Diuron	Pre- or postemergence	Velpar K and Advance
Metribuzin	Pre- or postemergence	Sencor
Simazine	Preemergence	Gesatop
Tebuthiuron	Preemergence	Combine, Tebuthiuron
Inhibitors of mitosis and early growth		
Alachlor	Preemergence	Laço
Pendimethalin	Preemergence	Herbadox
Trifluralin	Preemergence	Trifluralina Nortox, Treflan, and Premerlin
Inhibitor of respiration		
MSMA	Postemergence	Daconate, Dessecan
Inhibitors of amino acid synthesis		
Flazasulfuron	Pre- or postemergence	Katana
Glyphosate	Postemergence	Glifosato, Roundup, and Trop
Halosulfuron	Postemergence	Sempra
Imazapic	Pre- or postemergence	Plateau
Imazapyr	Pre- or postemergence	Arsenal, Contain e Chopper
Trifloxysulfuron +Ametryn	Postemergence	Krismar
Inhibitors of pigment synthesis		
Clomazone	Preemergence	Gamit
Isoxaflutole	Preemergence	Provence
Clomazone + Ametryn	Preemergence	Sinerge, Ranger
Growth regulators		
2,4-D	Pre- or postemergence	DMA
Dicamba	Pre- or postemergence	Banvel 480
Picloram +2,4-D	Pre- or postemergence	Dontor

Table 6.
Main herbicides used in sugarcane and application times in relation to weed emergence.

MSMA: recommended for late postemergence in direct spray application. It causes high toxicity to the sugarcane crop. It has been widely used in mixtures with other herbicides (Diuron, 2,4-D, Tebuthiuron, Ametryn, Velpar K, and others). It is effective in the control of several annual and perennial grasses, as well as nutsedge when it has about 4–8 leaves. Doses of 0.5–0.87 L/ha are recommended for MSMA 790.

Tebuthiuron: recommended for preemergence, including application to dry soils. It is not efficient in postemergence application. It persists in soil for a longer period, with residual effect of 12–15 months. It should not be used in areas intended for future renewal of the sugarcane plantation and should use crop rotation with soybean, peanut, and bean. It has a broad spectrum of control, and it is efficient in controlling several dicotyledons and grasses, including *Panicum maximum* jacq. Cv. Colonião, but not efficient for *Digitaria horizontalis*.

9. Chemical and organic fertilization of regrowth

The fertilization of the regrowth of sugarcane recommended for small farm-ers is based on the recovery of nitrogen and potassium removed by harvesting the previous cut. In one ton of natural matter of sugarcane biomass (industrializable culms + dry leaves + green leaves + buds), there is 850 kg of industrializable culms on average. Thus, the average index of industrializable culms in biomass is 0.85. For each ton of sugarcane biomass exported from the plantation, 1.2 kg of nitrogen and 1.5 kg of potassium are removed [1, 4]. Assuming that the yield of the plant-cane was 150 t of biomass (about 120 t of industrializable culms), the application of 180 kg of nitrogen and 225 kg of potassium is recommended. There is no need to split fertilization for fear of leaching losses of both nitrogen and potassium [1].

If sugarcane has been harvested for animal feed, the amount of dry leaves on the soil is small. Thus, burying the fertilizer between the crop rows is recommended. Urea can be used as the source of nitrogen in this case. If the fertilizer cannot be buried between the crop rows, ammoniacal or nitric sources should be used to fertilize the regrowth, because volatilization losses are very high when urea is applied over the straw. More detailed information on losses by volatilization and quantification of potassium in the profile of soil grown with sugarcane can be found in Ref. [1].

The application of the fertilizer in regrowth using a walk-behind spreader has been traditional in small properties. It is of low cost and precise, which greatly increases the workers' efficiency. A worker using a spreader covers more than 2.0 ha per day. The use of dairy cattle manure to fertilize sugarcane is a way to reduce production costs, recycle nutrients, and improve soil physical, chemical and biological properties. The efficiency of dairy cattle manure in the mineral nutrition of sugarcane depends on the chemical composition of the waste, the dose and environmental factors, especially temperature and soil moisture. The chemical analysis of cattle manure allows us to calculate the amounts that should be applied to the soil to restore the nutrients removed by harvesting. **Table 7** shows the results of the chemical analysis of manure of dairy cattle fed with different roughage and amounts of concentrated feed. In harvesting 150 tons of forage (culms + side branches + leaves), 150, 45, and 225 kg of N, P, and K, respectively, were removed. The K/N ratio of forage is 1.5, which is much higher than that of cattle manure. Thus, if the sugarcane plantation is fertilized with cattle manure, it is necessary to supplement fertilization with potassium to replace nutrients removed with the harvest.

Poultry litter is another alternative for the organic fertilization of sugarcane. In recent years, this waste has had its demand and price reduced because of a ban on its use in cattle feed. The concentration of nutrients in poultry litter is influenced by the bedding material. Most poultry farmers of the Zona da Mata region use rice husk, coffee husk, napier grass, shavings, and corn cobs. **Table 8** shows the nutrient con-tents of five poultry litters composed of different bedding materials collected after being used in a batch of broilers (48 days on average), at a density of 15 birds/ m^2.The authors evaluated the maturation and broth quality of the sugarcane variety RB867515, which were influenced by the fertilization with poultry litter. The study

Feed		% de DM in manure	N	P	K	Ca	M	S	K/N ratio
Roughage	Concentrad feed (kg/ cow/day)		. ------- g/kg of manure dry matter--------.						
Pasture	9.0	14.1	19.4	14.4	12.0	13.9	7.1	3.5	0,62
Corn silage	8.0	13.9	23.6	10.0	10.4	10.5	5.1	3.1	0,44
Corn silage	10.0	11.7	21.1	10.9	11.2	11.6	4.5	3.4	0,53
Corn silage	12.0	12.9	24.1	11.2	11.7	12.3	4.6	3.6	0,48
	Average	13.1	22.0	11.6	11.3	12.0	5.3	13.6	0.52

Table 7.
Percentage of dry matter in dairy cattle manure (% DM in manure) and nitrogen (N), phosphorus (P), potassium (K), calcium (Ca), magnesium (Mg), and sulfur (S) contents in manure of dairy cows fed with different roughage and quantities of concentrated feed.

Poultry litter	N	P	K	Ca	Mg	S
	.---------------------- g/kg of DM---------------------------.					
Rice husk	34.7 a	15.9 b	26.8 b	25.7 a	6.2 a	16 ab
Coffee husk	32.8 a	14.4 b	28.9 ab	25.0 a	5.5 b	15 b
Napier grass	34.8 a	15.1 b	23.3 c	25.5 a	6.0 a	15 b
Wood shavings	30.9 a	13.7 b	24.4 c	25.8 a	5.7 b	14 b
Corn cob	34.2 a	18.6 a	29.7 a	28.3 a	6.7 a	18 a
Average	33.5	15.5	26.6	26.1	6.0	15.0
CV (%)	10.6	12.2	10.0	6.6	8.2	10.1

Means followed by the same letter in the column do not differ statistically from one another by the Tukey test at 5%.
Source: Ref. [16].

Table 8.
Nutrients in dry matter (g/kg of DM) of poultry litter in which five different materials were used for bedding.

was conducted in three agricultural years: first, second, and third regrowth. The experiment was a randomized complete block design with four replicates. The treatments consisted of fertilization with three doses of poultry litter 7; 10 and 13 t of poultry litter dry matter/ha/yr), in addition to a control treatment (no chemical or organic fertilization), and chemical fertilization (180 kg N + 225 kg K/ha/yr). Sugarcane was harvested in early August in all 3 years. Fertilization of sugarcane with poultry litter did not interfere with maturation, nor did it affect broth quality, even at high doses. The difference in broth quality from 1 year to another was small. The average of the 3 years for soluble solid content, sucrose in the broth and broth purity were 22, 19, and 86%, respectively. Thus, the use of poultry litter is as an organic fertilizer is an alternative to chemical fertilization. Also, just as cattle manure, it should be supplemented with potassium, because the average K/N ration in poultry litter is 0.80.

10. Finalconsiderations

The technologies recommended to the small farmers for the implantation and management of sugarcane plantations have resulted in high yields in the plant-cane cycle with small decreases in the subsequent cycles. Furthermore, the techniques proposed by the authors maximized the use of inputs, land, and human resources, thus reducing the operating costs.

Author details

Mauro Wagner de Oliveira[1], Luís Cláudio Inácio da Silveira[2], Alexandre Bosco de Oliveira[3*], Márcio Henrique Pereira Barbosa[2], Manoel Gomes Pereira[4] and Terezinha Bezerra Albino Oliveira[1]

1 Federal University of Alagoas, Rio Largo, Brazil

2 Federal University of Viçosa, Viçosa, Brazil

3 Federal University of Ceará, Fortaleza, Brazil

4 Research and Development Sector, Triunfo Mill, Boca da Mata, Brazil

*Address all correspondence to: aleufc@gmail.com

References

[1] Oliveira MW, Macêdo GAR, Martins JA, Silva VSG, Oliveira AB. Mineral Nutrition and Fertilization of Sugarcane. Available from: https://www.intechopen.com/books/sugarcane-technology-and-research/mineral-nutrition-and-fertilization-of-sugarcane [Accessed: 23 October 18]

[2] Amabile RF, Fancelli AL, Carvalho AM. Comportamento de espécies de adubos verdes em diferentes épocas de semeadura e espaçamentos na região dos cerrados. Pesquisa Agropecuária Brasileira. 2000;**35**:47-54

[3] Duarte Júnior JB, Coelho FC. Adubos verdes e seus efeitos no rendimento da cana-de-açúcar em sistema de plantio direto. Pesquisa Agropecuária Brasileira. 2008;**67**:723-732

[4] Oliveira MW, Freire FM, Macêdo GAR, Ferreira JJ. Nutrição mineral e adubação da cana-de-açúcar. Informe Agropecuário. 2007;**28**:30-43

[5] Mascarenhas HAA, Tanaka RT, Costa AA, Rosa FV, Costa VF. Efeito residual de leguminosa sobre rendimento físico e econômico da cana-planta. Campinas: Instituto Agronômico; 1994. p. 15

[6] Wutke EB, Arévalo RA. Adubação verde com leguminosas no rendimento da cana-de-açúcar e no manejo de plantas infestantes, Boletim Técnico IAC. Vol. 198. Campinas: Instituto Agronômico de Campinas; 2006. p. 28

[7] Oliveira MW, Brighente IMC, Oliveira TBA, Gava GJC. Taxa de crescimento, acúmulo de nutrientes e atividade alelopática da crotalaria juncea. In: III Simpósio Brasileiro de Agricultura Sustentável. 2011. Viçosa: Anais do... III Simpósio Brasileiro de Agricultura Sustentável; 2011. CD ROM

[8] Perin A, Santos RHS, Urquiaga S, Guerra JGM, Cecon PR. Produção de fitomassa, acúmulo de nutrientes e fixação biológica de nitrogênio por adubos verdes em cultivo isolado e consorciado. Pesquisa Agropecuária Brasileira. 2004;**39**:35-40

[9] Chada SS, De-Polli H. Nodulação de leguminosas tropicais promissoras para a adubação verde em solo deficiente em fósforo. Pesquisa Agropecuária Brasileira. 1988;**23**:1197-1202

[10] Ribeiro Júnior WQ , Ramos MLG. Fixação biológica de nitrogênio em espécies para adubação verde. In: Carvalho AM, Amabile RF, editors. Cerrado: Adubação verde. Brasilia: Embrapa Cerrados; 2006. pp. 171-209

[11] Salcedo IH, Sampaio EVS, Alves GD. Mineralização do carbono e do nitrogênio em solo cultivado com cana-de-açúcar. Revista Brasileira de Ciência do Solo. 1985;**9**:33-38

[12] Rufty TW, MacKown CT Jr, Israel DW. Phosphorus stress effects on assimilation of nitrate. Plant Physiology. 1990;**94**:328-333

[13] Malavolta E, Vitti GC, Oliveira SA. Avaliação do estado nutricional das plantas. Piracicaba: Associação Brasileira para Pesquisa da Potassa e do Fosfato; 1989. p. 201

[14] Magalhães JV. Absorção e translocação de nitrogênio por plantas de milho (*Zea mays*, L.) submetidas a períodos crescentes de omissão de fósforo na solução nutritiva (Dissertation). Viçosa: Universidade Federal de Viçosa; 1996

[15] Procópio SO, Silva AA, Vargas SL, Ferreira FA. Manejo de plantas daninhas na cultura da cana-de-açúcar. Viçosa: Suprema; 2003. p. 150

[16] Souza CR. Disponibilidade e transformações de formas de fósforo em camas de aviário por meio de fracionamento químico e ressonância magnética nuclear do ^{31}P. Viçosa – UFV [Dissertação]. 2004. p. 76

8

Water Productivity Improvement Under Salinity Conditions: Case Study of the Saline Areas of Lower Karkheh River Basin, Iran

Nader Heydari

Abstract

Soil salinity and low drainage capacity are major factors limiting crop yield and low agricultural water productivity (WP) in the Lower Karkheh River Basin (L-KRB) in the southwest to south of Iran. The objective of this chapter is to provide and elaborate cost-effective and adaptive solutions and measures for the amelioration of the situation and for enhancing WP in this area. The suggested approaches and measures are based on field experiments conducted both under farmer's field and research station. The main cultivated crop in the region is wheat. For the first step, WPs of wheat were determined under some farmer's fields. Based on results, WP of wheat in the area varied between 0.24 and 0.77 kg/m^3. The study aimed to provide simple measures and management practices for reducing salinity and water-logging hazards and ultimately improving crop WP in the studied area. The focus of experiments was on the methods of land preparation for irrigation, sowing methods by the different adapted machines, and seeding rates. Changing of traditional basin irrigation to a modern mixed system of border-basin irrigation method could improve irrigation performance and will reduce applied water greatly and hence could improve WP. Construction and provision of fixed and low-cost water intake structures on farm ditches could have more impacts on the improvement of irriga-tion performance as well. Optimum (modern) border irrigation and optimum basin treatments had higher WPs (1.36 and 1.04 kg/m^3, respectively) than the traditional irrigation practiced by the farmers (0.61 kg/m^3). Although the WP of optimum border was highest, the basin irrigation method is suggested because this method is more adaptive and sustainable in terms of acceptance by the local farmers.

Keywords: water, salinity, productivity, basin, border, crop, improvement, Karkheh

1. Introduction

Agriculture plays an important role in Iran's economy. It accounts for 18% of gross domestic product (GDP), 25% of employment, 85% of food requirements, 25% of non-oil exports, and 90% of input materials for the local industries [1].

The climate of Iran is very diverse, and great extremes are common features of it due to its geographic location and highly varied topography. The major area of the country (almost 90%) could be classified as arid to semiarid. Of the characteristic of such a climate are hot summers with the temperatures reaching up to 55°C in the interior and southern parts. Evaporation demand is much higher than precipitation in most areas of the country. Consequently, water resource management under these conditions is an important issue and a great challenge especially in the agricul-tural sector.

Despite reliance of the country on agriculture, especially irrigated agriculture, water resources required for agricultural production are limited but provide a vital input to agricultural production in Iran. Currently more than 93% of water consumption (84 billion cubic meter; BCM) is used to irrigate 8.2 million hectares (Mha) of lands. Considering the growing demand for water in industrial and municipal sectors, combined with the environmental concerns, in the near future, there will be lesser freshwater resources available for agriculture in the country.

The latest agricultural statistics reveal that Iran produced 77 million tons of agricultural products from 84 BCM of water consumed. Therefore, currently the average water productivity (WP) in agriculture is almost 0.92 kg/m^3. This value is quite low and necessitates the use of appropriate approaches for its improvement [2].

Studies conducted on farmers' fields in five regions in the country (Kerman, Hamedan, Moghan, Golestan, and Khuzestan provinces) revealed that WP for the irrigated wheat varies in the range of 0.56–1.46, sugar beet (0.59–1.28), sugarcane (0.31), potato (1.45–3.0), silage corn (6.46), cotton (0.73), alfalfa (1.48), barley (0.56), and chickpea (0.18) kg/m^3 [3, 4]. Based on the review of 84 references on values of WP during the past 25 years, it was found out that the average WPs of wheat are about 1.09 kg/m^3 [5].

However, there is no information available or accessible that addresses assess-ment of WP in the Lower Karkheh River Basin (L-KRB). Preliminary estimates that are based on farmers' field visits and questionnaire on crop yield and applied water suggest WP of irrigated wheat to be about 0.6 kg/m^3 in this region.

Soil and water salinity are major threats and barriers to the optimal crop produc-tion systems in most parts of the world, especially in arid and semiarid regions, including Iran [6, 7]. The salinization of land and water resources has been the main consequence of two factors including naturally occurring phenomena and anthro-pogenic activities. The first factor has caused primary fossil salinity and/or sodicity, while the second factor, which is more prominent, has caused human-induced or secondary salinity and/or sodicity [8].

Soil and water resource salinities are the major threats to the crop production and sustainability of natural resources, especially in irrigated agriculture based on groundwater resources in Iran. Salt-affected soils are the major features of many parts of the country, particularly in the Central Plateau, which is surrounded by two main ranges of high mountains along with the northwest to northeast (Alborz range) and northwest to southern parts and southeast (Zagros range).

Irrigated agriculture is the main cropping system in the country, while out of which at least 4.1 Mha of 8.2 Mha of total irrigated lands (nearly 50%) suffer from different levels of salinity and sodicity [9]. This is under conditions that irrigated agriculture is the main focus of government plans for increasing agricultural products and food security, especially in the recent decades.

Irrigated agriculture with low irrigation efficiencies has been one of the causes of human-induced salinization of land and water resources. This phenomenon has occurred mostly in unique topographic conditions of semi-closed or closed intermountain basins where irrigated agriculture has been practiced for many years. Distribution of land salinity in Iran is diverse. The extent and characteris-tics of salt-affected soils in Iran have been investigated and are reported by several researches [10–14]. Overall, the slight to moderate salt-affected soils are mostly formed on the piedmonts at the foot of the Elburz Mountains in the northern part of the country, while the lands with severe to extreme degrees of salinity are mostly located in the Central Plateau, Khuzestan province plains, Southern Coastal Plains, and some parts of the Caspian Coastal Plains (mainly in Golestan province in the northeast) [7].

Average yields of the common crops vary highly depending on the climatic and soil and water conditions. However, because of many limiting factors, mainly water and salinity stresses, the achieved yields are generally suboptimal. Average crop yield losses due to salinity stress are estimated to be up to 50% in areas where salinity is present [15].

Large areas of Iran suffer from salinity and sodicity hazards. Wide distribution of such areas in the country reflects the fact that many factors are contribution to this phenomenon. Indeed, the causes of soil salinity could be divided into natural or primary causes and secondary or man-made causes. The natural causes include geological and physiographic conditions, climatic conditions, and salt loads by water. The man-made causes are mainly because of improper irrigation management followed by waterlogging problems. Waterlogging is mostly occurring in the irrigation networks developed under regulated waters, e.g., dams. Because of water scarcity, especially in recent decades, the increased use of marginal brackish waters for irrigation without required management also has worsened this problem and is sparking the soil's secondary salinization. Overall, the man-made causes of soil salinity in Iran could be nominated as poor water management, use of saline groundwater, over exploitation of groundwater, poor land preparation, fallowing and overgrazing of lands, improper cropping pattern, and sea-water intrusion into coastal areas.

Because of the importance of facing with salinity in agriculture of Iran, many research projects are conducted specifically on salinity issues in the country till date. These research projects cover some provinces or regions in the country facing with soil and water salinity hazard, e.g., Yazd, Golestan, Fars, Khurasan, Khuzestan, Markazi, Hormozgan (Bushehr), Moghan, Azerbaijan, Esfahan, and Qom. Very roughly it could be stated that till now the majority of research projects conducted in the country cover mainly the following areas including the extent of salt-affected soils, character-istics of salt-affected soils, methods of reclamation of salt-affected soils, crop fertility and productivity potentials of salt-affected soils, productivity of saline water used in different parts of the country, amount and distribution of saline water resources (as drainage water and groundwater resources), and quality of saline waters in regard to the salts and other contaminants. However, very limited salinity research projects have been conducted on the aspects of evaluation and improvement of WP under salinity conditions in the country in general and in the L-KRB in specific.

Karkheh River Basin (KRB) is a typical and important basin in Iran regarding the supply of water resources for the fertile plains and favorite climatic conditions for productive agriculture in the downstream basin, mainly located in the Khuzestan province. The other important characteristic of the KRB is that both dryland and irrigated agricultural production systems exist in the basin. Water in KRB is becoming scarcer because of climate change together with the population growth. Therefore water demand, especially for the irrigated areas on the downstream basin which are under intensive development, is increasing. The productivity of rainfed agriculture is low, conventional irrigation management is poor, cropping systems are suboptimal, and policies and institutions are weak [16]. Considering these inefficiencies, improvement of water productivity in agricultural sector is a recent and important policy of the country especially in KRB.

In the upper KRB, the dryland agriculture prevails. The challenges for the rural households in such areas are similar to the ones in other dryland areas, i.e., agricultural options are limited, and wheat, barely, and pulses are dominant cropping patterns in the landscape. Agricultural outputs are usually low and unstable, due mainly to the resource degradation, drought spells, and climate change impacts [16]. Irregular rainfall on poorly vegetated hill slopes results in severe soil ero-sion, downstream flooding, and sedimentation. Consequently, the lifetime of the Karkheh Reservoir Dam in the downstream basin is dwindling rapidly. These environmental constraints combined with their economic problems make this southwest corner of Iran one of the poor areas of the country with a high out-migration rate [16].

KRB had been selected as one of the nine benchmark basins of the CGIAR Challenge Program on Water and Food (CPWF[1]). One of the CPWF Phase 1 projects focuses on interventions for the improvement of on-farm agricultural WP in KRB. This project was carried out jointly by the International Center for Agricultural Research in the Dry Areas (ICARDA) and Agricultural Research, Education and Extension Organization (AREEO) of Iran. The objectives of the project were to develop biophysical interventions to improve the farm and basin level of WP and sustainable management of the natural resources and to develop appropriate policies and institutions supporting the project interventions to help the poor communities for the improvement of their income and livelihoods. Moreover, the project aimed at strengthening and enhancing the capacity of the National Agricultural Research and Extension Systems (NARES) of Iran.

KRB is becoming a water-scarce area, and droughts and climate change are becoming permanent features of this region. Because of water scarcity and degradation of land and water resources, livelihoods of rural communities are at stake. With the current rate of deterioration of natural resources if no remediation is taken, the situation will worsen in the near future. However, there are great potentials for the improvement of land and water productivities in the KRB. Therefore KRB was well adapted to be a pilot area for the development-oriented research activities to be implemented under Phase 1 of CGIAR Challenge Program on Water and Food (CPWF). KRB situation provided a unique opportunity for the CPWF to make an impact through improvements in land and water productivities, which in turn will improve the livelihoods of rural poor living in this basin. The issues of KRB have a great similarity with other basins located in the similar hydrological conditions, e.g., West Asia and North Africa (WANA) region.

This chapter provides an overview of the soil and water potential of the L-KRB and the salinity and waterlogging constraints to agricultural production and agricultural WP improvement under saline areas of L-KRB. The findings are mostly based on the research results conducted during the CPWF Phase 1 comprehensive project in KRB.

---

[1] The Challenge Program on Water and Food (CPWF) is one of the pilot programs designed to reinvent the business model for the CGIAR. The CPWF was launched in 2002 as a reform initiative of the CGIAR, the Consultative Group on International Agricultural Research. CPWF aims to increase the resilience of social and ecological systems through better water management for food production (crops, fisheries, and livestock). CPWF does this through an innovative research and development approach that brings together a broad range of scientists, development specialists, policymakers, and communities to address the challenges of food security, poverty, and water scarcity. CPWF Phase 1 worked in nine river basins globally: Andean system of basins, Indo-Gangetic, Limpopo, Mekong, Nile River, Yellow River, Sao Francisco, Volta, and Karkheh River Basin.

2. Karkheh River Basin characteristics

KRB is located in the west to southwest of Zagros ranges in Iran (**Figure 1**). It is located between 56°, 34′–58°, 30′ north latitude and 46°, 06′–49°, 10′ longitude. The area of the basin (inside Iran) is 50,764 square kilometers (km^2). Out of which 27,645 km^2 are mountains and 23,119 km^2 are plains and hills. The mountainous areas of KRB are mostly in the eastern and central parts. The plains are mostly in the northern and southern parts and cover almost 45% of the basin area. Hypsometric studies indicate that 75% of the basin is located in altitudes of 1000–2000 and 0.6% of the basin is above 2500 m altitude.

The Karkheh River arises from the confluence of numerous large and small tributaries including the three large rivers, namely, Gamasiyab, Ghareh-So, and Kashkan. The Karkheh River has various names along its route and is locally best known as the Saymareh River at the point where the Gamasiyab and Ghareh-So Rivers combine, and later the point where the Kashkan River flows into the main waterway is known as the Karkheh River. When approaching to the Khuzestan province hypsometric and slope of the basin decrees and gently the river ultimately flows into the Hawr-al-Azim (HAA) wetland at the basin outlet. Therefore KRB could be classified as a closed basin.

Based on general hydrological classification of basins in Iran, the KRB is considered as one of the sub-basins of the Persian Gulf Great Basin.

The pattern of precipitation in KRB is affected by Mediterranean regime. It means that the dry season is coinciding with summer and rainy season match with cold months. The rainfall distribution in the basin is very scatter, but most of the rain falls in winter and autumn seasons. The annual precipitation of the basin is 219 mm in Hamidieh (in L-KRB) to 765 mm in the northern dryland farming areas (in upper KRB).

Based on climatic maps, the hottest areas of the basin are located in its southern parts (L-KRB) and are surrounded by the 25°C iso-temperature (isohyets) contours. The coldest areas of the basin are located in altitude higher than 3000 m and are mostly located in the north and northeast of the basin and are surrounded by the 5°C isohyets' contour map.

Evaporation in KRB varies between 1800 and 3600 mm depending on the altitude. For example, it is around 3561 mm in Abdul-Khan Station in an altitude of 40 m in L-KRB. Almost 79% of annual evaporation occurs from May to September.

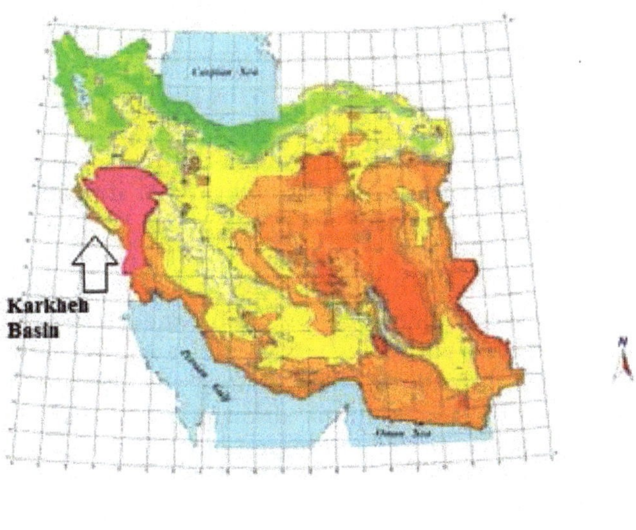

Figure 1.
Geographical location and boundaries of KRB.

In KRB both surface and groundwater resources are used. Based on data, in 1994, the share of total agricultural water uses from both resources was 3.92 BCM. The agricultural water consumption till completion of the ongoing water works could be increased to 7.43 BCM (90% increase).

The water resources of the basin in general have a good quality. However, the quality of groundwater in the Southern plains deteriorates to some extent. The potential of surface water resources of KRB is 7.4 BCM. In wet years it can be doubled, and in dry years, it can be reduced to half. Agricultural water withdrawn in KRB is 3.96 BCM (in year 1994). Out of which 36.8% is supplied from groundwater and 63.2% is supplied from surface water resources.

Plains of L-KRB are the highest consumer of surface water resources. Based on year 1994 statistics, out of different plains of KRB, Azadegan plain (Dasht-e Azadegan, DA) with 662 MCM water consumption is the greatest consumer of water in the basin. This plain also is the greatest consumer (660.2 MCM) of surface water in the basin.

Based on year 1994 statistics, out of 4157.4 MCM consumed water resources, 2504.6 MCM (60.2%) were from surface water, and 1653 MCM (39.8%) from groundwater resources. Out of the total consumption, share of rural, urban, industry and mining, and fishery consumptions was 1.23, 3.93, 0.32, and 0.35 BCM, respectively. The share of agricultural water uses this year was 94.2%. The share of agricultural water consumption in the basin is the highest. The sums of indus-trial and mining consumptions are very low and just consist 0.32% of total water consumptions. Therefore, from the aspects of water resources' uses, the KRB could be defined as an agricultural basin.

Two major agricultural production systems prevail in the KRB. The dryland system prevails in the upstream areas (upper KRB), while the fully irrigated areas are located in some part of upstream and mainly in downstream of the KRB (L-KRB). The dryland areas are well established and cover most of the basin agricultural lands, occupying 894,125 ha, whereas irrigated lands occupy 578,862 ha but are expected to expand up to 340,000 ha following the completion of irrigation and drainage networks under Karkheh reservoir Dam [16].

Owing to the different flowing rivers, abundant water resources, fertile lands, and sufficient extraterrestrial energy, Khuzestan province in the southwest Iran is one of the potentially most suitable regions for agricultural production. However, salinization of land and water resources has become a serious threat to the efficient use of agricultural lands. It is estimated that out of the total 6.7 Mha of the province, 1.2–1.5 Mha (18–22% of total area) are faced with the conjunctive problems of soil salinization and waterlogging [16].

The next agricultural production system in the KRB is irrigated agriculture. It is estimated that about 1 Mha of land are irrigable in KRB. Out of which about 380,000 ha are currently under cultivation [16]. About 340,000 ha of additional available arable lands will be brought under irrigation following the completion of irrigation and drainage networks under Karkheh Reservoir in L-KRB [16].

The drainage outlet of the KRB is the HAA wetland along with the Iran-Iraq border (**Figure 2**). At present, there are very limited modern irrigation and drainage networks under operation within the L-KRB. However, the networks are under completion, and irrigated agriculture is developing gradually [17]. The government has started construction of irrigation and drainage networks, especially on the tertiary canal level, with the goal of improving on-farm water management and modernization of traditional irrigation networks. The focuses of these activities are the arable lands under the Karkheh Reservoir and in pilot projects such as DA plains in the southern parts of the L-KRB (**Figure 2**).

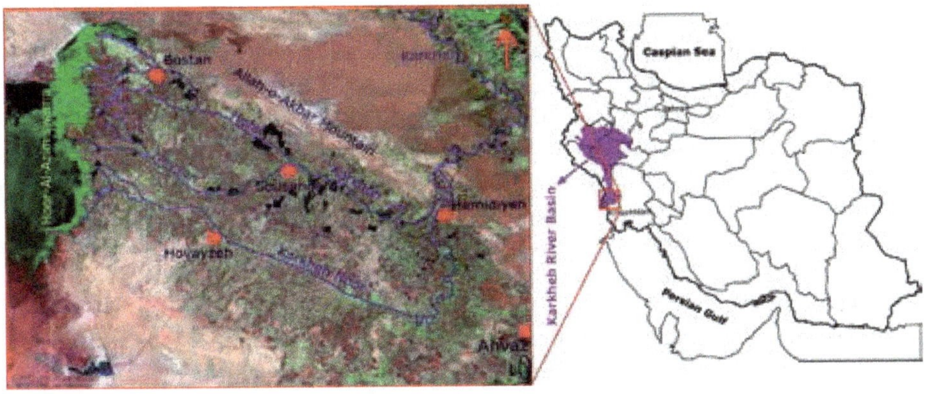

Figure 2.
Karkheh River Basin (KRB) and Dasht-e-Azadegan (DA) region in the L-KRB.

In the L-KRB because of the differences in factors affecting agricultural WP in the northern and southern parts, two distinct regions can be identified. In the northern part, there are no serious limiting factors in regard to soil and water quality. In this part improving farmer's skills and use of appropriate farming systems can improve WP sufficiently. Shortfalls in water supply and excess irrigation water losses, mainly in earthen canals, cause low WPs of cultivated crops in this area. Extension of new farming systems, e.g., pressurized irrigation and land preparation methods (e.g., raised-bed and double-row cropping systems, etc.), could be some useful approaches for improving WP in the area. Overall in this part, successful introduction and implementation of new farming systems and technologies in accordance with other agricultural services could be effective ways for crop WP improvement.

In the southern parts of L-KRB, heavy soil texture and subsurface water recharge from upstream areas cause natural condition for waterlogging. This situation is more induced by deep percolation losses resulting from low irrigation efficiencies of irrigated lands in the area.

The available soil data indicate that the majority of arable lands in L-KRB possess with various degrees of limitations. However, soil salinity, waterlogging, lack of soil organic matter, soil structural deterioration, intrinsic low permeability, and low infiltration rate caused by soil compaction are the main factors limiting economic and sustainable crop production of the irrigated lands in this area [16].

Waterlogging and soil salinity are the main causes of inefficiencies in achieving high WPs and are threats to sustainable agricultural production in the L-KRB. Major factors causing soil salinization in the L-KRB could be classified as follows [18–20]: shallow water table, existence of salt containing soil layers, inadequate natural drainage, inadequate artificial drainage networks, high evaporation demand of the climate, salt intrusions by wind, and salty sediment transport during flood periods.

In the southern parts of L-KRB, mainly DA plain, available data and surveys show that the problem of soil salinity is intensified because of deficiencies in farmers' knowledge and skills and lack of new and improved farming practices.

In general, the main cause of soil salinity in the L-KRB is high water table which is often less than below 2.0 m from soil surface and usually varies between 1.2 and 3.0 m below the soil surface. If sufficient developments of drainage networks are not provided, the problem will be worsened considering the coming development plans with the aim of expansion of new irrigation networks.

The total rainfall in Susangerd and Bostan towns' climatological stations are 180 and 200 millimeter, respectively. The agricultural service centers also are equipped with rain gauges.

Current crops in Azadegan Plain in southern L-KRB include cereals (such as wheat, barely, rice), vegetables (such as melon, watermelon, tomato, cucumber, eggplant, okra, lettuce, cabbage, carrot, radish, onion, etc.), grains such as beans, and fodder crops (such as alfalfa, barely, maize, and sorghum). More than 78% of agricultural production in Azadegan Plain is dominated by grains, mainly wheat and barley [16, 21]. This is because of soil salinity and sodicity with high toxic elements which makes serious limitation for cultivation of other crops. Currently water supply limitations, agriculture economy (guaranteed purchase of with by the government), and security problems in the region (wheat need less labor, less irrigation, and in overall less need for the stay of farmer in his land) are some other reasons for the farmers' higher interests on wheat cultivation.

The main challenge of agriculture in this region is waterlogging and soil salinity. Waterlogging and secondary soil salinization occur in a certain period of the year. For example, early November is planting date of wheat cultivation system in DA. Late November is the first irrigation for land preparation, and the harvest time is in late May. Deep percolation losses of irrigation during this period cause rises in water table. The peak of water table rise is in February. The salinity [Electrical conductivity (EC)] of shallow groundwater and EC of irrigation water in this area are about 6–9 and 3 dS/m, respectively. The highest depth of water table depth is between 0 and 1.2 m. Operation of main drains has started in recent decade (in 2003), and their outlet is HAA Wetland in the border of Iran-Iraq [16].

3. Amelioration and management approaches for improving WP in the saline areas of lower KRB

There is no doubt that one of the most important requirements for the reclamation of lands in the L-KRB is installation of adequate drainage network for the entire irrigated area. Installation of drainage network is a fundamental solution to improve the quality of salt-affected soils in the L-KRB. Drainage system will reduce the adverse effects of shal-low water table and waterlogging issues in the agricultural lands. Hence it will contribute to the improvement of crop production and crop WP. Promising efforts have been initi-ated by the government in this regard, but still the progress is low and very costly.

To avoid further salinization of agricultural lands and to ameliorate the current situation, the communities and agricultural agencies are called to apply sound management practices until adequate drainage systems are installed.

One of the most important prerequisites to enable sustainable crop production in the area is the development of a monitoring network for observing the effect of different management practices on the salt content of groundwater as well as the salt and water balance of the crop's root zone. The regular monitoring and data acquisition will provide the database required for providing the best measures to prevent restoration of soil and water salinity and secondary salinization of the crop root zone. Moreover, water and salt balance studies at the watershed level will increase the capability to predict the role of any hydrological impacts on the fate and behavior of catchments' salinity.

Salinity and depth to shallow water table in DA were monitored in observation wells during November 2003 to April 2004 [18, 19]. There was a large variation in salinity of groundwater ranging between 4 and 100 dS/m. No trend of salinity changes throughout the study area was found. However, trend of groundwater salinity changes may partly be explained in regard to the soil texture variation in a manner that was lower in light-textured soils than that of heavier textured ones.

Also salinity was lower in the vicinity of the river tributary than those further away from the river. The depth to water table was lowest in April as a result of deep percolation from winter rains, excess irrigation of fields, river floods, and seepage from earthen channels. The depth to water table reaches its maximum in September due to high evaporation during the hot dry summer. This pattern seems to repeat itself throughout the years resulting in accumulation of salt in the top surface soil layer.

Generally saying, the agricultural cropping systems and practices in the area are suboptimal. But there are great scopes and opportunities for their improvement. At present, the crop varieties used by the farmers are not adapted to the prevailing soil conditions, and significant improvements in crop production could be realized by introducing or applying of salt-tolerant crop varieties and species. As already noted the majority of the cultivated areas (almost 90%) are allocated to the winter wheat. The average yields of its two cultivated varieties grown, i.e., Chamran and Verinak, are low and about 2 tons/ha. Introducing and testing of high-yielding varieties to the area available in the country may respond well and be promising. Some salt-tolerant varieties such as Kavir, Bam, and Sistan were tested as part of the CPWF projects for improving WP in the area, and the results were promising.

Appropriate irrigation schedules based on soil moisture depletion or climatic data, and improvement of irrigation efficiency, would prevent excess losses of irrigation water into the subsoil or groundwater. Land leveling could improve water distribution in the field and prevent waterlogging problems. More attention should be paid to the irrigation systems. Efficiency of irrigation water application should be increased, and water should be applied more uniformly along the field [18, 19]. Some of these measures are tested in the field as part of the CPWF project, and the obtained results are given and elaborated in the next sections.

Another important issue limiting crop production in the area is the accumulation of salt in the top soil. This phenomenon mainly occurs during the fallow period, when the soil is uncultivated and the farm is left for the next cropping season. Conducting soil leaching activities prior to the sowing and land preparation activities could reduce the soil surface salinity. Therefore, it will ameliorate the adverse effects of salt stress on crop establishment especially in the early stages of crop growth. Other suitable practices are mulching with crop residues, adding organic matters to the soil, selection of suitable crop rotations, and implementation of proper cropping patterns [18, 19].

4. Determination and evaluation of water productivity

As improvement of WP and identification of its sources of inefficiencies are set as one of the top priorities in Iran, especially in KRB, some studies were conducted in the downstream areas of L-KRB located in the DA plain in the Khuzestan province (**Figure 3**).

The main objectives of these studies were to determine and evaluate WP of irrigated wheat, as a major cultivated crop in DA. Moreover some recommendations and simply applicable management approaches for the better management of irrigation practices and the amelioration of salinity-waterlogging hazards on crop yield and WP were suggested.

The researches were conducted in seven farmers' fields, typical of the farms in the region and during cropping season of 2006–2007. In **Figure 3** location of the selected fields is demonstrated.

The measured parameters were irrigation water inflows and runoffs; soil texture; soil and water salinity; soil and water pHs; soil organic matter; the P, K, Fe, Mn, Zn, and Cu elements of the soil profile; depth and quality (EC) of groundwater during growth season; and finally crop yield.

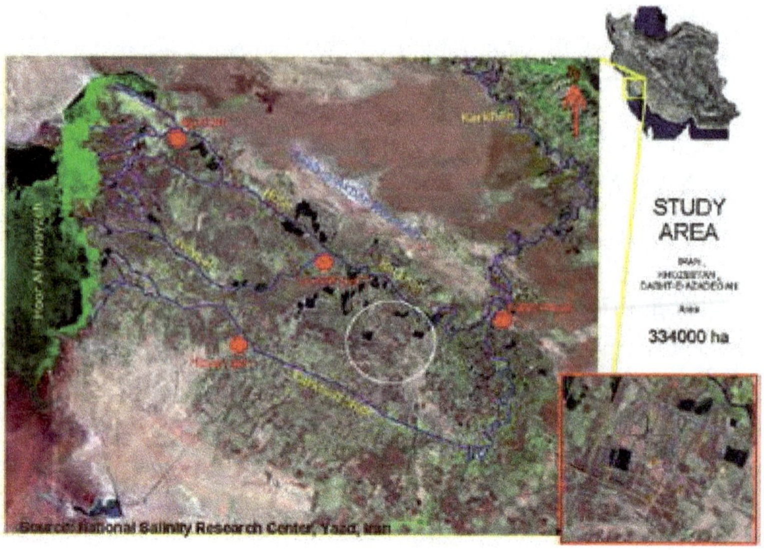

Figure 3.
The study area and location of the selected fields.

Table 1 shows some soil and water characteristics of the studied farms measured prior to the planting stage. Crop yield and yield components were measured through 20 field samples before harvest. The amount of applied irrigation water was measured by using Washington State College flume (WSC) flumes of different types. The irrigation intervals were the same as practiced by the farmers in the selected area. WPs of wheat crop were calculated using the measured total applied waters and measured crop yields. The results are shown in **Table 2**.

The range of WP in the country is generally wide, and for the wheat crop, it is between 0.56 and 1.46 kg/m^3 [3, 4]. Analysis of measured WPs in the DA area also indicates that the range of WP values is relatively high and varies between 0.24 and 0.86 kg/m^3 (**Table 2**).

Results indicated that in general by increasing the farm sizes the amount of water consumed per hectare increases. This fact indicates the higher problems associated with irrigation water management in larger field sizes. The lack of required equipment and facilities, lack of farmer's skills, and shortfalls in proper land leveling have led to higher water losses and hence higher water applications (even three times more) in the larger farm sizes.

Farm code	Area (ha)	Soil texture	EC (dS/m) (30 cm depth)	Depth of water table (cm)	EC of ground water (dS/m)
F1	1.05	SiL	26.4	105	8.8
F2	1.47	SiCL	10	205	39
F3	4.49	CL	52.6	180	71.5
F4	3.44	C	17	195	31
F5	1.73	C	21.5	182	48
F6	0.46	SiC	21.3	173	46
F7	5.24	C	10.5	213	8.7

Table 1.
Some soil and water characteristics of the selected fields.

Farm code	Water applied (m^3/ha)	ET (mm)	Yield (kg/ha)	WUE (kg/m^3)
F1	3109	517	2392	0.77
F2	3460	522	1022	0.30
F3	2062	477	1336	0.65
F4	3792	505	1453	0.38
F5	3527	553	3032	0.86
F6	2311	553	4851	2.09
F7	5933	517	1431	0.24

Table 2.
Applied waters, crop yields, and WPs of wheat in different fields.

Evaluation of the relationships between WP and applied water, yield, initial soil salinity, initial groundwater salinity, groundwater depth, and farm sizes of the selected fields indicated that there was no clear and distinct correlation between WP and these factors in all cases. In other words, a combination of these factors is affecting WP, and meanwhile managerial factors are more prominent than the basic physical factors. Consequently, sources of inefficiencies and the limiting factors affecting WP in south-ern part of L-KRB are complex and can be categorized into four main factors as follows:

- Sociocultural problems governing the area and causing low motivation for investment in irrigation management and on-farm improvement activities by the farmers
- Hindering factors that are out of the farmer's management control and authority (e.g., irrigation intervals and rationing) and shortage of agricultural inputs (e.g., fertilizers, other agrochemicals, equipment and machineries, etc.)
- Infrastructure limitations and lack of technical supports (e.g., inadequate drainage, no reclamation activities, and incomplete irrigation and drainage networks) that need extensive planning and investments and should be supported more by the government
- Managerial issues and limitations whose solutions are simple and do not need much investments and could be accomplished easily

The results indicated that the issues and challenges hindering improvement of WP in all the selected fields are not the same and vary depending on the farmer's characteristics, the farmer's management, and the location of the farms. Some of these limitations and issues are elaborated below:

Traditional common irrigation practice in the area is a combination of border-basin irrigation methods. It consists of the long borders (till 400 m) which are divided into different basins (12–15 in wide). Every basin receives its own water from the previous basin. The applied water remains for a long time in the first basins before flowing into the next one (**Figure 4**). This causes stagnation of water in the basin for a long period and stuffiness of the cultivated seeds. As usual the inflow rates to the plots are too high, and there are soil erosions, soil movements, and washing off of the cultivated seeds.

As there is not enough control on irrigation cutoff time, large amounts of outflows concentrate in the lower parts of the plots and create surface

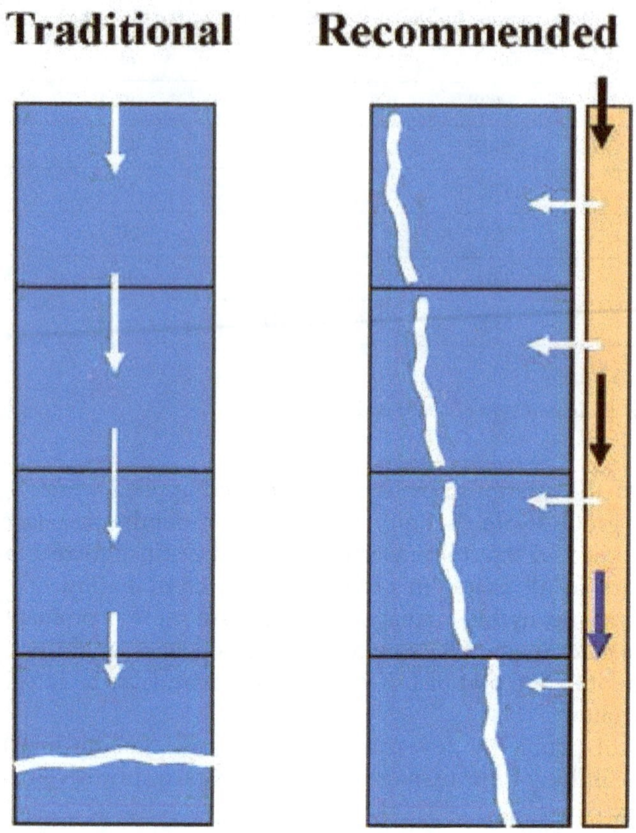

Figure 4.
The traditional (left) and recommended optimum border-basin irrigation (right) methods.

waterlogging problems. Recommend that by using a farm ditch alongside the border and construction of proper intakes, each basin to receive its inflow water individually (**Figure 4**).

Water intake and proper conduct of water into the irrigation plots is another issue. Farmers should pay high efforts to control the inflow, and this makes waste of the irrigation time. Consequently it causes poor water management and waste of water. Recommend that by construction of temporary and low-cost intake structures (gates), water intake and hence water management to be facilitated and improved.

Improper shaping of the plots in accordance with the land slope causes uneven water distribution in the basins.

Improper land preparation and agronomic practices (weed control, planting date, etc.) are some inefficiencies and shortfalls in regard to the crop production and improvement of WP in the studied area.

Considering the above limitations and issues, the following solutions and measures are recommended for improving WP in the saline areas of L-KRB:

- Conversion of traditional common irrigation practices to proper modern basin-border irrigation methods
- Construction of fixed and low-cost water intake structures on farm ditches
- Proper land leveling and bedding according to the farm slope

- Application of on-farm management improvement instructions provided by the rural extension services
- Conduct of training programs for the farmers and supervision of the farms by irrigation experts in order to provide required guidance for the upgrading of irrigation management and performances
- Provision of the required structures and enabling conditions for the volumetric allocation of water to the farmers and implementation of proper cropping patterns

5. Evaluation of the best management practices for improving WP in the salt-prone areas of lower KRB

The main objective of this section is to find out cost-effective and short-term solutions for enhancing WP under salinity conditions. According to this necessity, the following targets were identified for the saline areas of L-KRB:

- Identification of simple management practices for reducing soil salinity stress and improving agricultural WP
- Determination and comparison of WP values under different irrigation managements, i.e., traditional vs. improved border-basin irrigation methods (**Figure 4**)
- Recognition of the effects of different cultivation/sowing methods on wheat's WP in the area

The experiments were conducted during cropping season of 2006–2007 in DA plain in L-KRB. The experimental area was located between 47° 55′ to 48° 30′ E longitude and 31° 15′ to 31° 45′ N latitude, and it is about 3–12 m above the mean sea level. Soil texture was silty clay loam (SCL) to clay loam (CL). Soil's average pH was 7.8, and average soil salinity at the depth of 0–90 cm was 10.5 dS/m. Sowing was done in November and the crop was harvested in May.

The source of irrigation water was Karkheh River. The EC of groundwater and irrigation waters were 11.3 and 1.4 dS/m, respectively.

Groundwater depth at the early stages of the growth season (in winter) and before the start of rainfalls and irrigation season was at the depth of 2.4 m. It was gradually raised by the start of irrigation events and changed from 35 to 98 cm from soil surface during the growth season.

The experimental treatments were as follows [19]:

T_1 = Border irrigation + sowing by centrifugal broadcaster + one pass disk.
T_2 = Border irrigation + sowing by seed drill (Taka type).
T_3 = Border irrigation + sowing by three-row bed seeder (Barzegar-e Hamedani type).
T_4 = Basin irrigation + sowing by centrifugal broadcaster + one pass disk.
T_5 = Basin irrigation + sowing by seed drill machine (Taka type).
T_6 = Basin irrigation + sowing by three-row bed seeder (Barzegar-e Hamedani type).
T_c = Traditional irrigation and sowing method by farmer (as control).

Dimensions of plots for the T1, T2, treatments (border irrigation) were 160 m x 10 m, while the plot dimensions of plots for the T4, T5, treatments (basin irrigation) were 40 m x 10 m. The selected dimensions were optimal sizes and were selected based on US Soil Conservation Service (SCS) criteria.

The control treatment, i.e., traditional irrigation method, was a combination of basin-border irrigation method practices by the local farmers (**Figure 4**). Local farmers choose the farm borders' length according to their farm dimensions which are usually between 100 and 400 m. The border's width usually ranges between 5 and 14 m. The farmers divide borders to several small basins with 30–70 m length, depending on their farm topography (**Figure 4**). In every irrigation event, they fill the first basin and then water transfers to the next one, and this process continues until filling of the last basin and completion of the irrigation in the irrigation border (**Figure 4**).

All the treatments were sown by the Chamran wheat seed variety. In treatments T_1 and T_4, the seeds were sown by centrifugal broadcaster machine, and the seeding rate was 240 kg/ha. In other treatments (T_2, T_3, T_5, T_6), two seeding machine types, i.e., seed drill machine (Taka type) sowed the seeds for the T_2 and T_5, and the three-row bed seeder (Barzegar-e Hamedani type) sowed the seed for the T_3 and T_6 treatments. The seeding rate for these treatments was 180 kg/ha. In the control treatment (T_c), which was sown by centrifugal broadcaster and managed by the farmer, the seed rate was 350 kg/ha.

In all treatments except control, the optimized irrigation management (**Figure 4**) was practiced. Other farming practices were the same for all the treatments.

In **Table 3** and **Figure 5**, some soil chemical characteristics measured prior to the planting date and fluctuations of water table depths (average of three points) during the cropping season are presented, respectively.

Crop yield and yield components were measured through sampling methods. Prior to harvest 20 soil samples were taken from the field. Volume of applied irrigation water was measured using WSC flumes. Interval and the number of irrigation events for the farmer's managed and optimum irrigation management treatments (**Figure 4**) were the same. In fact, the difference was in how to manage irrigation water and the methods of water application, which directly affected volume of consumed water.

The results indicated that the border irrigation with centrifugal sowing method (T1) provided the highest WP (1.6 kg/m^3) (**Table 4**).

The optimum border irrigation had the maximum WP (1.36 kg/m^3), while the control treatment (traditional border-basin irrigation method under centrifugal sowing with 350 kg seed used) provided the minimum one, i.e., WP equal to 0.61 kg/m^3 (**Table 4**).

There was no significant difference ($\alpha = 0.05$) in yields between the treat-ments with the control. Although the consumption of seed used in both Taka and Barzegar-e Hamedani sowing methods was 50% less than the centrifugal broadcast-ing method, the seed germination percentage was more in these sowing methods (**Table 5**).

Either "improved basin" or "border irrigation" methods could be recommended for the improvements of water management and WP in the L-KRB area. However, the basin irrigation method (**Figure 4**) is more adaptive and sustainable in terms of acceptance by the local farmers for the following reasons:

Soil depth	EC	pH	OC	P	K	Fe	Zn	Cu	Mg
(cm)	(dS/m)		(%)	(ppm)	(ppm)	(ppm)	(ppm)	(ppm)	(ppm)
0–30	11.1	7.7	0.4	5.1	166	3.8	1.3	0.6	5.2
30–60	9	7.9	0.1	2.2	88.7	2.5	0.1	0.4	1.2
60–90	11.4	7.9	0.1	1.8	59	3.2	0.1	0.5	1.5

Table 3.
Some soil chemical characteristics measured prior to the planting date.

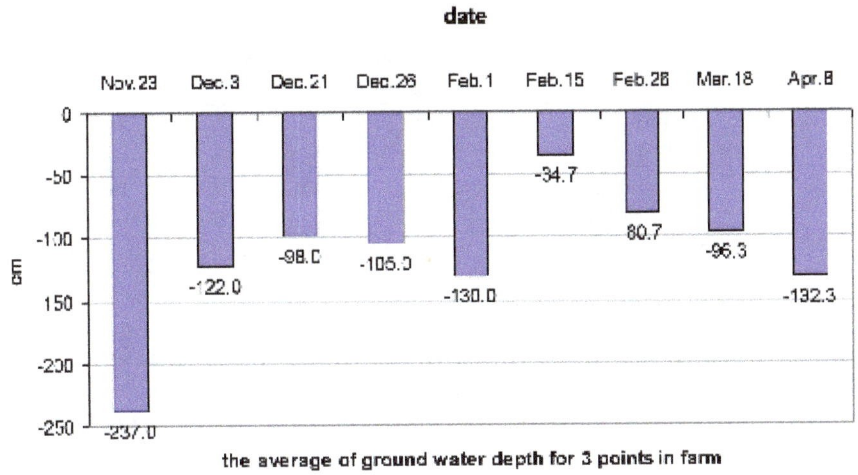

Figure 5.
Variation of the groundwater depth during growth season.

Irrigation method	Sowing method	Yield (kg/ha)	Applied water (m^3/ha)	WP (kg/m^3)	WP (avg. of irrigation treatments) (kg/m^3)
Basin-border (farmer)	Centrifugal (350 kg seed/ha)	1953	3205	0.61	0.61
Optimum border	Centrifugal (250 kg seed/ha)	2590	1618	1.60	1.36
	Taka (180 kg/ha)	2434	1774	1.37	
	Barzegar-e Hamedani (180 kg/ha)	1901	1729	1.10	
Optimum basin	Centrifugal (250 kg seed/ha)	2730	2394	1.14	1.04
	Taka (180 kg seed/ha)	2521	2417	1.04	
	Barzegar-e Hamedani (180 kg seed/ha)	2198	2344	0.94	

Table 4.
Yield, water applied, and WP of the selected treatments [19].

- Basin irrigation method requires low levels of land leveling and uniform slope along the irrigation dimensions of plots for the T1, T2.
- Basin irrigation method is more adaptive to the farm micro relief caused by common cultivation practices.
- Basin irrigation method is more adaptive to the sociocultural conditions of the area.
- Basin irrigation method requires less labor (considering shortages in agricultural labor in the area).
- Considering shortfalls and/or lack of land leveling and low levels of on-farm improvement activities in the area at present situation, the basin irrigation method is the most adapted method to this condition.

Irrigation method	Sowing method	Seed consumption rate (kg/ha)	Number of shrub in m^2	Sprouting percentage (%)	Yield (kg/ha)
Basin-border (farmer)	Centrifugal (350 kg/ha)	350	247	34	1953
Optimum border	Centrifugal (250 kg/ha)	250	341	56	2308
	Taka (180 kg/ha)	180	262	60	
	Barzegar-e Hamedani (180 kg/ha)	180	286	65	
Optimum basin	Centrifugal (250 kg/ha)	250	387	63	2483
	Taka (180 kg/ha)	180	332	75	
	Barzegar-e Hamedani (180 kg/ha)	180	353	80	

Table 5.
Seed consumption, number of shrub, and sprouting percentage of the treatments [19].

- Basin irrigation method requires less control on flow considering high rate of variation in channel discharges and high rate of flow control requirements.
- Basin irrigation method has the advantage of providing pre-cultivation leaching opportunities, considering high levels of salinity and its variation in the wheat farms of the area.

Because of 50% reduction in seed consumption, the high rate of seed germination, and better flow of water, the use of seed drill machine (Taka type) or the three-row bed seeder (Barzegar-e Hamedani type) is recommended.

6. Summary and conclusions

Soil and water salinization are the most serious hindering factors on the improvement of WP, enhancement of agricultural production, and sustainability of natural resources in Iran and especially in L-KRB. The existence of vast areas with saline sodic conditions in various parts of the country especially in L-KRB reflects the fact that there are many factors affecting this phenomenon, and it should pay enough attention.

The soil texture in the area is mainly heavy textured and with low hydraulic conductivity. Natural drainage is very low, and the potentials for soil waterlog-ging followed by salinity hazard are high. Consequently agricultural production is suboptimal and crop WPs are low. Under these conditions installation of drainage networks may seem as a rapid solution for the removal of salinity and waterlogging problems, but simple and cost-effective water and crop managerial measures also have their own importance and merits.

The government's huge investments in the expansion of irrigation networks in the L-KRB and future increase on applied water for agricultural activities will cause extreme changes in the surface and groundwater hydrology and in overall climatic parameters of the region. It is expected that without consideration of required arrange-ments, e.g., development of drainage networks and on-farm improvements activities, the problems of land salinity and waterlogging will be intensified in the future.

However, the trend of previous activities in this regard indicates that the rate of expansion of drainage networks is low. Therefore immediate actions are required for the mitigation of salinity and waterlogging hazards and for the improvement of crop WP in the L-KRB. At current condition this could be tackled by soil, water, and crop management activities such as assessment of current WP values, recognition of sources of inefficiencies on improvement of crop WPs, improvement of irrigation efficiencies, more use of salt-tolerant crop varieties, land leveling, changing of irrigation methods, changes in cropping patterns, and other on-farm improvement activities.

Waterlogging and soil salinity are the major threats to WP and sustainable agricultural production in the L-KRB. These problems are somehow because of physical characteristics of the region (heavy soil texture, high evaporation demand, low soil hydraulic gradient, etc.) but are mainly man-made problems. These challenging issues could be managed by proper measures and approaches including infrastructure activities (hardware) and to a greater extent by proper on-farm water management (software) measures.

Hardware measures may include completion of drainage networks and completion of farm canals (tertiary-level networks). The Ministry of Energy recently has emphasized this critical need in the government body in an attempt to obtain enough authority and resources to complete the irrigation and drainage networks in the country and in parallel to the Ministry of Agriculture.

Software measures include application of new approaches and tools such as the use of proper models relevant to the study of the issues at plant, farm, system, and basin levels, preparation and implementation of comprehensive plans at the basin level, conducting detailed or semi-detailed studies at the basin level and development of well-defined strategies and policies for water management in the Khuzestan province, and especially in the L-KRB, where two thirds of the country's water recourses flows at this province.

Recently the government has received the importance of water users and basin stakeholder participation in better management of water. Hence Water User Association (WUA) is developing irrigation networks, and the work is in process. However, the levels of their success to achieve their goals and their effects on WP improvement and performance of the irrigation networks vary much and need more research and evaluations in this regard.

Water scarcity and soil and water salinities in the arid regions such as L-KRB and all of the aforementioned issues suggest that the water in the agricultural sector should be consumed efficiently. The most important way to achieve this objective, especially in the L-KRB, is the improvement of agricultural WP index.

Research studies related to water table management, soil salinity control, irrigation water management, selection of suitable crop varieties, and improved agronomic practices will help to improve agricultural WP and farmer's livelihoods in the L-KRB. Waterlogging and resource salinity are major threats to WP enhancement and sustainable agricultural production in the L-KRB. Therefore sound and adaptive measures and solutions are required for this region. The KRB reflects in many aspects the problems and challenges associated with the water management and crop production in other arid to semiarid regions of the world. Therefore, it was intended by the CPWF to link the work studies in KRB with the Euphrates and Amu Darya river basins, which have been postponed to the next phase of the CGIAR Challenge Program on Water and Food (CPWF).

Overall, soil salinity and waterlogging, in addition to the other sources of inefficiencies in agricultural WP improvements, are the major limiting factors in the L-KRB. The causes to these hindering factors are somewhat because of physical characteristics of the region, but they are mainly man-made problems which could be managed easily with low costs using proper measures and approaches including infrastructure activities (hardware) and to a greater extent by the water and crop management (software) measures.

Author details

Nader Heydari

Iranian Agricultural Engineering Research Institute (AERI), Agricultural Research, Education and Extension Organization (AREEO), Karaj, Iran

*Address all correspondence to: nrheydari@yahoo.com

References

[1] Keshavarz A, Heydari N, Ashrafi S. Management of agricultural water consumption, drought, and supply of water for future demand. In: 7th Int'l Conference on Sustainable Development and Management of Drylands in 21 Century, Tehran, Iran. 2003

[2] Heydari N, Keshavarz A, Dehghanisanij H. Agricultural water consumption management in Iran considering aridity and drought incidences. In: Proceedings of National Conference on Prevention of Losses from National Resources, Academy of Science of Iran, June 15-16, 2005, Tehran, Iran (in Persian). 2005

[3] Heydari N, Eslami A, Ghadami A, Kanoni A, Asadi ME, Khajehabdollahi MH. Determination of water productivity of some major crops in different regions of Iran. Res. Rep. No. 84/988, Iranian Agr. Eng. Res. Ins., Karaj, Iran (in Persian). 2006

[4] Heydari N. Determination and evaluation of water use efficiency of some major crops under farmer's management in Iran. Journal of Water and Irrigation Management. 2011;**1**(2, autumn):2011 (in Persian)

[5] Zwart SJ, Bastiaanssen WGM. Review of measured crop water productivity values for irrigated wheat, rice, cotton, and maize. Journal of Agricultural Water Management. 2004;**69**:115-133

[6] Alizadeh A, Malak-Mohammadi I, Kamali GA. Sustainable utilization of saline water in agriculture based on indigenous knowledge. In: Taha FK, Ismail S, Jaradat A, editors. Prospects of Saline Agriculture in the Arabian Peninsula: Proceedings of the International Seminar on Prospects of Saline Agriculture in the GCC Countries, 18-20 March 2001, Dubai, United Arab Emirates. 2004. pp. 483-490

[7] Koocheki A, Moghaddam PR. History of research on salt-affected lands of Iran: Present status and future prospects. In: Taha FK, Ismail S, Jaradat A, editors. Prospects of Saline Agriculture in the Arabian Peninsula: Proceedings of the International Seminar on Prospects of Saline Agriculture in the GCC Countries, 18-20 March 2004, Dubai, United Arab Emirates. 2004. pp. 507-520

[8] Ghassemi F, Jakeman AJ, Nix HA. Salinization of Land and Water Resources. Australian National University. Centre for Resource and Environmental Studies: CAB International; 1995. 526 p. ISBN: 0851989063, 9780851989068

[9] Cheraghi SAM. Institutional and scientific profiles of organizations working on saline agriculture in Iran. In: Taha FK, Ismail S, Jaradat A, editors. Prospects of Saline Agriculture in the Arabian Peninsula: Proceedings of the International Seminar on Prospects of Saline Agriculture in the GCC Countries, 18-20 March 2001, Dubai, United Arab Emirates. 2004. pp. 399-412

[10] Abtahi A, Sys C, Stoops G, Eswaran H. Soil forming processes under the influence of saline and alkaline groundwater in the Sarvestan Basin (Iran). Pédologie. 1979;**29**:325-357

[11] Banie MH. Soils Map of IRAN: Land Resources and Potentialities. Tehran: Soil and Water Research Institute (SWRI); 2001

[12] Dewan ML, Famouri J. The Soils of Iran. Rome: Food and Agriculture Organization of the United Nations; 1964

[13] Mahjoory RA. The nature and genesis of salt-affected soils in Iran. Soil Science Society of America Journal. 1979;**43**:1019-1024

[14] Matsumoto S, Cho T. Field investigation on the agricultural development of arid regions in Iran, III soil profile investigations and salt accumulation related to the depth of ground water level. Journal of the Faculty of Agriculture, Tottori University (Japan). 1985;**20**:86-97

[15] Siadat H. Salt-affected soils of Iran. In: Country Report Presented in Seminar on Salt-affected Soils, 1997, Cairo, Egypt. 1997

[16] Anonymous. Some information on soil, water, and agronomic practices in the lower Karkheh River Basin. Hand notes of basin coordinator and other expert members of the CPWF-KRB project team which were collected during different field visits of the basin (in Persian). 2007

[17] Anonymous. Soil Survey of West Dasht-e-Azadegan. Tehran, Iran: Ministry of Energy; 1989. 186 p (in Persian)

[18] Cheraghi SAM, Hasheminejhad Y, Heydari N. Causes and management of salt-prone land degradation in lower Karkheh River Basin. In: Oweis T et al., editors. A Compendium of Review Papers. 2008. pp. 81-91

[19] Cheraghi SAM, Heydari N, Qadir M, Oweis T. Improving crop growth and water productivity on salt-affected soils in the lower Karkheh River Basin. In: Final Research Report No. 4 CPWF Project: Improving On-farm Agricultural Water Productivity in the Karkheh River Basin (PN8), ICARDA, November 2008. 2008

[20] Ghobadian A. Causes Leading to Soil Salinity in Khuzestan and Their Control. Ahvaz, Iran: College of Agriculture, Jundi Shapur University; 1969. Publication Number 1, 67 p

[21] Mahab-e Ghods. Studies of Agricultural Development Plan of Dasht-e Azadegan, Irrigation and Drainage Plan of Karkheh. Tehran, Iran: Mahab-e Ghods Consulting Engineers; 1992 (in Persian)

9

Current Approaches to Pesticide Use and Glyphosate-Resistant Weeds in Brazilian Agriculture

Kassio Ferreira Mendes, Rodrigo Nogueira de Sousa and Ana Flávia Souza Laube

Abstract

The aim of this chapter is to show a general vision about the pesticides use in Brazil. Pesticides are chemical products that contribute to agricultural produc-tion processes, mainly large scale, as agents of chemical, physical and biological processes. The glyphosate is the most widely used pesticide in Brazil to vast area cultivated with genetically modified glyphosate-resistant crops. Also, this herbicide is the most widely used in the world to control weeds in various crops, making chemical control cheap, easy and efficient. However, with the advance in the cultivation of glyphosate-resistant crops and the intensive use of this herbicide associated with the non-use of other herbicides, glyphosate-resistant weed biotypes are emerging very quickly. In this way, its use must be performed consciously in order not to occur significant increase in the amount of weeds resistant to this herbicide. Despite its great use in the country, it presents few records of resistant weeds when compared to other herbicide action mechanisms. Thus, good agricultural practices are indispensable and more innovations in technologies are necessary for the future. Therefore, adopting a long-term weed management perspective and integration systems for all agricultural practices is of paramount importance to farmers.

Keywords: herbicide resistance, mode of action, chemical product, conventional farming

1. Introduction

The domestication of plants over a long time came with several challenges to maintain their sustainability. It is known that cultivated crops would suffer attacks from pests and diseases, causing great yield losses with the ever-present possibility of hunger for the population, mainly due to the lack of resources and knowledge. Even today with advances in technologies to control invaders, food losses due to pests and diseases range from 10 to 90%, with an average of 35–40%, for all potential crops of food and fiber [1]. Pesticides are chemical products that contribute to agricultural production processes, mainly large scale, as agents of chemical, physical, and biological processes [2]. In the south of Brazil, the monoculture of soybeans, wheat, and rice was associated with the mandatory use of pesticides for those who intended to use government rural credit. Today, pesticides are disseminated in con-ventional agriculture, as a short-term solution for pest and disease infestation [3].

Brazil is one of the largest agricultural producers in the world and the second country that exports these products, playing an important role in the local economy. To maintain such production, this sector intensively uses transgenic seeds and chemical inputs, such as fertilizers and pesticides. Brazil is the largest consumer of pesticides in the world, with an extensive area of planting [4]. The consumption of herbicides in Brazil was about 540,000 tons of formulated (commercial) products in 2017 [5]. Glyphosate is the most widely used pesticide in Brazil, with 173,150.75 tons of acid equivalent marketed in 2017 [6]. One of the main consequences of weed resistance to herbicides is the increase in weed control costs, which is hardly addressed in scientific publications, but of great importance for the produc-tive sector.

Glyphosate (*N*-phosphonomethyl glycine) is a nonselective and systemic herbicide applied in postemergence, which belongs to a chemical group of replaced glycines. It presents a wide spectrum of actions, enabling the control of annual and perennial weeds with broad and narrow leaves. Due to excellent weed control along with its ease of handling, low cost, and increased productivity, glyphosate has become the most widely used herbicide in the world [7]. This herbicide is registered in Brazil for the following crops: cotton, rice, plum, banana, cocoa, coffee, sugarcane, citrus, coconut, eucalyptus, beans, tobacco, apple, papaya, corn, nectarine, pear, peach, pine, rubber tree, soybean, wheat, grape, pastures, forage ryegrass, and black oats [7]. Glyphosate acts by inhibiting the enzyme 5-enolpyr-uvylshikimate-3-phosphate synthase (EPSPs) and, consequently, the biosynthesis of aromatic amino acids, lignins, flavones, isoflavones, anthocyanins, and several indispensable components for the plant, leading to plant death [7].

It is of paramount importance to understand the behavior of the herbicide in the plants. Therefore, the absorption of glyphosate takes place in the aerial part of the plants, having a maximum absorption 96 h after application, translocated by simplasto with photoassimilates of the leaves for meristematic tissue reaching the target site. In water it presents a weak acid behavior and presents four variable dis-sociation constants (pKa between 2.6 and 10.3), in which it presents cell absorption facilitated by phosphate carriers that are in the cell membranes [8].

Glyphosate is the most widely used pesticide in the world due mainly to the large number of genetically modified crops resistant to this product [9]. However, with the increase in the number of agricultural areas with transgenic crops (glyphosate-resistant), mainly soybean, cotton, and corn, together with the high use and incor-rect application of this herbicide, new cases of resistant weeds appear [8]. In the world, 47 species of glyphosate-resistant weeds are already reported, and 9 of them are in Brazil [10]. **Figure 1** presents the number of weed species resistant to various herbicides of different resistance mechanisms reported worldwide.

Acetolactate synthase (ALS) inhibitor herbicides have the highest number of resistant species (162 species) (**Figure 1**). 4-Hydroxyphenylpyruvate dioxygen-ase (HPPD) inhibitor herbicides have the smallest number of resistant species (two species). **Figure 2** shows that the United States is the country with the most cases of unique resistance. Brazil has 51 cases of herbicide-resistant weeds already recorded.

Figure 3 presents the amount of herbicide-resistant weed species within weed families. The family that has the most cases of herbicide resistance is the Poaceae family with 82 registered cases. The Caryophyllaceae family presents fewer cases of resistant weed species (six species).

Figure 4 presents the amount of weed species that have simple resistance to each herbicide. Atrazine presents the largest number of weed species with simple resistance, with 66 species registered. For glyphosate there are reported 43 species of weed resistance, the second being herbicide with the highest number reported.

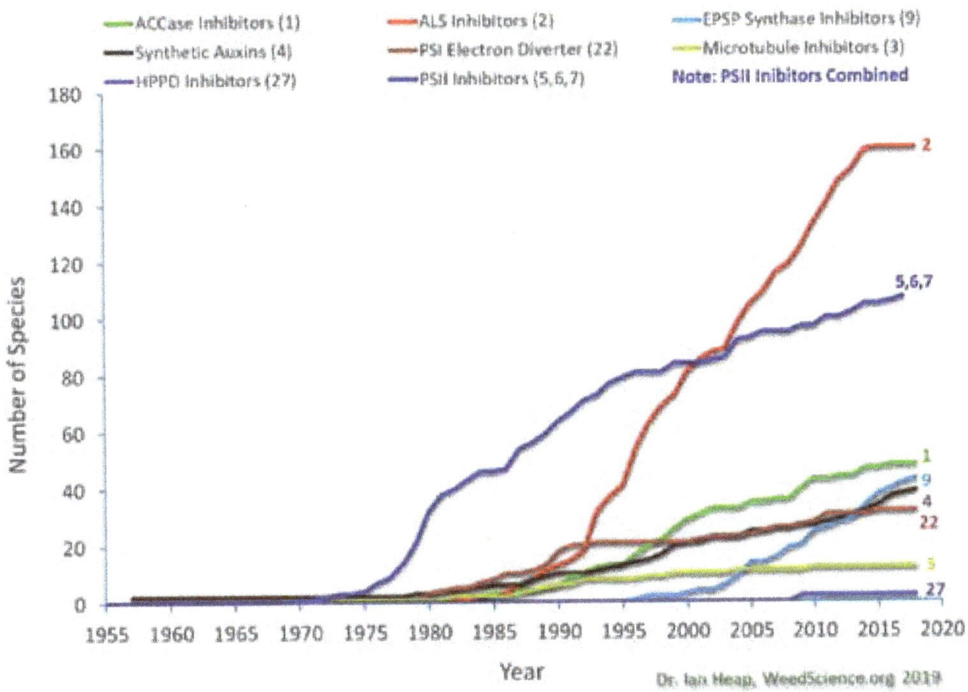

Figure 1.
Number of weed species resistant to various herbicides of different resistance mechanisms reported worldwide. Source: Heap [10].

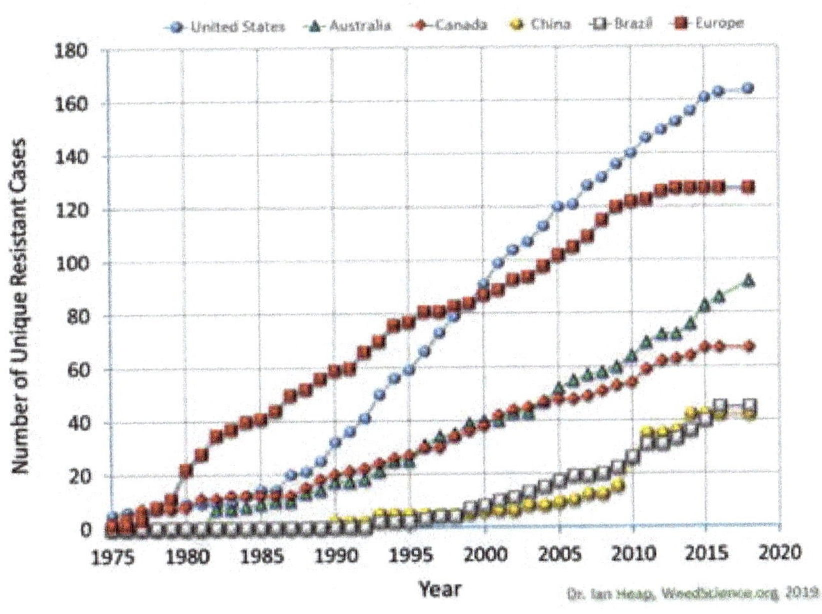

Figure 2.
Increased cases of resistant weeds reported in various countries and Europe. Source: Heap [10].

The mechanisms that generate resistance to herbicides in weeds can be separated into the following: (i) related to the site of action (target-site resistance, TSR) and/or (ii) not related to the site of action (nontarget-site resistance, NTSR) [9]. Mutation of the gene encoding enzyme EPSPS and amplification of this gene are

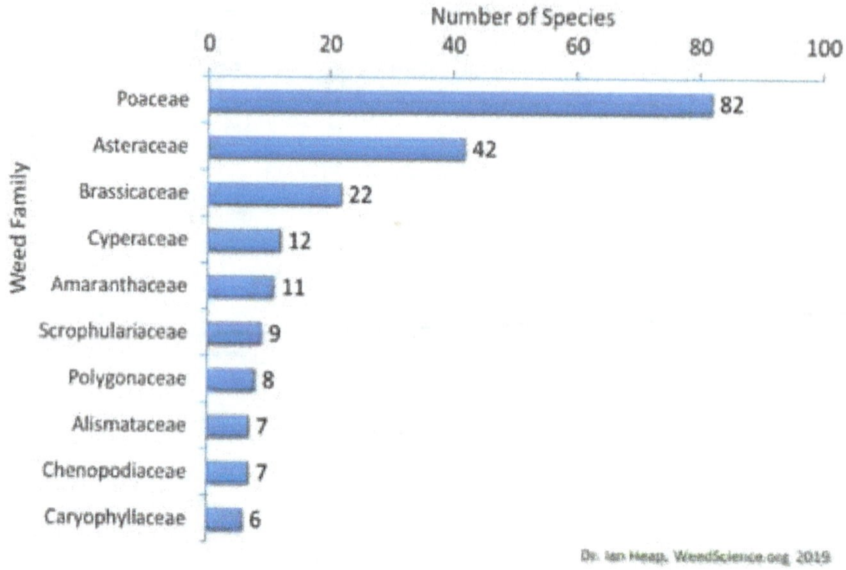

Figure 3.
Number of herbicide-resistant weed species per weed family. Source: Heap [10].

examples of TSR mechanism, while reduced absorption, differential translocation, high metabolism, and glyphosate sequestration by the vacuoles are examples of NTSR mechanism [11]. Thus, it is necessary to know about the mechanisms of weed resistance to herbicides to perform good management practices for the prevention of the occurrence of new resistant biotypes in other areas and, mainly, for the determination of preventive management programs to the selection of resistant biotypes and also for the determination of the practices of weed control already selected [9].

The aim of the authors in this chapter was to present pesticide use and general characteristics of glyphosate- and herbicide-resistant weeds in Brazil.

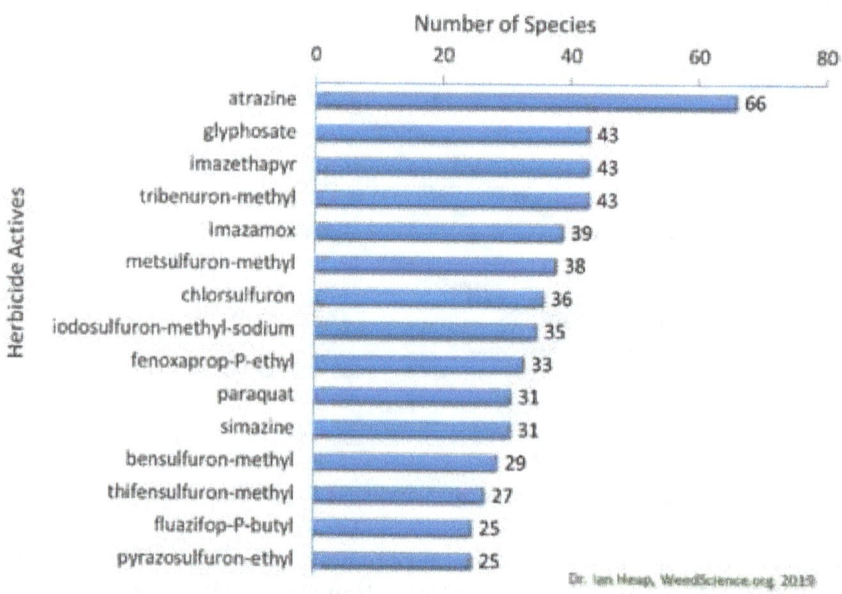

Figure 4.
Number of weed species with simple resistance to herbicides. Source: Heap [10].

2. Pesticide use in Brazilian agriculture

The agricultural production in Brazil plays an important role in the Brazilian economy, thanks to which this country is one of the world's leading producers of agricultural commodities. To keep up with production, this sector uses intensively transgenic seeds and chemical inputs, such as fertilizers and pesticides—chemical or biological substances used to protect crops against the introduction and spread of pests such as insects, fungi, bacteria, viruses, mites, nematodes, and weeds [4, 12]. Regarding the function of pesticides, they all have the same common action, which is to block the vital metabolic processes of the organisms in which they are toxic.

Currently, the total amount of pesticide commercialized in Brazil is US$10,522 billion per year, 14% less than 2014 (**Figure 5**), or 21% in a global market estimated to be worth US$50 billion. In a country with a high pest index due to tropical climate, the farmers' challenge is to reduce pesticide application (which is nowadays the main pest management), as well as to reduce the cost of production and the associated risks to human health and natural resources.

Among the several alternatives for pest control in crops, the chemical method is still the most widely used, due to its practicality, efficiency, and speed. However, if pests are not controlled, they can drastically reduce the crop productivity. Among the pesticide classes, herbicides (selective and nonselective) used for weed control and also applied for crop desiccation represent 33% of consumption in the country, followed by insecticides (29%) and fungicides (28%) (**Figure 6**).

Due to the high total amount of pesticides used, some agricultural crops deserve attention, not because these products are intensively applied per unit of cultivated area but because these crops occupy large areas in Brazil. Half of the pesticides commercialized in the country are used in soybean crop, followed by the main crops such as sugarcane (12%), corn (11%), and cotton (9%) (**Figure 7**).

Pesticide use differs in the various regions of the country, where intensive and traditional agricultural (not use chemical product intensively) activities are mixed. Located in Midwest, the Mato Grosso state is the one that uses the most pesticides (24%) in the country, and the second is São Paulo state, located in Southeast

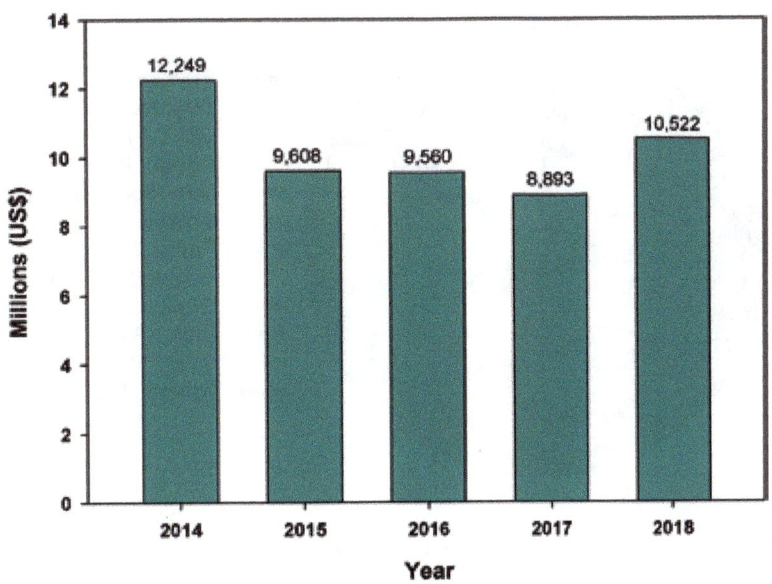

Figure 5.
Total commercialization of pesticides in Brazil from 2014 to 2018. Source: SINDIVEG [13].

(**Figure 8**). The consumption of pesticides in the Midwest increased in the 1970s and 1980s due to the occupation of the Cerrados and the cultivation of soybean, cotton, corn, and sugarcane continues to increase in this region. The South region represents 26% of pesticide consumption, while in the Northeast region, it is only 9%.

Empty pesticide packaging, unlike any plastic packaging, cannot be reused for domestic uses. This is because the products are aggressive, i.e., harmful to human and animal health, and can cause contamination if reused. And due to the toxicity of pesticides, their handling requires extreme care, attention, and personal protective equipment, and empty containers cannot be disposed in the dumping ground due to the aforementioned eminent risks. Therefore, all pesticides that are marketed in Brazil have empty packaging collected by the National Institute for Empty Packaging Processing (inpEV), which is responsible for the final destination of this material.

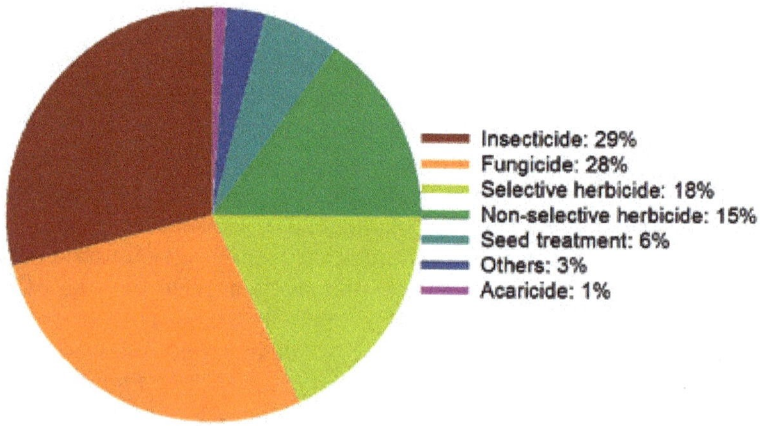

Figure 6.
Commercialization of classes of pesticides used in Brazil. Source: SINDIVEG [13].

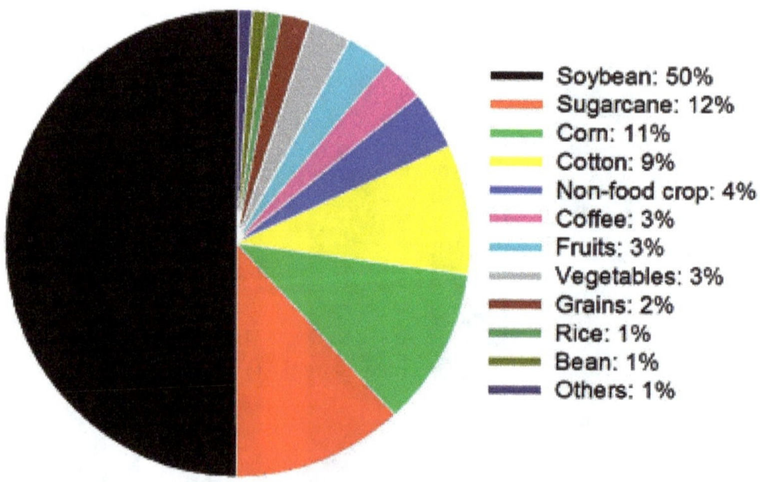

Figure 7.
Commercialization of pesticides used by crops in Brazil. Non-food crops: reforestation, pasture, floriculture, and tobacco. Fruits: citrus, apple, grape, melon and watermelon, banana, and others. Vegetables: potatoes, tomatoes, onions, garlic, and others. Grains: wheat, oats, rye, barley, and peanuts. Others: stored grains and others. Source: SINDIVEG [13].

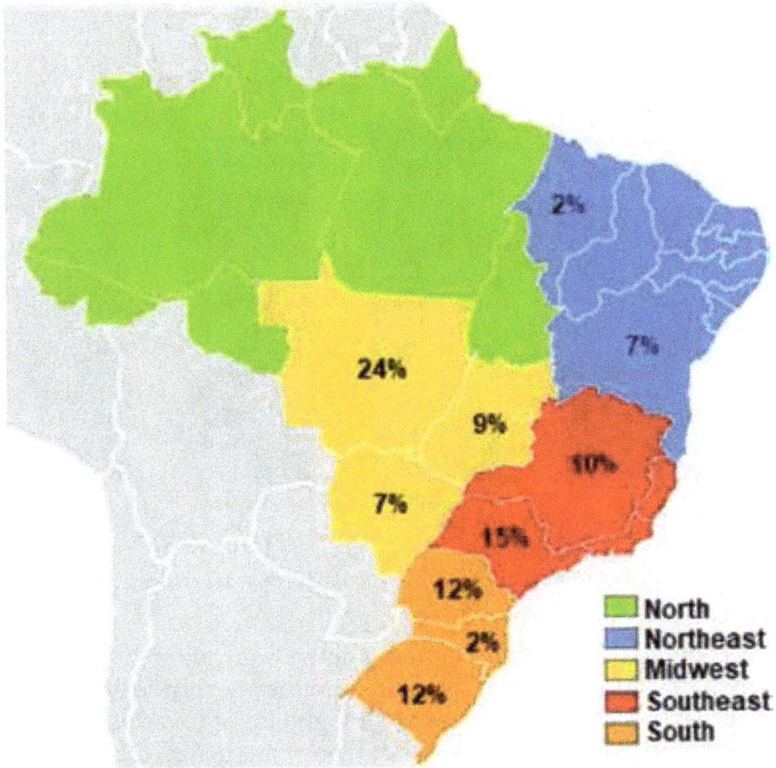

Figure 8.
Commercialization of pesticides for use in agriculture by Brazilian states. Source: SINDIVEG [13].

3. Glyphosate: sales in Brazil and worldwide, origin and mode of action

Currently, the most efficient method used for weed control is the use of herbicides, mainly in large areas of cultivation, for which its rapid action added more viable costs. Among the herbicides used, glyphosate is the most marketed worldwide in more than 119 countries with about 150 trademarks for this product [14]. **Figure 9** presents the commercialized quantities of glyphosate and its channels in Brazil.

According to the ABRASCO [16], 110 products with glyphosate alone have been sold in Brazil, in 29 different companies, and 173,150.75 tons sold in 2017, the amount being 3 times more than the second most commercialized herbicide, the 2,4-D. This considerable increase in sales was due to the production of corn and cotton after the development of transgenic soybeans, from 40,000 tons of products marketed to 300,000 per year in Brazil. In 2013, Asian countries, especially China and India, were the ones that consumed glyphosate-based products the most. At the same time, the United States accounted for more than 25% of all glyphosates marketed. The estimate is that in 2020, the demand for this herbicide is worldwide, which exceeds one ton [17].

The discovery of glyphosate occurred in 1950, and this acid was an interesting complexion agent, a pH reducer, a detergent, and several other applications [18]. The glyphosate molecule was invented by the Cilag/Ciba industry in Switzerland, during the process of selection of chelating compounds for paints. In the mid- 1960s, some scientists at Stauffer discovered other chelating properties of glypho-sate. However, only in the early 1970s did Monsanto scientists discover the herbicide properties of glyphosate. Two decades after it began to be marketed, there were

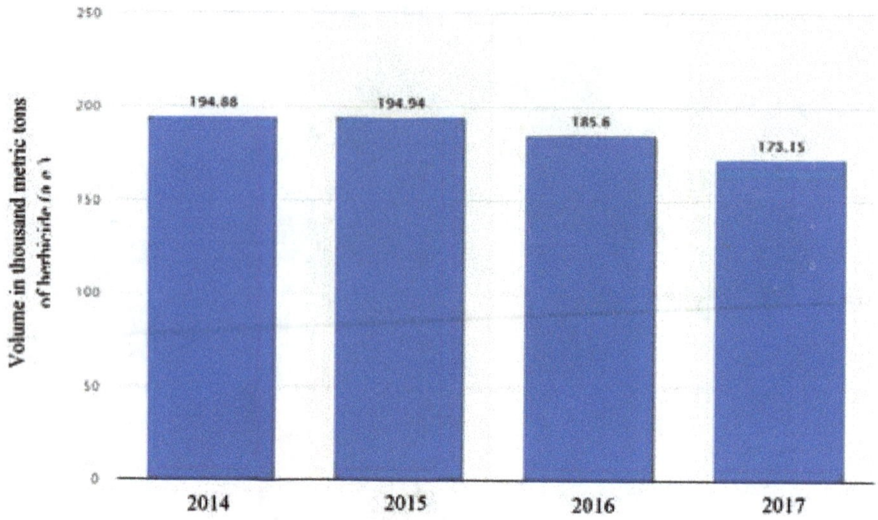

Figure 9.
Annual distribution of the quantity, in tons, of glyphosate in Brazil, from 2014 to 2017. Source: IBAMA [15].

more than 90 commercial products with this active equivalent [19]. Today, glyphosate is widely used and has become the most marketed herbicide in the world.

Due to its lack of selectivity, the use of glyphosate was initially limited to preplanting, directed jet, pre-harvest, and postemergence of weeds. With the introduction of glyphosate-resistant crops in the mid-1990s, it is now used for weed control in resistant crops without concerns about crop damage. Currently, glyphosate-resistant crops are grown in several countries, with great adoption in the United States, Canada, Argentina, and Brazil. The wide adoption of glyphosate-resistant crops has caused changes in weed species in these crops and resulted in the evolution of resistant weeds [20].

Glyphosate is a nonselective herbicide (affecting all "natural" or non-transgenic plants), which has a broad spectrum, is systemic, and is applied in postemergence, belonging to the glycine-derived chemical group that has been widely used in the world in the last four decades [20].

The mechanism of action of glyphosate is the inhibition of the enzyme EPSPS and, consequently, the biosynthesis of aromatic amino acids, tryptophan, phe-nylalanine, and tyrosine [18, 20], and precursors of compounds such as lignins, flavonoids, and benzoic acids [21]. This leads to several metabolic disorders, inhibiting the biosynthesis of proteins and secondary products and generating a significant increase in the concentration of shikimate, a common precursor in the metabolic route of the three amino acids (**Figure 10**) [22]. Glyphosate inhibits the enzyme EPSPS by competing with the phosphoenolpyruvate (PEP) substrate, preventing shikimate from being transformed into chorismate. The synthesis of the enzyme EPSPS occurs in the cytoplasm, which is transported to the chloroplast where it operates; glyphosate binds to it by glutamic acid carboxylic (glutamine) at position 418 of the amino acid sequence. The final action of the herbicide is not explained by the reduction of amino acids and the accumulation of shikimate. It is believed that the deregulation of the shikimic acid route causes the loss of carbons available for other cellular reactions in the plant, once 20% of plant carbon is used in this metabolic route, because tryptophan, phenylalanine, and tyrosine are fore-runners of most aromatic compounds in plants. Glyphosate causes the reduction of phytoalexin synthesis. There is an increased concentration in toxic levels of nitrate, ethylene, kinetic acid, and other compounds that accelerate plant death [22].

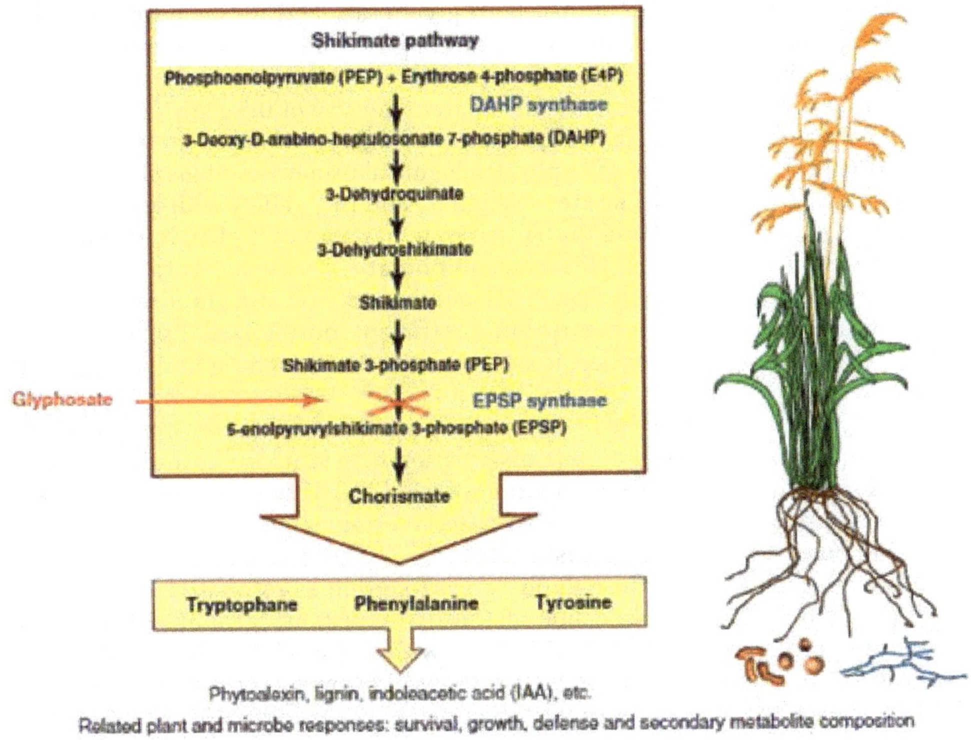

Figure 10.
Glyphosate acts by inhibiting the enzyme 5-enolpyruvylshikimate-3-phosphate synthase in the shikimate pathway, blocking the production of tryptophan, phenylalanine, or tyrosine. Source: Adapted from Helander et al. [23].

The dose of glyphosate used depends on the species that will be controlled and can range from 0.18 to 2.16 kg a.e. ha^{-1}. After application of this a perherbicide, iod of 4–6 h without rain is necessary to increase the efficiency. After being treated with this herbicide, the plants die between 7 and 14 days. For absorption to be facilitated, it should be used in low flow and larger drops. The yellowing of meristems is a symptom in plants that can lead to necrosis and then to death in days or weeks [21].

4. Physicochemical properties of glyphosate

Glyphosate presents the molecular formula $C_3H_8NO_5P$ (molecular weight = g mol169.1^{-}) ^{1} [24]. This herbicide can be formulated as isopropylamine salt, ammonium salt, or trimethylsulfonic salt [25] (sulfate) [19]. The chemical glyphosate group is a replaced glycine. The glyphosate structure is presented in **Figure 11**.

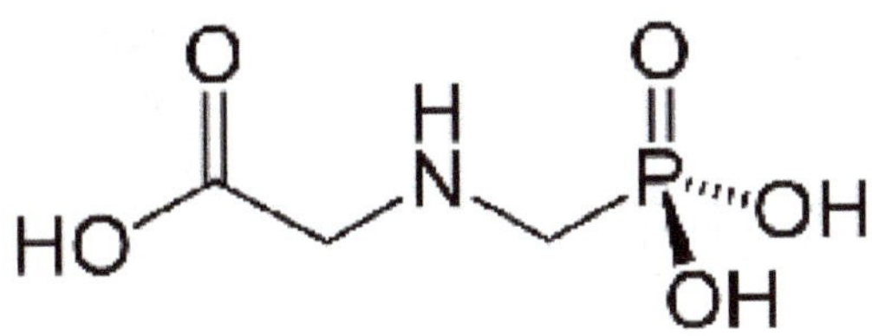

Figure 11.
Chemical structure of glyphosate.

Under environmental conditions, both glyphosate and its salts are crystalline solids, which have high solubility in water (12 g L^{-1} to 25°C, for glyphosate) and are practically insoluble in common organic solvents such as acetone and ethanol. Glyphosate melts at 189.5°C and has an apparent density of 0.5 g cm^{-3} and presents significant water solubility in the presence of light, including at temperatures above 60°C [24]. **Table 1** shows the physicochemical properties of glyphosate.

As shown in **Table 1**, glyphosate has high S_W and pK_a values with acid character and a low K_{OW} value, indicating that glyphosate has a great hydrophilic tendency, which decreases soil sorption. However, glyphosate quickly binds to positive soil charges (mainly in clays) such as in soils abundant in iron and aluminum oxides. This indicates high values of sorption coefficient normalized by soil organic carbon (OC) (K_{oc}) than other herbicides, limiting their leaching in the soil profile [27].

The pK_a values found in the literature for glyphosate are pKa1, 0.8; pKa2, 2.16; pKa3, 5.46; and pKa4, 10.14. These dissociation constants indicate the degree of dissociation of the herbicide as a function of pH [24]. This shows the relationship between the amount of matter that exists after a certain reagent has been consumed and the amount of material that exists initially.

At pH values below 0.8, most glyphosate is found in a protonation on the amine site. In a pH of 0.8, being the value of the first constant, 50% of the molecules present this protonation and the other 50% with a dissociation in the phosphate group. From this value up to pH of 2.2, the molecular formula is predominant, with a dissociation ($-PO_2H-$) and a protonation ($-NH_2^+-$), and in a pH of 2.2, 50% of the compound will have dissociation despite maintaining protonation in the amine group. Among the pH values of 2.2 and 5.4, the predominant form of the herbicide

Properties	Values
Chemical name (IUPAC)	*N*-(phosphonomethyl) glycine
Common name	Glyphosate
CAS number	1071-83-6
Molecular formula	$C_3H_8NO_5P$
Molecular weight	169 g mol^{-1}
Class	Herbicide
Group	Replaced glycine
Melting point	189.5°C
Boiling point	Decomposes before boiling
Degradation point	200°C
Vapor pressure (PV)	1.31×10^{-5} Pa (25°C, acid)
Henry's law constant to 25°C (H)	2.10×10^{-07} (Pa m^3 mol^{-1})
Solubility in water (S_W)	10.5 g L^{-1} (20°C)
Acid partition coefficient (pK_a)	2.34 (at 25°C)
Octanol-water coefficient (K_{ow})	6.31×10^{-4} (pH 7, 20°C)
Sorption coefficient (K_d)	209.4 mg L^{-1}
Half-life time degradation in soil (DT50)	15 days

Source: Adapted from PPDB [26].

Table 1.
Physicochemical properties of glyphosate.

is with two dissociations, thus having 50% of the molecules with three pH dissociations of 5.5. In a pH ranging from 5.5 to 10.2, there are three and four dissociations of glyphosate. Above pH = 11, the glyphosate is fully dissociated [24].

Amino acids and their derivatives present a zwitterionic behavior, that is, in its structure the carboxylic acid has a more acidic characteristics than the ammonium group. In glyphosate, phosphate and carboxylic groups have a greater acidic characteristics than the ammonium group. Cikalo et al. [28] observed the zwitterionic behavior of glyphosate when describing its dissociation. Thus, in the first dissociation of glyphosate, it would lose hydrogen linked to oxygen and only in the last dissociation that hydrogen linked to nitrogen.

5. Glyphosate-resistant weeds in Brazil

The occurrence of weed resistance to herbicides is a natural and inheritable capacity of certain biotypes within a given population to develop and reproduce after being exposed to herbicide doses that would be lethal to a normal population of the same species. This resistance is from an evolutionary process, occurring naturally at low frequency, and the selection pressure exerted by repetitive application of some herbicide or different types of herbicides that have the same mechanism of action increases the number of resistant individuals in the population. Herbicide resistance is identified when, generally, 30% of the plants are resistant [29].

Several weed species are inherently more resistant to glyphosate than others. A biotype of glyphosate resistance that occurred naturally was the *Convolvulus arvensis* without reporting the use of glyphosate [20]. A biotype of *Lotus corniculatus* resistant to glyphosate doses was identified by Boerboom et al. [30]. The natural resistance to this herbicide, these and other species, was not a problem until the emergence of glyphosate-resistant crops. With the adoption of these crops, many species became problematic because they occupied places where other weed species did not inhabit and glyphosate-resistant crops were cultivated [20].

Biotypes that have glyphosate resistance have been selected in crops such as corn, soybeans, and various orchards. In Brazil, glyphosate-resistant biotypes of *Conyza bonariensis* [31], *Conyza canadensis* [32, 33], *Conyza sumatrensis* [34], *Lolium multiflorum* [35, 36], *Digitaria insularis* [9, 37], *Chloris elata* [38], *Eleusine indica* [39], *Amaranthus palmeri* [40], and more recently *Amaranthus hybridus* [41] and *Euphorbia heterophylla* [42] were identified.

Due to the increase in the adoption of glyphosate-resistant crops, there was a great difficulty in selecting the herbicide to be used in weed populations in the past decade. The alternating herbicides that have different modes of action or herbicides mixed in tanks are recommended in resistance management programs; however, this is often ignored by farmers, because the cost to control weeds only with glypho-sate is much cheaper.

It is recommended that a rotation be made between cultivating glyphosate-resistant crops and nonresistant crops, so that the development of glyphosate resistance in weeds is delayed. However, the correct use of glyphosate with other herbicides, a survey of the weed population, extension of the area, and economy of producers are important factors in the management of weeds in glyphosate-resistant crops [20].

Figure 12 shows all the species of glyphosate-resistant weeds worldwide. The first case of reported resistance was in 1996 of the species *Lolium rigidum* in Victoria, Australia. In Brazil, the first case of glyphosate-resistant weed identified was in 2003, which was the species *Lolium perenne* ssp. *multiflorum*. In 2019, four

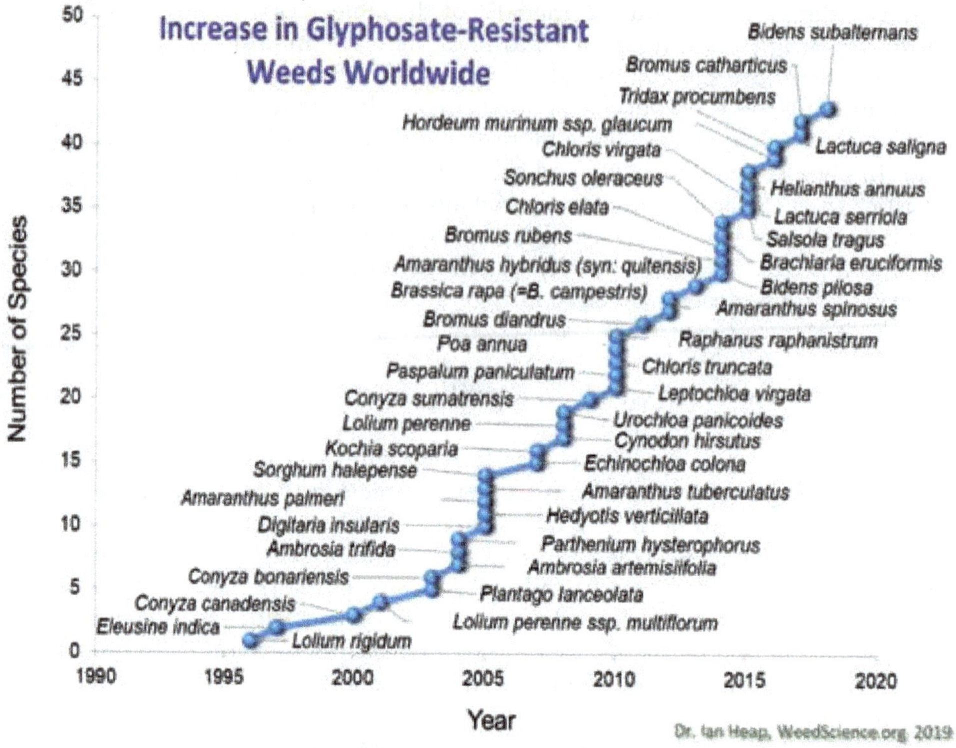

Figure 12.
Glyphosate-resistant weeds worldwide. Source: Heap [10].

cases of new glyphosate-resistant weeds were reported: in Australia, the spe-cies *Avena sterilis* ssp. *ludoviciana*; in Argentina, the species *Carduus acanthoides* which presented multiple resistance to 2,4-D and glyphosate; in Colombia, the species *Chloris radiata* which showed simple resistance to glyphosate; and also in Argentina, resistance of the species *Echinochloa crus-galli* var. *crus-galli* which was reported [10].

6. Conclusions

Among the pesticides, glyphosate is widely used in Brazil and in the world to control weeds in various crops. However, its use should be performed consciously, since there is no significant increase in the amount of weeds resistant to this herbicide. Despite its great use in the country, it presents few reports of resistant weeds than another mode of action of herbicides, such as acetolactate synthase and photosystem II (PSII) inhibitors.

Knowledge of glyphosate characteristics, weed biology, resistance mechanisms, and the production system used that favors the emergence of herbicide-resistant weed biotypes is important to appropriately manage and prevent or delay new cases of resistant weeds in the field.

Crops with glyphosate-resistant transgenic technologies will continue to be important in the future for weed management, and resistant biotypes will con-tinue to be selected. Thus, good agricultural practices are indispensable, and more innovations in technologies are necessary for the future. Therefore, adopting a long-term weed management perspective and integration system for all agricultural practices is of paramount importance to farmers.

Author details

Kassio Ferreira Mendes[1*], Rodrigo Nogueira de Sousa[2] and Ana Flávia Souza Laube[3]

1 Department of Agronomy, Federal University of Viçosa, Viçosa, MG, Brazil

2 Department of Soil Science, "Luiz de Queiroz" College of Agriculture, University of São Paulo, Piracicaba, SP, Brazil

3 Department of Chemistry, Federal University of Viçosa, Viçosa, MG, Brazil

*Address all correspondence to: kfmendes@ufv.br

References

[1] Dekker M. Economic benefits of pest management. In: Peshin R, editor. Encyclopedia of Pest Management. New York, USA: Springer; 2002. pp. 224-227. ISBN: 0618249060

[2] BRASIL. Ministério do Meio Ambiente. 2019. Available from: http://www.mma.gov.br/seguranca-quimica/agrotoxicos [Accessed: 13 December 2019]

[3] Portal São Francisco. História dos Agrotóxicos. 2019. Available from: https://www.portalsaofrancisco.com.br/biologia/historia-dos-agrotoxicos [Accessed: 13 December 2019]

[4] Pignati WA, Lima FANS, Lara SS, Correa MLM, Barbosa JR, Leão LHC, et al. Spatial distribution of pesticide use in Brazil: A strategy for Health Surveillance. Ciência & Saúde Coletiva. 2017;**22**(10):3281-3293. ISSN: 1413-812

[5] IBAMA—Instituto Brasileiro do Meio Ambiente e dos Recursos Naturais Renováveis. Consolidação de dados fornecidos pelas empresas registrantes de produtos técnicos, agrotóxicos e afins, conforme art. 41 do Decreto n° 4.074/2002. Vendas de Ingredientes Ativos por UF. 2018. Available from: https://www.ibama.gov.br/agrotoxicos/relatorios-de-comercializacao-de-agrotoxicos [Accessed: 13 December 2019]

[6] ANVISA—Agencia nacional de vigilância sanitária. Nota Técnica N° 23/2018/SEI/CREAV/GEMAR/GGTOX/DIRE3/ANVISA. 2018. Available from: http://portal.anvisa.gov.br/documents/111215/117833/Nota+t%C3%A9cnica+23+de+2018+-+Glifosato/faac89d6-d8b6-4d8c-8460-90889819aaf7 [Accessed: 13 December 2019]

[7] Galli AJB. A molécula glyphosate e a agricultura brasileira. In: Velini ED, Meschede DK, Carbonari CA, Trindade MLB, editors. Glyphosate. Botucatu: FEPAF; 2009. pp. 4-493. ISBN: 978-85-98187-09-9

[8] Ferreira SD. Resistência ao glyphosate em biótipos de *Digitaria insularis* e nível de dano econômico em soja e milho. 2019. 178 f. Tese (Doutorado em Agronomia), Universidade Estadual do Oeste do Paraná, Marechal Cândido Rondon; 2019

[9] Brunharo CACG. Resistência da planta daninha capim-branco (*Chloris polydactyla*) ao herbicida glyphosate. 2014. 153 f. Dissertação (Mestrado em Ciências—Fitotecnia). Piracicaba: Escola Superior de Agricultura "Luiz de Queiroz"—Universidade de São Paulo; 2014

[10] Heap I. The International Survey of Herbicide Resistant Weeds. Available from: http://www.weedscience.org [Accessed: 13 December 2019]

[11] Kaspary TE, Lamego FP, Langaro AC, Ruchel Q, Agostinetto D. Investigation of the mechanism of resistance to glyphosate herbicide in hairy fleabane. Planta Daninha. 2016;**34**(3):555-564. ISSN: 1806-9681

[12] Vasconcelos Y. Pesticides in the Balance. São Paulo, Brazil: FAPESP; 2018. Available from: https://revistapesquisa.fapesp.br/en/2019/02/25/pesticides-in-the-balance/ [Accessed: 25 July 2019]

[13] SINDIVEG—Sindicato Nacional da Indústria de Produtos para Defesa Vegetal. Estatísticas do Setor. 2018. Available from: https://sindiveg.org.br/estatisticas-do-setor/ [Accessed: 25 July 2019]

[14] Mesquita HC, Rodrigues AP, MendonçaJúnior AF. Riscostoxicológicos

do herbicida glyphosate. Agropecuária Científica no Semiárido. 2011;**7**(2):1-5. ISSN: 1808-6845

[15] IBAMA—Instituto Brasileiro do Meio Ambiente e dos Recursos Naturais Renováveis. Relatórios de comercialização de agrotóxicos. 2017. Available from: https://www.ibama.gov.br/agrotoxicos/relatorios-de-comercializacao-de-agrotoxicos [Accessed: 14 December 2019]

[16] ABRASCO—Associação Brasileira de Saúde Coletiva. Entenda o que é glifosato, o agrotóxico mais vendido do mundo. Available from: https://www.abrasco.org.br/site/outras-noticias/movimentos-sociais/entenda-o-que-e-o-glifosato-o-agrotoxico-mais-vendido-do-mundo/40996/ [Accessed: 14 December 2019]

[17] CONITEC—Comissão Nacional de Incorporação de Tecnologias no SUS. Abordagem do Paciente Intoxicado por Produtos Comerciais Formulados à base de Glifosato. Available from: http://conitec.gov.br/images/Protocolos DiretrizesBrasileiras_Agrotoxico_Cap3 .pdf [Accessed: 14 December 2019]

[18] Oliveira Júnior RS. Mecanismos de ação de herbicidas. In: Oliveira Júnior RS, Constantin J, Inoue MH, editors. Biologia e manejo de plantas daninhas. Vol. 1. Curitiba: Omnipax; 2011. pp. 141-192. ISBN: 978-85-64619-02-9

[19] Kruse ND, Trezzi MM, Vidal RA. Herbicidas inibidores da EPSPS: Revisão de literatura. Revista Brasileira de Herbicidas. 2000;**1**(2): 139-146. ISSN: 2236-1065

[20] Nandula VK, Reddy KN, Duke SO, Poston DH. Glyphosate-resistant weeds: Current status and future outlook. Outlooks on Pest Management. 2005;**16**(4):183-187. ISSN: 14658933

[21] EMBRAPA—Empresa Brasileira de Pesquisa Agropecuária. Principais herbicidas indicados para cultura de milho no preparo convencional do solo e plantio direto, para controle total da vegetação. 2006. Available from: http://www.cnpt.embrapa.br/biblio/do/p_do61_15.htm [Accessed: 14 December 2019]

[22] Ferreira FA. Mecanismos de ação dos herbicidas. 2005. Available from: https://www.cnpa.embrapa. br/produtos/algodao/publicacoes/trabalhos_cba5/336.pdf [Accessed: 14 December 2019]

[23] Helander M, Saloniemi I, Saikkonen K. Glyphosate in northern ecosystems. Trends Plant Science. 2012;**17**(10):569-574. ISSN: 1360-1385

[24] Amarante Júnior OP, Santos TCR, Brito NM, Ribeiro ML. Glifosato: Propriedades, toxicidade, usos e legislação. Química Nova. 2002;**25**:589-593. ISSN: 0100-4042

[25] Roman ES, Vargas L, Rizzardi MA, Hall L, Beckie H, Wolf TM. Como funcionam os herbicidas: Da biologia à aplicação. Passo Fundo: Berthier; 2007. p. 158. ISBN: 9788589873512

[26] PPDB—Pesticide Properties Data Base. Glyphosate (Ref: MON 0573). Available from: https://sitem.herts.ac.uk/aeru/ppdb/en/Reports/373.htm [Accessed: 14 December 2019]

[27] Sprankle P, Meggitt WF, Penner D. Adsorption, mobility and microbial degradation of glyphosate in the soil. Weed Research. 1975;**23**(3): 229-234. ISSN: 1365-3180

[28] Cikalo MG, Goodall DM, Matthews W. Analysis of glyphosate using capillary electrophoresis with indirect detection. Journal of Chromatography A. 1996;**745**(1-2): 189-200. ISSN: 0021-9673

[29] Christoffoleti PJ, López-Ovejero R. Principais aspectos da resistência

de plantas daninhas ao herbicida glyphosate. Planta Daninha. 2003;**21**(3):507-515. ISSN: 1806-9681

[30] Boerboom CM, Wyse DL, Somers DA. Mechanism of glyphosate tolerance in birdsfoot trefoil (*Lotus corniculatus*). Weed Science. 1990;**38**(6):463-467. ISSN: 1550-2759

[31] Vargas L, Bianchi MA, Rizzardi MA, Agostinetto D, Dal MT. *Conyza bonariensis* biotypes resistant to the glyphosate in Southern Brazil. Planta Daninha. 2007;**25**(3):573-578. ISSN: 0100-8358

[32] Moreira MS, Nicolai M, Carvalho SJP, Christoffoleti PJ. Resistência de *Conyza canadenses* e *C. bonariensis* ao herbicida glyphosate. Planta Daninha. 2007;**25**(1):157-164. ISSN: 0100-8358

[33] Lamego FP, Vidal RA. Resistance to glyphosate in *Conyza bonariensis* and *Conyza canadensis* biotypes in Rio Grande do Sul, Brazil. Planta Daninha. 2008;**26**(2):467-471. ISSN: 0100-8358

[34] Santos G, Oliveira Júnior RS, Constantin J, Francischini AC, Osipe JB. Resistência múltipla de *Conyza sumatrensis* ao chlorimuron-ethyl e ao glyphosate. Planta Daninha. 2014;**32**(2):409-416. ISSN: 0100-8358

[35] Roman ES, Vargas L, Rizzardi MA, Mattei RW. Resistance of Italian ryegrass (*Lolium multiflorum*) to glyphosate. Planta Daninha. 2004;**22**(2):301-306. ISSN: 0100-8358

[36] Vargas L, Roman ES, Rizzardi MA, Silva VC. Identification of glyphosate-resistant ryegrass (*Lolium multiflorum*) biotypes in apple orchards. Planta Daninha. 2014;**22**(4):617-622. ISSN: 0100-8358

[37] Carvalho LB, Cruz-Hipolito H, González-Torralva F, Alves PLCA, Christoffoleti PJ, Prado R. Detection of sourgrass (*Digitaria insularis*) biotypes resistant to glyphosate in Brazil. Weed Science. 2011;**59**(2):171-176. ISSN: 1550-2759

[38] Brunharo CACG, Patterson EL, Carrijo DR, Melo MS, Nicolai M, Gaines TA, et al. Confirmation and mechanism of glyphosate resistance in tall windmill grass (*Chloris elata*) from Brazil. Pest Management Science. 2016;**72**(9):1758-1764. ISSN: 1526-4998

[39] Takano HK, Oliveira Júnior RS, Constantin J, Braz GBP, Gheno EA. Goosegrass resistant to glyphosate in Brazil. Planta Daninha. 2017;**35**:e017163071. ISSN: 0100-8358

[40] Gonçalves Netto AG, Nicolai M, Carvalho SJP, Borgato EA, Christoffoleti PJ. Multiple resistance of *Amaranthus palmeri* to ALS and EPSPs inhibiting herbicides in the State of Mato Grosso, Brazil. Planta Daninha. 2016;**34**(3):581-587. ISSN: 0100-8358

[41] Oliveira C, Mathioni SM, Lemes L, Ozório E, Jauer A, Altmann T, et al. População de caruru (*Amaranthus hybridus*) resistente ao glyphosate são encontradas no Rio Grande do Sul. Boletim Informativo—SBCPD; 2019. pp. 23-25

[42] EMBRAPA – Empresa Brasileira de Pesquisa Agropecuária. Mais uma planta daninha resiste ao glifosato no Brasil. 2020. Available from: https://www.embrapa.br/busca-de-noticias/-/noticia/50622096/mais-uma-planta-daninha-resiste-ao-glifosato-no-brasil [Accessed: 18 March 2020]

10

Agricultural Homoeopathy: A New Insight into Organics

Mazón-Suástegui José Manuel, Ojeda-Silvera Carlos Michel, García-Bernal Milagro, Avilés-Quevedo María Araceli, Abasolo-Pacheco Fernando, Batista-Sánchez Daulemys, Tovar-Ramírez Dariel, Arcos-Ortega Fabiola, Murillo-Amador Bernardo, Nieto-Garibay Alejandra, Ferrer-Sánchez Yarelys, Morelos-Castro Rosa María, Alvarado-Mendoza Alex, Díaz-Díaz Miriam and Bonilla-Montalvan Boris

Abstract

At present, agricultural homoeopathy is being increasingly implemented worldwide to mitigate the negative effects caused by the indiscriminate use of chemical products in conventional agricultural practices. It is a viable alternative to improve organic agriculture, since homoeopathic medicines are innocuous substances with a capability to activate measurable response mechanisms when used in plants, animals and humans. Experimental research results allow us to conclude in this chapter that agricultural homoeopathy is able to stimulate favourable biological and even genetic responses in plants (basil *Ocimum basilicum* L., bean *Phaseolus vulgaris* L., cucumber *Cucumis sativus* L., tomato *Solanum lycopersicum* L.), which shows a novelty insight for organic agriculture.

Keywords: agricultural homoeopathy, organic production

1. Introduction

According to the four principles of organic agriculture, health, ecology, care and fairness, all materials used as fertiliser to control pests must be innocuous for people, animals and environment. Those principles are mainly focused to respecting and emulating tha natural biological cycles in order to leave for future generations the same possibility of using natural resources of the planet. All materials or supplies must come preferably from the same farm to avoid introducing external risk agents and to reduce costs [1]. In this context, homoeopathy should be an alternative resource to help crop plants improve their production under several biotic and abiotic stress conditions.

Homoeopathy is a branch of universal medicine based on the 'principle of the like' (*Similia Similibus Curentur* = Like treats Like) and in minimum doses. It means that a substance in a massive dose generates pathological symptomatology;

symptomatology, has the possibility of cure it, if applied in the minimum doses obtained by dilution and intense agitation, in other words, homoeopathic succussion [2]. Homoeopathy is derived from a Hippocratic concept developed by the German physician, Samuel Hahnemann (1755–1843), with serially diluted medicines (1:9, 1:99, 1:999 and others) in water-ethanol vehicle alternating dilution with succussion [3]. Through this process, medicinal products were obtained, called potencies, dilutions or dynamisations, in decimal, centesimal or thousands, and others [4].

The starting point in obtaining a certain boosting curative capacity is a concentrate or mother tincture (MT), which is an alcoholic extract prepared from plants, animals, minerals and even inert materials as nanoparticulate metals [4]. Despite its high dilution, it is possible to detect molecules or nanoparticles of the 'ponderable active principle' (MT) in dynamisations, even in high centesimal dilutions (12CH, 30CH, 200CH) although according to Avogadro's theory, the dynamisation 12CH should have a single molecule of the original substance contained in the MT [5].

The principle of similarity should be understood as the parallelism between the toxic effect of a substance and its therapeutic use for the treatment of similar, but not necessarily identical, symptomatology. Some drugs used for the treatment of depression, anxiety and panic can induce panic and depressive symptoms in a healthy person when they have been homoeopathically energised by means of a dilution-succussion process [6]. The most paradigmatic and controversial aspect of homoeopathy is that these serial dilutions have a measurable effect even when they are given to an animal or a plant in infinitesimal doses, which leave no possibility of suggestion and placebo effect.

Formerly, the somewhat ethereal or subjective concept of 'vital force' was handled, which has fallen into disuse because it has been found that a large part of the effects of homoeopathic medicine are mediated by cells of the immune system [7, 8]. This is the case of ultradilutions of Aspirin® that have a platelet pro-aggregating effect, just opposite to the effects of aspirin in a ponderable dose [9], and whose action would be mediated by the COX-2 coenzyme [10].

Animal model research has its advantages, and possibly the most robust, reproduced and tested research topic in the world is the effect of thyroxine on the induction of frog metamorphosis. The effect of thyroxine in massive dose is just the opposite to the effect of the same product in homoeopathic dose [11], which has been proven by different authors in different frog species [12]. In veterinary medicine, homoeopathy is used, among others, for the treatment of foetal death in pigs (Day, 1984); control of mastitis in cattle [13] and sheep (Day, 1986); and elimination of ticks in bovines [14], salmonellosis in chickens [15] and gastroenteritis in dogs [16]. In aquaculture systems, it can be administered in food or added directly to culture water [17] to promote resistance to pathogens in stress situations and a better postinfection recovery [2, 18]. Freshwater fish, when raised in low stress conditions, have greater survival, production potential and meat quality [19]. Other relevant research topics are gonadic development and sexual maturation [20], stress response [21], physiological and neuroendocrine changes [20, 22], and growth/survival [20]. Recently published results [23, 24] have shown that aquacultural homoeopathy is able to produce measurable biological effects in marine molluscs, shrimp and fish, which trigger physiological, humoral, genomic, metagenomic and transcriptomic responses [17, 25, 26] to recover the internal homoeostasis of the treated individual, which is a synonym of health.

In the plant model, the most replicated study is the protective effect of homoeopathically ultradiluted arsenic versus the effect of arsenic itself in ponderable doses (pre- and post-treatment) on maize seeds [27]. Positive results have been achieved in prevention and control of plague organisms in the *Solanum quitoense* Lam [28] crop, in germination and initial growth of *Hancornia speciosa* Gomes [29] and in remediation of soils affected by heavy metals [30]. The homoeopathic medicine *Natrum muriaticum* has been used successfully to increase resistance to salt stress in *Solanum lycopersicum* [31]. Those previous and other published results, suggest that aquacultural and agricultural homoeopathy are viable alternatives from the economic, ecological and social points of view to contribute to the environment and extinguish the harmful footprint left on the planet by the indiscriminate use of various toxic agrochemicals [32].

The homoeopathic medical therapy establishes a principle that there are no illnesses but sick people and that every disease is only the reflection of a disturbance of the dynamic balance between the organism and biotic and abiotic elements of their environments. It is officially recognised as an alternative, holistic or integra-tive medicine therapy in various countries of the world where medical professionals are also trained at bachelor, master and doctoral degrees. Particularly in Mexico, the practice of homoeopathy was authorised by a presidential decree on July 31, 1895, and its study and practice are now officially recognised in the General Health Law (2015), and only health professionals can prescribe homoeopathic medicines. Homoeopathic medicines must have an official code and registration, so they differ from other products that are not medicines but herbs for infusion or herbal reme-dies. In countries such as Brazil, there are homoeopathic medicines for exclusive use in veterinary medicine and aquaculture of marine and freshwater fishes registered with the Ministry of Agriculture [17, 24].

Homoeopathy is an emerging holistic therapy whose application continues to grow throughout the world and is gaining ground thanks to rigorous scientific research in human, animal and plant models. Due to new discoveries in the field of quantum physics, it has even been proposed to rename homoeopathy as 'adaptive network nanomedicine' [33]. As a counterpart, the official medicine known as 'allopathy' is derived from a Galenic concept: the 'principle of opposites' based on the application of massive doses of various chemotherapeutic agents. These drugs, generically synthetic, are officially classified as antimicrobial, antiviral, anti-inflammatory, antispasmodic, antihistamine, anti-fever, etc., which, although they can alleviate, eliminate, or conceal symptoms, invariably have side effects and contraindications. Because it can affect health, pharmaceutical laboratories must communicate and describe these risks to the consumer on the label of the drug, following legal provisions.

This chapter describes recent research results about the use of homoeopathic medicines (HOM) and homoeopathic treatments (HOM treatments) in vegetable model. For the purposes of this chapter, commercial homoeopathic medicines for human use have been utilised in the form of liquid hydro-alcoholic dynamisations (Similia® Laboratories, Mexico) and injectable aqueous dynamisations (Rubio Pharma®, Mexico). They were considered 'stock dynamisations' from which the respective 'study dynamisations' were obtained through a serial process of decimal or centesimal dilution-succussion. The following HOM treatments were used for plants in controlled laboratory conditions and semicontrolled conditions in the field, as dynamisations of *Natrum muriaticum* (NaM), *Silicea terra* (SiT), *Magnesia phosphorica* (MaP), *Arsenicum album* (ArA), *Zincum phosphoricum* (ZiP) and *Phosphoric acid* (PhA) and *Magnesium metallicum* (MgM) (Provider: Similia®, CDMX, México) and *Magnesium Manganum phosphoricum* (MaMnP) (Provider: Rubiopharma®). These and other nosode-type HOM products have been designed by CIBNOR, which is processing the respective industrial property titles (Office for industrial protection and technology transfer, OTT-CEPAT/CIBNOR, www.cibnor.gob.mx). Distilled water (DW) was used as control treatments. HOM treatments and controls are used to soak seeds and sprinkle leaf area or added directly to substrate for crop species. It is very important to consider that homoeopathy could be used in fulfilling the organic agriculture principles. Nowadays authors of this chapter are studying the effects and substance homoeopathy in organic agriculture principles.

2. Effect of homoeopathy on plant species

2.1 Promotion of plant growth health, nutrition and performance

The application of homoeopathy in agriculture is known as agricultural homoeopathy, which offers an ecologically and economically viable model with the potential to reduce the use of agrochemicals in world agriculture. Homoeopathy in plants contributes to the improvement of internal processes to optimise their growth and development [34]. Scientifically proven results in crops have validated their ability to modify the physiological response of the plant, abundance of foliage and amount of fruit [30]. Recently the homoeopathic medicines *Sulphur*, *Silicea terra* and *Nux vomica* have been assessed on different plants of commercial interest, including corn [35], while the use of other HOM treatments is recommended, such as *Calcarea carbonica*, *Carbo vegetabilis* and *Magnesia carbonica*, because their active ingredients can provoke favourable responses in plants [34].

2.1.1 Germination and emergence of cucumber (Cucumis sativus) and tomato plants (Solanum lycopersicum)

Horticultural fruit crops are of great interest in agricultural industry due to their high consumption worldwide either fresh or canned, so it is important to increase their production. At the Universidad Técnica Estatal de Quevedo (UTEQ, Ecuador, FOCICYT Project), the effect of HOM treatments on tomato and cucumber was studied during germination and emergence from certified seeds of tomato and cucumber (Floradade and Marketmore varieties, respectively). The bioassays were carried out in the experimental area of the Plant Biotechnology Laboratory (germination) and in 'La María' farm (emergence) at UTEQ.

For germination and emergence, a completely randomised design was applied with a 2 axis bifactorial arrangement where factor A was dynamisations (7CH and 13CH) and B was homoeopathic medicines. HOM treatments T1 (NaM 7CH), T2 (SiT 7CH), T3 (MaP 7CH), T4 (ArA 7CH), T5 (NaM 13CH), T6 (SiT 13CH), T7 (MaP 13CH) and T8 (ArA 13CH) were applied in cucumber and T1 (NaM 7CH), T2 (SiT 7CH), T3 (ZiP 7CH), T4 (PhA 7CH), T5 (NaM 13CH), T6 (SiT 13CH), T7 (ZiP 13CH) and T8 (PhA 13CH) in tomato. For both species, distilled water (DW) was applied as a control. Each experimental treatments included six repetitions, each one with 30 seeds. The seed were previously disinfected and washed and then submerged for 20 min in each of the corresponding HOM treatments. The germination tests were carried out in sterilised Petri dishes (150 × 15 mm) placing filter paper as a substrate on the dish bottom. In each dish, 5 ml of the corresponding HOM treatment was initially added, and humidity of the paper substrate kept adding distilled water daily. The germination tests were carried out under controlled conditions with a 12:12 h photoperiod and 27 ± 1°C temperature. The seeds were considered germinated when the radicle is measured around 2 mm in length. To evaluate emergence, the seeds were also previously submerged in the HOM treatments and then planted in 200-well polystyrene trays with commercial substrate (Novarbo®). The seeds were considered emerged when the seedling broke the surface and emerged through the substrate.

Germination was recorded daily, determining the final percentage at 24 h in cucumber and 7 days in tomato. Emergence was also recorded daily, and the final percentage was determined at 4 days for cucumber and at 15 days for tomato. The germination and emergence rates were calculated using Maguire's equation [36]: ($M = n1/t1 + n2/t2 + \dots n30/tn$ (1)), where n1, n2, ... n30 are the number of seeds germinated and emerged at times t1, t2, ... tn. From each treatment/repetition, ten

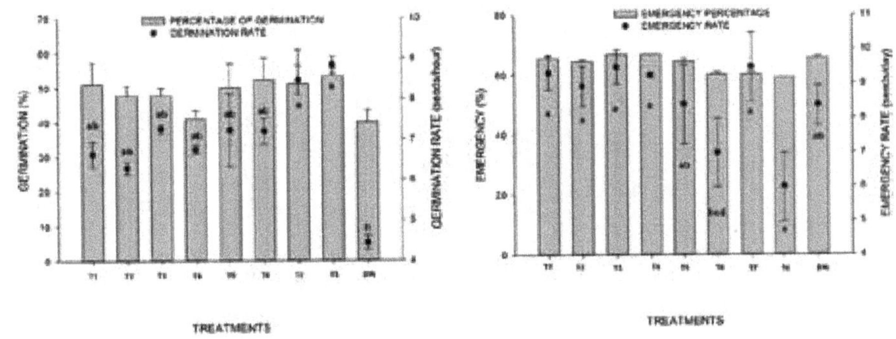

Figure 1.
Effect of HOM treatments on cucumber (Cucumis sativus) germination (% and rate; left) and emergence (% and rate; right); average values with different literals correspond to statistically different treatments (p ≤ 0.05).

random seedlings were taken for measurement of the morphometric variables at 14 days after emergence. After separating the tissues (radicle, stem and leaves), the length of the stem from the base to the apical part was measured with a graduated ruler (mm). *Idem*, after having washed the roots with drinking water and finally with distilled water, the length of the radicle was measured from the base of the stem where root hairs start up to where the main root ends. To determine fresh and dry biomass of the aerial part (stems + leaves) and radicle, each one was weighed separately in analytical balance (Mettler Toledo®, model AG204). Then aerial part and radicle were placed separately in paper bags in a drying oven (80°C) during 72 h until complete dehydration and then weighed in the same analytical balance. The data were expressed in grams of plant material (fresh or dry).

Test 1. In cucumber the results revealed that germination rate and percentage reached the highest value with T8 (**Figure 1**) with 53% in comparison to the control (40%). The highest germination rate (9 seeds/h) was also obtained with this HOM treatment, compared to control (4 seeds/h). The emergence percentage was similar for T1 and for the control (65%), which was lower than in other treatments. The emergence rate for T1 was higher than the other treatments and without difference with the control (**Figure 1**). During germination, the plants treated with T7 reached the longest stem length (6.5 cm) without statistical difference from other HOM treatments: T1, T2, T4, T5, T6 and T8. In those cases, significant statistical differences were found with the control (4.5 cm). The longest radicle was obtained with T7 (10.9 cm), with respect to the control (8.4 cm). Regarding biomass, the aerial part with the highest fresh biomass was recorded with T2 (0.18 g), which was statistically different from the control (0.12 g). The plants treated with T8 had the highest fresh radicle biomass (0.08 g), compared to the control (0.06 g). Also with T8 treatment, the highest dry radicle biomass (0.0032 g) was recorded with respect to the control (0.0018 g). During the emergence stage using T3, the longest stem length (12.3 cm) was reached, compared to control (9 cm). The greatest radicle length (6.0 cm) was obtained with T4, and no statistical difference was recorded with the rest of the treatments (5.0–5.9 cm). The highest yield in fresh biomass production of the aerial part was obtained with T2 (0.79 g) with respect to the control (0.47 g). Seedlings with the highest fresh radicle biomass (0.12 g) were obtained with T7, showing statistical differences compared to other experimental interactions (0.04–0.07 g). Finally, the highest radicle dry weight (0.0049 g) was attained with HOM treatment T5, which was statistically higher than that of the control (0.0017 g).

Test 2. In tomato, no significant differences were recorded regarding germination rate and percentage. However, with HOM treatment T2, the highest rate and percentage of emergence were obtained (3 and 26%), unlike the control group

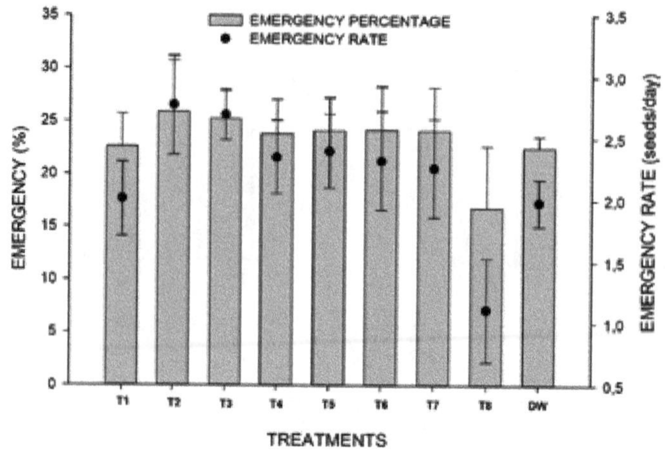

Figure 2.
Effect of HOM treatments on tomato (Solanum lycopersicum) emergence (% and rate).

(≤2 and ≤23%, respectively). In general, most of the HOM treatments had a stimulating effect during the emergence stage (**Figure 2**) since significant differences were recorded in the morphometric variables evaluated. During germination, the length of stem (SL) was greater using T3 (5.5 cm) with respect to the control (4.3 cm). With HOM treatment T2, the highest root development was obtained with a length of 8.2 cm compared to the control (5.6 cm). Significant differences were observed in dry biomass of stem and root obtaining higher dry stem biomass with T3 and T5 (0.01 g) than that obtained with control and other HOM treatments (0.0027 g). When T1 was applied, a higher dry root biomass (0.002 g) was observed with respect to control (0.0005 g), and other HOM treatments (≤0.0013 g) were assessed. During the emergence stage, significant differences were found regarding stem length with HOM treatment T2 (6.6 cm), and with T3 a similar growth was obtained (5.9 cm). The length of the stem was smaller with the rest of the treatments (≤5.8 cm) but, even so, higher than the control (4.6 cm). Regarding root length, no significant differences were found; however, the seedlings treated with T4 and T7 had the highest root growth (4.5 and 4.6 cm, respectively) beyond the control (3.3 cm). Finally, with regard to biomass, no significant differences were observed between HOM treatments and the control.

2.1.2 Initial plant growth of common bean (Phaseolus vulgaris L.) variety Quivicán plants

The common bean (*Phaseolus vulgaris* L.) is one of the edible legumes of greater consumption worldwide, providing an important source of proteins, vitamins and minerals to the diet of populations in developing countries in the Americas [37]. For this reason, research development has focused on improving crop response variables to increase their agricultural production. Thus, the objective of this study was to assess homoeopathic medicines to promote the overall performance and productivity of *P. vulgaris* L. (white-tinted Quivicán variety) during the stages of initial plant growth in pots with substrate, within cultivation houses with shadow mesh at Centro de Investigaciones Biológicas del Noroeste (CIBNOR), La Paz, Mexico.

Test 1. Certified seeds of this Quivicán variety (Seed Company, Villa Clara, Cuba) were used; a completely randomised design was used with three HOM treatments, T1 (MgM-31CH), T2 (MgMnP-3CH) and T3 (MgM-31CH + MgMnP-3CH), and a control with distilled water (DW), each one with four replicates. This

Treatments	Morphological parameters										
	SL (cm)	RL (cm)	FRB (mg)	FLB (mg)	FSB (mg)	DRB (mg)	DLB (mg)	DSB (mg)	FA (cm^2)	SD (mm)	L No.
T1	13.4^{ab}	20.6^{b}	12.25^{ab}	13.75^{b}	5.25^{b}	0.71^{b}	1.43^{b}	0.67^{a}	476.15^{b}	2.07^{b}	5.75^{a}
T2	13.4^{ab}	20.05^{ab}	8.88^{b}	14.75^{b}	7.0^{b}	0.58^{b}	1.50^{b}	0.72^{a}	435.14^{ba}	2.92^{b}	5.75^{a}
T3	15.4^{a}	27.97^{a}	23.75^{a}	21.12^{a}	8.2^{a}	1.52^{a}	2.05^{a}	0.85^{a}	629.37^{a}	3.07^{a}	6.25^{a}
DW	10.5^{b}	21.17^{b}	11.12^{b}	7.87^{c}	3.37^{c}	1.01^{ab}	0.72^{c}	0.30^{b}	233.83^{c}	2.2^{c}	3.75^{b}

Average values with different literals in the same column differ statistically (Tukey HSD, $p \le 0.05$)

Table 1.
Effect of HOM treatments on morphometric parameters during growth stage of the common bean (Phaseolus vulgaris L.) variety Quivicán.

experiment was developed to evaluate initial growth phase. Seeds were planted (five replications per treatment) in plastic pots (3 seeds/pot) with 5 kg of commercial substrate (Sogemix PM®). Emergence evaluation was done daily; when 50% + 1 of the seeds emerged, 1 ml of the respective HOM treatment and water for the control (NT) was applied around the stems of the plants on alternate days. After 35 days, we proceeded to measure stem length (SL), root length (RL), fresh biomass (g) of root, stem and leaves (FRB, FSB, FLB), foliar area (FA) (cm^2), stem diameter (SD) (mm) and number of leaves (L No). The yields corresponding to dry biomass (DRB, DLB, DSB) were also measured. The most relevant result of the study was highly favourable and statistically significant effect of HOM treatment T3 with respect to the untreated control group (**Table 1**).

The cause-effect results in the plant model demonstrated a synergistic effect of the homoeopathic medicines included in HOM treatment T3, which evidently could not have been a placebo effect because it clearly favoured the growth of *P. vulgaris* variety Quivicán. Magnesium and manganese were components of T3; both are essential for the growth of any living cell and necessary secondary macronutrients for plant growth and development. Around 75% of foliar magnesium was involved in protein synthesis, and 15–20% of total magnesium was associated with pigments, a constituent element of the chlorophyll molecule and extremely important in photosynthesis [38]. Magnesium acts mainly as a cofactor of several enzymes involved in photosynthetic carbon fixation and also in basic metabolism [39]. Both magnesium and manganese play an important role in plant general nutrition and enhance or reinforce their resistance to diseases [40]. The results obtained in this study suggest that agricultural homoeopathy has application in the cultivation of the common bean (*Phaseolus vulgaris* L.) variety Quivicán, during the stages of initial plant growth.

2.2 Attenuation of the effects of abiotic stress

Currently, numerous studies have focused on elucidating the negative effects of abiotic stress on agricultural crops. Saline stress associated with high tempera-tures and solar irradiation is the most important environmental process that stops cultivated plant growth, development and survival, decreasing productivity [41]. Worldwide, millions of hectares have a high degree of aridity and the presence of salts [42]. In Mexico, arid and semiarid regions constitute more than 50% of the national territory [43], so the study of these factors is important due to the negative impact they have on the agricultural sector. Salinity has environmental, social and economic consequences because sustainability and yield of the cultivated spe-cies decrease in the affected areas. Salinity affects plant metabolism and growth, causing a decrease in biomass production [44]. Global losses due to salt stress are estimated at 12 billion dollars per year and affect a fifth of the farmland [45]. Among the main harmful effects, decrease in water absorption, ion assimilation that can cause toxicity and nutritional imbalance and physiological changes, such as the reduction of the photosynthetic rate due to a lower leaf area, which reduces crop viability [46] are included.

Several alternatives have been studied to mitigate the effect of salinity in agriculture, such as genetic improvement, selection of tolerant varieties and physical treatments to seeds to induce tolerance or agrochemicals to stimulate plant growth. In these sense, HOM treatments imply the application of ultra-diluted substances, which are an eco-friendly and economic variant that could be used effectively in any condition and circumstance [47]. Unfortunately, this organic-like alternative has been little studied worldwide.

2.2.1 Salinity stress by NaCl in the bean crop (Phaseolus vulgaris L.)

The common bean *Phaseolus vulgaris* L. is a key product in world food security [48], but the adverse environmental conditions, mainly drought and salinity in soils, affect their general performance reducing productivity and harvest [49]. Saline stress occurs due to high concentrations of sodium (Na^+) and chlorine (Cl^-) that seri-ously alter the plant metabolism, affecting its growth and development [50]. Water deficit is an osmotic stressing agent, specifically associated with salinisation, which reduces the rate of fixation of carbon dioxide (CO2) and affects processes associ-ated with photosynthesis [51]. Abscisic acid (ABA) is a key hormone that regulates the responses of plants to abiotic stress [52] and initiates the activation of stomatal closure when facing salt stress. When it happens, CO2 levels decrease and conse-quently photosynthesis, causing oxidative stress [53]. The objective of this study was to evaluate HOM treatment NaM (7CH) as attenuator of the harmful effect of salinity induced by NaCl application. The experiment was developed at CIBNOR to assess the initial growth stage of the species and study the expression of the genes associated with physiological response of *P. vulgaris* against HOM treatments.

A completely randomised design was applied with HOM treatment T1 (NaM 7CH) and control (DW) each one with five replicates and two concentrations of NaCl (0 and 75 mM). Seeds of the variety white-tinted Quivicán (Empresa de Semillas, Villa Clara, Cuba) were used.

Test 1. The seeds were disinfected, then imbibed for 30 min in HOM treatment NaM 7CH or in DW (control treatment), planted in 5 kg plastic pots (3 seeds/pot) with commercial substrate (Sogemix PM®) and grown for 35 days. Once the seeds germinated and the plants emerged, NaM 7CH or DW (15 ml) was applied on alternate days near the stem of each plant. The addition of NaCl began 15 days after germination and is applied gradually until reaching 75 mM to avoid osmotic shock.

The photosynthesis rate (A, $\mu mol\ m^{-2}\ s^{-1}$) was measured with a LCpro-SD portable computer with a wide-blade camera (ADC, Hoddesdon, Herts, United Kingdom). Three measurements were made during the fourth week of the trial, on healthy leaves and on completely sunny days. At the begining of frlowering, a sample was taken to perform the expression analysis of the 9-cis-epoxycarotenoid gene (PvNCDE1) [54]. Total RNA isolation, cDNA synthesis and real-time PCR amplification were performed following the methodology reported by Morelos et al. [55], and the relative expression of PvNCDE1 was estimated following the model proposed by Hellemans et al. [56]. One-way ANOVA followed by Tukey's exact test was performed using STATISTICA 8.0 (StatSoft, Inc.). A p-value <0.05 was considered significant.

At the end of the study, photosynthesis showed a considerable decrease in the control (DW) when the plants underwent saline stress (**Figure 3**). In these stressed plants, stomatal closure was produced by ABA, generating a direct blockage of photosynthesis due to the limited uptake of CO2, an indispensable substrate for the Calvin cycle. When it happens, a blockage of the photo phase occurs due to the null

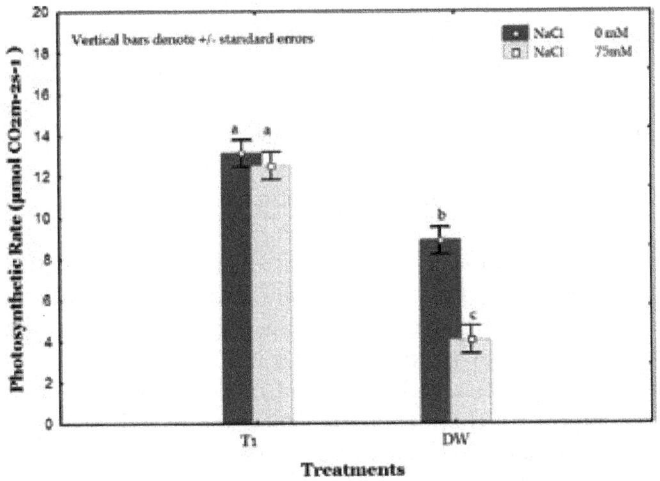

Figure 3.
Photosynthetic rate is recorded in Phaseolus vulgaris treated and non-treated with NaM-7CH and subjected to 0 and 75 mM NaCl. Average values with different literals in the same treatment or control differ statistically ($p \leq 0.05$).

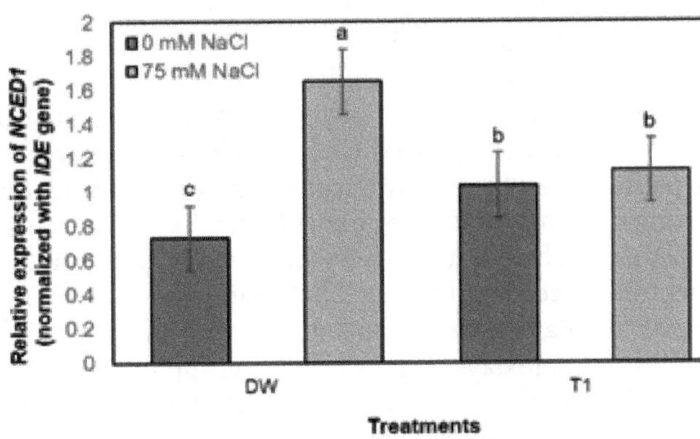

Figure 4.
Expression of the pvNCED1 gene of HOM-treated (NaM-7CH) bean Phaseolus vulgaris L., exposed to NaCl stress. Average values with different literals correspond to statistically different treatments ($p \leq 0.05$).

demand of the reducing power and ATP, resulting in oxygen production with consequent damage to the chloroplast membranes and their subsequent disintegration [57]. Contrary to the above, in HOM-treated plants that were subjected in parallel to salt stress, photosynthesis increased and no significant difference ($p > 0.05$) was observed with respect to plants not subjected to salt stress by the addition of NaCl. This result suggests that plants receiving HOM treatment did not become stressed, despite having been exposed to saline conditions (**Figure 3**).

The 9-cis-epoxycarotenoid gene (NCED) is known to be overexpressed under drought stress conditions in *P. vulgaris* bean, and it is considered fundamental for the regulation of ABA biosynthesis [58]. The increase of ABA under stress conditions causes a change in gene expression and adaptive physiological responses of plants [59]. As a result of this study, the relative expression of the pvNCED1 gene (**Figure 4**) was higher ($p < 0.05$) in the plants without HOM treatment of the control group (DW) than that were exposed to 75 mM of NaCl with respect to those not exposed to NaCl. The response of the HOM-treated plants did not show significant differences ($p > 0.05$) under normal conditions or in high salinity (75 mM NaCl), which means that the plants receiving homoeopathy were not stressed despite also receiving 75 mM NaCl since a lower expression of the pvNCED1 gene was found in the gene expression study. This result confirms a cause-effect relationship of T1 and the impossibility of a placebo effect in a plant model investigation where there is no possibility of suggestion of the treated individual.

2.2.2 Salinity stress by NaCl in basil (Ocimum basilicum L.)

Basil is an important aromatic species for its use as flavouring and dry and fresh seasoning in the food industry; as stimulant, antispasmodic and antialopecic in pharmacy; and as aromatising cosmetics in the perfume industry [60]. The growing interest of consumers for products of natural origin stimulates the market of aromatic plants, making them a viable option for the organic agricultural sector and the possibility of exporting them fresh or processed in extracts, essences and oils used in culinary industries, cosmetics and pharmaceuticals [61]. This work aimed to assess the effect of NaM as an attenuator of NaCl stress in plant photosynthesis and biomass production of two varieties of basil, grown under hydroponic system.

Test 2. This experiment was developed at CIBNOR, and seedlings were obtained from certified organic seeds of basil *O. basilicum* L., Emily and Napoletano varieties (Vis Seed Company, USA). A completely randomised experimental design with factorial arrange-ment (2A × 2B × 3C) was used, considering the Napoletano and Emily as factor A, the concentration of NaCl (0 and 75 mM) as factor B and HOM treatments NaM 7CH and NaM 13CH as factor C. The study included a total of 12 treatments each with 4 replica-tions. The bioassay was carried out in expanded polyurethane boxes of 69 × 38.5 × 25 cm and 38 L capacity, gauged with potable water (electrical conductivity 0.22 dS m^{-1}). Six plastic pots (150 ml) with an experimental plant inside were fixed above each box letting the roots inside the box with liquid-enriched media pass through 1 inch holes. The plants received a n Tutritioushe a sopplicatlutionion aof the dapted forHOM btreatmentsasil according t and oco Sntrolamperio (D [W) 62].began after a period of acclimatisation 7 days after transplant, spraying the aerial part of the plants with 150 ml $plant^{-1}$ on alternate days. After 15 days, the application of saline treatments began gradually to avoid osmotic shock until a concentration of 75 mM was reached.

The photosynthetic rate (A, µmol m^{-2} s^{-1}) was measured with (IRGA) LCpro-SD portable photosynthesis system equipment with a broad-leaf leaf chamber (ADC, Hoddesdon, Herts, UK) in a completely turgid and healthy leaf on completely sunny days (three measurements, 1 week before cutting); 45 days after transplant, the leaf area was determined (LA, cm^2) by integrating leaf area metre (Li-Cor®, model-LI-3000A, series PAM 1701) and biomass using fresh weight of the aerial part (BFAP, g); an analytical balance was used (Mettler Toledo®, model AG204).

The cause-effect results of this study revealed significant differences between varieties × NaCl × HOM treatment for A ($F_{2,60} = 4.14$, $p \leq 0.020$), FA ($F_{2,36} = 2.87$; $p \leq 0.01$) and BFAP ($F_{2,36} = 11.1$; $p \leq 0.0001$). Napoletano showed highest fresh weight of aerial part without the addition of NaCl and the HOM treatment NaM 7CH (**Table 2**). When both basil varieties were subjected to 75 mM NaCl, the photosynthetic rate decreased 53.6 and 63.6% with respect to the control in Napoletano and Emily, respectively. However, this result was reversed with the application of HOM treatments (NaM 7CH or NaM 13CH) in all variables and both varieties. At the end of the study, increases greater than 50% were obtained with respect to the plants that received 75 mM NaCl, but they did not receive a 'similar' HOM treatment.

These results suggest that *O. basilicum* plants naturally and positively responded to saline stress with both HOM treatments without the possibility of a placebo effect, both at cellular level and in the tissues as a whole, increasing growth. According to Zhu [63], stress agents offer environmental signals that are perceived and recognised by plants,

Varieties	NaCl (mM)	*NaM*	A (μmol m⁻² s⁻¹)	BFAP (g)	FA (cm²)
Napoletano	0	0	15.46^{bc}	174.6^{b}	4768.9^{a}
	0	7CH	22.7^{a}	190.9^{a}	5013.1^{a}
	0	13CH	15.68^{bc}	164.0^{c}	4304.6^{b}
	75	0	7.17^{g}	61.7^{h}	846.3^{g}
	75	7CH	17.02^{b}	140.7^{e}	3060.9^{d}
	75	13CH	13.15^{de}	110.7^{f}	2471.9^{e}
Emily	0	0	12.68^{ef}	143.2^{de}	3389.2^{cd}
	0	7CH	15.39^{bcd}	150.2^{d}	3501.9^{c}
	0	13CH	12.55^{ef}	142.4^{de}	3133^{cd}
	75	0	4.62^{i}	57.2^{h}	747.5^{g}
	75	7CH	13.47^{cde}	108.2^{f}	2195.4^{ef}
	75	13CH	10.69^{f}	93.4^{g}	1951.5^{f}

Average values with different literals in the same column differ statistically (Tukey HSD, $p \leq 0.05$)

Table 2.
Effect of the interaction of varieties × NaCl × NaM in the variables evaluated in two varieties of basil (O. basilicum) subjected to saline stress (NaCl).

transduced into cells and re-transmitted, which generate a cascade of biochemical, physi-ological and genetic responses that allow the plant to adapt to change when it is gradual, but this response depends on each species and variety. It is possible to hypothesise that the presence of magnesium nanoparticles initially bioavailable in sea salt (NaM origin) has contributed to the formation of chlorophyll molecules and photosynthesis, which is the main process of plant biomass production. Additionally, Magnesium has a predomi-nant role in enzymatic activity related to carbohydrate metabolism [64]. The results of this study suggest that agricultural homoeopathy besides being an organic-like treatment can increase general overall performance and productivity of *O. basilicum*, strengthening its ability to tolerate saline stress conditions without the need of using agrochemicals.

2.3 Attenuation of the effects of biotic stress

With the indiscriminate use of pesticides in agriculture, the resistance of pathogenic microorganisms to the chemicals used has increased, which has negatively affected the environment and therefore the consumption of products with high levels of toxicity. Therefore, technological alternatives have been generated to replace conventional ones. Recently the application of safe products has increased in agriculture, many of which are framed in novel agricultural homoeopathy. Agricultural homoeopathy is an alternative for agricultural farmers, compatible with traditional, organic, ecological, biodynamic and even conventional agriculture, capable of influencing the biological processes of plants by controlling health problems caused by fungi, viruses and bacteria. It contributes to pest control and influences crop growth and development [34].

Test 1. The experiment was developed at UTEQ, Ecuador, to assess in vitro activity of homoeopathic medicines against *Fusarium oxysporum* f. sp. lycopersici. One of the most important diseases affecting tomato crop is vascular wilt caused by *Fusarium oxysporum* f. sp. lycopersici (Fol) (Sacc.) Snyder and Hansen [65]. The control of fungal diseases in agriculture is controlled by agrochemicals including toxic pesticides whose excessive and indiscriminate use has caused the decrease or loss of the fertile soil layer and death of microorganisms in soil. Additionally, the pathogens have developed resistance to the active ingredient of the agrochemical that generally has a high cost [34]. Therefore, it is necessary to search for ecological and less polluting alternatives for the control of pests and diseases in agriculture, ensuring the safety and future of the agro-food industry. The objective of this study was to evaluate the in vitro effect of homoeopathic medicines on the pathogenic fungus *Fusarium oxysporum* f. sp. lycopersici, highly damaging various agricultural crops [66].

An experimental design which is completely randomised was applied with six homoeopathic medicines, each in two dynamisations (7 CH and 13 CH, Similia® CDMX, Mexico): T1 (MaP 7CH), T2 (ZiP 7CH), T3 (PhA 7CH), T4 (SiT 7CH), T5 (NaM 7CH), T6 (ArA 7CH), T7 (MaP 13CH), T8 (ZiP 13CH), T9 (PhA 13CH), T10 (SiT 13CH), T11 (NaM 13CH) and T12 (ArA 13CH) and one control (DW) to measure antifungal activity against *F. oxysporum*, using the method of the poisoned medium [67]. The diameter of the mycelium was measured daily, and the percentage of growth inhibition was determined by the formula [%inhibition = mycelial growth of the control – mycelial growth of the treatment/mycelial growth of the control × 100].

No significant differences were observed in radial growth of the phytopathogenic fungus in the HOM treatments with respect to the untreated control (NT). On the other hand, significant differences were observed between HOM treatments and their dynamisations (7 CH and 13 CH) with respect to NT. The dynamisation 13 CH increased the percentage of inhibition of the phytopathogenic fungus. These results confirmed the variability of the response induced by homoeopathic medicines in plant model, whose response depends on the dynamisation used [68]. These results are in agreement with those reported by Narváez-Martínez et al. [28] who used a homoeopathic treatment developed from a pathogen (nosode); when they applied different dynamics in tomato *Solanum quitoense* Lam, they found different effects against a pest caused by *Neoleucinodes elegantalis*. HOM treatments T2, T7 and T12 offered a greater inhibition percentage against the pathogen (70, 65 and 51%, respectively). These results agree with Tichavsky [69], who stated that *Phosphorus* homoeopathic medicine helped to control diseases caused by fungi, and two of these treatments contained phosphorus. According to Casali et al. [70], this result was due to the production of secondary metabolites (essential oils). However, during this study, HOM treatments T3 and T9 favoured the growth and reproduction of the fungus. Our results coincide with Damin et al. [71] who evaluated nine homoeopathic medicines against the pathogen *Metarhizium anisopliae* and obtained stimulation in the production of conidia by this fungus. It is necessary to conceptualise that homoeopathy acts on living beings; therefore, fungus can also be favoured with a HOM treatment.

The attained results revealed that the HOM treatments showed activity against *F. oxysporum*, highlighting *Zincum phosphoricum* and *Magnesium phosphoricum*. These cause-effect results in the plant model demonstrated that a placebo effect is not only absent in homoeopathic medicine but also supports agricultural homoeopathy. These results contribute to search for alternatives to control diseases caused by this phytopathogen in tomato plants, by using effective, innocuous and more eco-friendly tools to substitute the use of agrochemicals.

3. Conclusions

A lot of experimental scientific results related with the use of homoeopathic medicines in plants, besides new results and insights discussed along this chapter, eliminate the principal argument of homoeopathy detractors. Actually, they cannot sustain their arguments in the sense that the suggestion and the placebo effect are the only mechanisms of action of this old, and at the same time new, alternative for organic agriculture.

As shown throughout the text, there is sufficient scientific evidence that homoeopathy strengthens the energy and vitality of plants even under conditions of abiotic stress and promotes a dynamic balance of the plant with soil, water and the environment. Homoeopathic dilutions from different origins can be applied by soaking seeds and spraying leaves or directly to the soil or substrate. These ultra-diluted and innocuous treatments have the ability to initiate cascade responses, promoting favourable physiological reactions in the plant, with a systemic approach against the symptoms associated with an infection or a stressor. This is possible because, as in any living organism, plants have a genetic memory, which is continuously enriched. The response of plants treated with homoeopathy is natural, lasting and without the negative side effects that some agrochemicals have, which can even accumulate in their tissues and affect the safety of the harvested product, which makes it unfit for human consumption. Undoubtedly, homoeopathy has great potential, not only in human public health and aquaculture and veterinary production but also in organic, ecological and sustainable agriculture, which will be essential for the future development of humanity.

Acknowledgements

This study was financed by the Sectorial Fund for Education of Mexico: Project CB SEP-CONACYT-258282 'Experimental evaluation of homoeopathy and new pro-biotics in the culture of molluscs, crustaceans and fish of commercial interest' and Project PROINNOVA-CONACYT-241777 'Innovation and continuous improvement of products and processes to optimise hatchery seed production of the American oyster *C. virginica*', under academic responsibility of the first author (JMMS); the authors thank Diana Fischer for editorial services.

Author details

Mazón-Suástegui José Manuel[1], Ojeda-Silvera Carlos Michel[1*],
García-Bernal Milagro[2], Avilés-Quevedo María Araceli[3], Abasolo-Pacheco Fernando[4],
Batista-Sánchez Daulemys[1], Tovar-Ramírez Dariel[1], Arcos-Ortega Fabiola[1],
Murillo-Amador Bernardo[1], Nieto-Garibay Alejandra[1],
Ferrer-Sánchez Yarelys[4], Morelos-Castro Rosa María[1], Alvarado-Mendoza Alex[4],
Díaz-Díaz Miriam[2] and Bonilla-Montalvan Boris[4]

1 Centro de Investigaciones Biológicas del Noroeste S.C. (CIBNOR), La Paz, B.C.S., Mexico

2 Universidad Central de Las Villas (CBQ), Santa Clara, Villa Clara, Cuba

3 Freelance Services for R&D in Marine Aquaculture, La Paz, B.C.S., Mexico

4 Universidad Técnica Estatal de Quevedo (FCA), Quevedo, Ecuador

*Address all correspondence to: cojedas1979@gmail.com

References

[1] International Federation of Organic Agriculture Movements (IFOAM). The IFOAM norms for organic production and processing. Germany: IFOAM; 2014. p. 132

[2] Toledo M, Stangarlin J, Bonato C. Homeopathy for the control of plant pathogens. Physiology. 2011;**19**(20):21

[3] Silva LN, Moletta JL, Minho AP, Filippsen LF. Use of biotherapic in the control of the natural infestation by *Boophilus microplus*: Pilot study. International Journal of High Dilution Research. 2008;7(22):36-38

[4] Teixeira MZ. "New Homeopathic Medicines" database: A project to employ conventional drugs according to the homeopathic method of treatment. European Journal of Integrative Medicine. 2013;**5**:270-278

[5] Chikramane PS, Suresh AK, Bellare JR, Kane SG. Extreme homeopathic dilutions retain starting materials: A nanoparticulate perspective. Homeopathy. 2010;**99**(4):231-242

[6] Serretti A, Calati R, Goracci A, Di Simplicio M, Castrogiovanni P, De Ronchi D. Antidepressants in healthy subjects: What are the psychotropic/psychological effects? European Neuropsychopharmacology. 2010;**20**(7):433-453

[7] Ramachandran C, Nair PK, Clement RT, Melnick SJ. Investigation of cytokine expression in human leukocyte cultures with two immune-modulatory homeopathic preparations. Journal of Alternative and Complementary Medicine. 2007;**13**:403-407

[8] De Oliveira CC, de Oliveira SM, Goes VM, Probst CM, Krieger MA, Buchi DD. Gene expression profiling of macrophages following mice treatment with an immunomodulator medication. Journal of Cellular Biochemistry. 2008;**104**:1364-1377

[9] Eizayaga FX, Aguejouf O, Desplat V, Belon P, Doutremepuich C. Modifications produced by indomethacin and L-NAME in the effect of ultralow-dose aspirin on platelet activity in portal hypertension. Pathophysiology of Haemostasis and Thrombosis. 2007;**35**:357-363

[10] Aguejouf O, Eizayaga F, Desplat V, Belon P, Doutremepuich C. Prothrombotic and hemorrhagic effects of aspirin. Clinical and Applied Thrombosis/Hemostasis. 2009;**15**(5):523-528

[11] Endler PC, Pongratz W, Van Wijk R, Kastberger G, Haidvogl M. Effects of highly diluted succussed thyroxine on metamorphosis of highland frogs. The Berlin Journal on Research in Homoeopathy. 1991;**1**:151-160. Biological Models for Research in Homeopathy

[12] Welles S, Suanjak-Traidl E, Weber S, Scherer-Pongratz W, Frass M, Endler PC, et al. Pretreatment with thyroxin 10(−8) and the effect of homeopathically prepared thyroxin 10(−30) on highland frogs—a multi-researcher study. Research in Complementary Medicine. 2007;**14**:353-357

[13] Varshney JP, Naresh R. Comparative efficacy of homeopathic and allopathic systems of medicine in the management of clinical mastitis of Indian dairy cows. Homeopathy. 2005;**94**(2):81-85

[14] Figueiredo A, Fantatto RR, Agnolon IC, Lopes LG, Oliveira PRD, Mathias MIC, et al. *In vivo* study of a homeopathic medicine against Rhipicephalus (Boophilus) microplus in dairy cow. Revista Brasileira de Farmacognosia. 2018;**28**(2):207-213

[15] Berchieri JA, Turco WCP, Paiva JB, Oliveira GH, Sterzo EV. Evaluation of isopathic treatment of Salmonella enteritidis in poultry. Homeopathy. 2006;**95**(2):94-97

[16] Epstein SR, Bell IR. Homeopathy in the treatment of gastrointestinal conditions in animals: Part 1—What is homeopathy? Management. 2016;**9**:10

[17] Mazón-Suástegui JM, Tovar-Ramírez D, Salas-Leiva JS, Arcos-Ortega GF, García-Bernal M, Avilés-Quevedo MA, et al. Aquacultural Homoeopathy: A Focus on Marine Species. Aquaculture —Plants and Invertebrates. Rijeka: IntechOpen; 2018. In Press

[18] Bellavite P, Conforti A, Ortolani R. Immunology and homeopathy. Experimental studies on animal models. Evidence-based Complementary and Alternative Medicine. 2006;**3**:71-186

[19] Andretto AP, Fuzinatto MM, Bonafe EG, Braccini GL, Mori RH, Pereira RR. Effect of an homeopathic complex on fatty acids in muscle and performance of the Nile tilapia (*Oreochromis niloticus*). Homeopathy. 2014;**103**(3):178-185

[20] Junior RP, Vargas L, Valentim-Zabott M, Ribeiro RP, da Silva AV, Otutumi LK. Morphometry of white muscle fibers and performance of Nile tilapia (*Oreochromis niloticus*) fingerlings treated with methyltestosterone or a homeopathic complex. Homeopathy. 2012;**101**:154-158

[21] De Oliveira Feitosa KC, Povh JA, de Abreu JS. Physiological responses of pacu (*Piaractus mesopotamicus*) treated with homeopathic product and submitted to transport stress. Homeopathy. 2013;**102**(4):268-273

[22] Braccini GL, Natali MRM, Ribeiro RP, Mori RH, Riggo R, Oliveira CA, et al. Morphofunctional response of Nile tilapia (*Oreochromis niloticus*) to a homeopathic complex. Homeopathy. 2013;**102**:233-241

[23] Mazón-Suástegui JM, García-Bernal M, Saucedo PE, Campa-Córdova A, Abasolo-Pacheco F. Homeopathy outperforms antibiotics treatments in juvenile scallop *Argopecten ventricosus*: Effects on growth, survival, and immune response. Homeopathy. 2017;**106**:18-26

[24] Ortiz-Cornejo NL, Tovar-Ramírez D, Abasolo-Pacheco F, Mazón-Suástegui JM. Homeopatía, una alternativa para la acuicultura. Revista Médica de Homeopatía. 2017;**10**:28-34

[25] Rosero García AP. Evaluación de fármacos homeopáticos sobre el estado fisiológico de salud y nutrición en juveniles de pargo lunarejo *Lutjanus guttatus*. [thesis]. México: Centro Interdisciplinario de Ciencias Marinas, I.P.N.; 2017

[26] García-Corona JL. Análisis fisiológico y transcriptómico del efecto de medicamentos homeopáticos en la maduración gonádica del mejillón café *Modiolus capax* (Conrad, 1837). [thesis]. S.C. México: Centro de Investigaciones Biológicas del Noroeste; 2018

[27] Brizzi M, Lazzarato L, Nani D, Borghini F, Peruzzi M, Betti L. A biostatistical insight into the As(2)O(3) high dilution effects on the rate and variability of wheat seedling growth. Forschende Komplementärmedizin und Klassische Naturheilkunde. 2005;**12**(5):277-283

[28] Narváez-Martínez E, Toro H, León J, Bacca T. Evaluación de soluciones homeopáticas para controlar *Neoleucinodes elegantalis* guenée (Lepidóptera: Crambidae) en cultivo de Lulo. Biotecnología en el Sector Agropecuario y Agroindustrial. 2014;**12**(1):115-123

[29] Pinto JG, Nilbe CM, Cassiano C, Frazão da Silva E. Germinação

e crescimento inicial de mangaba (*Hancornia speciosa* Gomes) em função de preparados homeopáticos Carbo vegetabilis e dias após o despolpamento para semeadura. Revista Agrarian. 2014;7(24):244-250

[30] Dos Santos MG, Gonçalves AC Jr, Junior DLB, Schwantes D, Campagnolo MA, Coelho GF, et al. Utilizacão de medicamentos homeópaticos para remediacão de solos contaminados por metáis tóxicos. Journal of Agronomic Sciences. 2016;**5**:26-45

[31] Giardini-Bonfim FP, Dias-Casali VW, Ronie-Martins E. Germinacão e vigor de sementes de tomate (*Lycopersicon esculentum* Mill), peletizadas com preparados homeopáticos de *Natrum muriaticum*, submetidas a estresse salino. Enciclopédia Biosfera, Centro Científico Conhecer-Goiânia. 2012;**8**(14):625-633

[32] Mazón-Suástegui JM, Murillo Amador B , Batista-Sánchez D, Agüero-Fernández Y, García-Bernal MR y Ojeda-Silvera CM. *Natrum muriaticum* como atenuante de la salinidad (NaCl) en albahaca (*Ocimum basilicum* L.). Nova Scientia 2018;21(10):148-164

[33] Bell IR, Schwartz GE. Enhancement of adaptive biological effects by nanotechnology preparation methods in homeopathic medicines. Homeopathy. 2015;**104**(2):123-138

[34] Meneses MN. Agrohomeopatía como alternativa a los agroquímicos. Revista Médica de Homeopatía. 2017;**10**(1):9-13

[35] Modolon TA, Pietrowski V, Alves IFA, Guimarães ATB, Pizzato M. Efeito de dinamizações seqüenciais do preparado homeopático *Nux vomica* no desenvolvimento inicial de plantas de milho submetido ao percevejo barriga-verde *Dichelops melacanthus* Dallas (Heteroptera: Pentatomidae). Cadernos de Agroecologia. 2016;**10**(3):2236-7934. Available in: <http://revistas.aba-agroecologia.org.br/index.php/cad/article/view/16997>. Access in: 13 feb. 2019

[36] Maguire JD. Speed of germination —aid in selection and evaluation for seedling emergence and vigor. Crop Science. 1962;**2**:176-177

[37] Suárez-Martínez SE, Ferriz-Martínez RA, Campos-Vega R, Elton-Puente JE, de la Torre CK, García-Gasca T. Bean seeds: Leading nutraceutical source for human health. CyTA Journal of Food. 20016;**14**(1):131-137

[38] White PJ, Broadley MR. Biofortification of crops with seven mineral elements often lacking in human diets—iron, zinc, copper, calcium, magnesium, selenium and iodine. New Phytologist. 2009;**182**:49-84

[39] Hermans C, Conn SJ, Chen J, Xiao Q , Verbruggen N. An update on magnesium homeostasis mechanisms in plants. Metallomics. 2013;**5**:1170-1183

[40] Huber DM. The role of mineral nutrition in defense. In: Horsfall JG, Cowling EB, editors. Plant Disease: *An Advanced Treatise*. New York: Academic Press; 1980. pp. 381-406

[41] Ojeda-Silvera CM, Murillo-Amador B, Nieto-Garibay A, Troyo-Diéguez E, Reynaldo-Escobar MI, Ruíz-Espinoza FH, et al. Emergencia y crecimiento de plántulas de variedades de albahaca (*Ocimum basilicum* L.) sometidas a estrés hídrico. Ecosistemas y Recursos Agropecuarios. 2015;**2**:151-160

[42] Frahm MA, Rosas JC, Mayek N, López E. Breeding beans for resistance to terminal drought in the lowland tropics. Euphytica. 2004;**136**:223-232

[43] Fereres E, Soriano MA. Deficit irrigation for reducing agricultural water

use. Journal of Experimental Botany. 2007;**58**:147-159

[44] Kandil AA, Shareif AE, Gad MA. Effect of salinity on germination and seeding parameters of forage cowpea seed. Research. Journal of Seed Science. 2017;**10**:17-26

[45] Khaliq S, Ullah Z, Rehman A, Khaliq R. Physiological and biochemical basis of salt tolerance in *Ocimun basilicum* L. Journal of Medicinal Plants Studies. 2014;**2**:18-27

[46] Abbas G, Saqib M, Akhtar J, ulHaq MA. Interactive effects of salinity and iron deficiency on different rice genotypes. Journal of Plant Nutrition and Soil Science. 2015;**178**:306-311

[47] Ruiz Espinoza FDJ. Agrohomeopatía: Una opción ecológica para el campo mexicano. La Homeopatía de México. 2001;**70**(613):110-116

[48] Delgado H, Pinzón EH, Blair M, Izquierdo PC. Evaluación de líneas de fríjol (*Phaseolus vulgaris* L.) de retrocruce avanzado entre una accesión silvestre y radical cerinza. Revista U.D.C.A Actualidad & Divulgación Científica. 2013;**16**(1):79-86

[49] Kaymakanova M, Stoeva N. Physiological reaction of bean plants (*Phaseolus vulgaris* L.) to salt stress. General and Applied Plant Physiology. 2008;**34**:177-188

[50] Hasegawa PM, Bressan R, Zhu J, Bohnert H. Plant celular and molecular responses to high salinity. Annual Review of Plant Biology. 2002;**51**:463-499

[51] Sudhir P, Murthy S. Effects of salt on basic process of photosynthesis. Photosynthetica. 2004;**42**(4):481-486

[52] Kazan K. Diverse roles of jasmonates and ethylene in abiotic stress tolerance. Trends in Plant Science. 2015;**20**:219-229

[53] Véry AA, Robinson MF, Mansfield TA, Sanders D. Guard cell cation channels are involved in NaCl–induced stomatal closure in a halophyte. The Plant Journal. 1998;**14**(5):509-521

[54] Hiz MC, Canher B, Niron H, Turet M. Transcriptome analysis of salt tolerant common bean (*Phaseolus vulgaris* L.) under saline conditions. PLoS One. 2014;**9**(3):e92598

[55] Morelos RM, Ramírez JL, García-Gasca A, Ibarra AM. Expression of the myostatin gene in the adductor muscle of the Pacific lion-paw scallop Nodipecten subnodosus in association with growth and environmental conditions. Journal of Experimental Zoology Part A: Ecological Genetics and Physiology. 2015;**323**(4):239-255

[56] Hellemans J, Mortier G, De Paepe A, Speleman F, Vandesompele J. qBase relative quantification framework and software for management and automated analysis of real-time quantitative PCR data. Genome Biology. 2007;**8**(2):R19

[57] Appels A, Lagudah HE. Manipulation of chromosomal segments from wild wheat for the improvement of bread wheat. Australian Journal of Plant Physiology. 1990;**17**:253-266

[58] Qin X, Zeevaart JAD. The 9-*cis*-epoxycarotenoid cleavage reaction is the key regulatory step of abscisic acid biosynthesis in water-stressed bean. Proceedings of the National Academy of Sciences of the United States of America. 1999;**96**:15354-15361

[59] Schroeder JI, Kwak JM, Allen GJ. Guard cell abscisic acid signalling and engineering drought hardiness in plants. Nature. 2001;**410**:327-330

[60] Batista Sánchez D, Murillo Amador B, Nieto Garibay A, Alcaraz Meléndez L, Troyo Diéguez E, Hernández Montiel L, et al. Mitigación de NaCl por efecto

de un bioestimulante en la germinación de *Ocimum basilicum* L. Terra Latinoamericana. 2017;**35**:309-320

[61] Méndez R. Cultivos orgánicos: Su control biológico en plantas medicinales y aromáticas. Segunda Edición ed. Vol. 168. Bogotá: ECOE; 2008

[62] Samperio RG. Hidroponía Básica. 10a ed. Editorial Diana. México; 1997. p. 176

[63] Zhu JK. Salt and drought stress signal transduction in plants. Annual Review of Plant Biology. 2002;**53**:247-273

[64] Xiao JX, Hu CY, Chen YY, Yang B, Hua J. Effects of low magnesium and an arbuscular mycorrhizal fungus on the growth, magnesium distribution and photosynthesis of two citrus cultivars. Scientia Horticulturae. 2014;**177**:14-20

[65] Agrios GN. Plant Pathology. Fourth ed. New York, USA: Academic Press; 2004. p. 635

[66] Ramaiah AK, Raj Kumar H. *In Vitro* Antifungal activity of some plant extracts against *Fusarium oxysporum* f. sp. Lycopersici. Asian Journal of Plant Science and Research. 2015;**5**(1):22-27

[67] Ochoa Fuentes YM, Cerna Chávez E, Landeros Flores J, Hernández Camacho S, Delgado Ortiz JC. Evaluación *in vitro* de la actividad antifúngica de cuatro extractos vegetales metanólicos para el control de tres especies de *Fusarium* spp. Revista Mexicana de Ciencias Agrícolas. 2012;***81***(1):69-73

[68] Damin S, Alves L, Alexandre TM, Bonini AK, Bonato CM. Preparados homeopáticos sobre a atividade do fungo entomopatogênico *Beauveria bassiana* (Bals.) Vuill. (Ascomycota: Cordycipitaceae). Revista Brasileira de Agroecologia. 2014;**9**(3):41-53

[69] Tichavsky R. Origen y tipos de medicamentos homeopáticos. In: Manual de Agrohomeopatía. Primera editor. Vol. 1. Instituto Comenius. Monterrey. Nuevo león, México; 2007. pp. 31-32

[70] Casali VWD, Andrade FMC, Duarte ESM. Acologia de Altas Diluições. Viçosa: Departamento de Fitotecnia da Universidade Federal de Viçosa; 2009. p. 537

[71] Damin S, Alves LFA, Bonini AK, Alexandre TM. In vitro assay on homeopathic solutions on Metarhizium anisopliae (Metsch) Sorok (Ascomycota: Clavicipitaceae). Arquivos do Instituto Biológico. 2015;**82**:1-8

11

Nonchemical Weed Control in Winter Oilseed Rape Crop in the Organic Farming System

Aušra Marcinkevičienė, Marina Keidan, Rita Pupalienė, Rimantas Velička, Zita Kriaučiūnienė, Lina Marija Butkevičienė and Robertas Kosteckas

Abstract

A field experiment was conducted during the 2014–2017 period at Aleksandras Stulginskis University (now—Vytautas Magnus University Agriculture Academy) on a *Endocalcaric Endogleyic Luvisol* (*LV-can.gln*) according to the WRB 2014. The three nonchemical weed control methods were explored: (1) thermal (using wet water steam), (2) mechanical (interrow loosening), and (3) self-regulation (smothering). In the thermal and mechanical weed control treatments, winter oilseed rape was grown with an interrow spacing of 48.0 cm and in weed smothering (self-regulation) treatment with an interrow spacing of 12.0 cm. Winter oilseed rape was grown in the soil with a regular humus layer (23–25 cm) and with a thickened humus layer (45–50 cm). Annual weeds predominated in the winter oilseed rape crop. In the soil with both humus layers, regular and thickened, the most efficient weed control method was mechanical weed management both during the autumn (efficacy 26.7–75.1%) and spring (efficacy 37.1–76.7%) growing seasons. Thermal and mechanical weed control in combination with the bio-preparations in droughty years significantly reduced the number of weed seedlings. Dry matter mass of weeds most markedly decreased through the application of the mechanical weed management method.

Keywords: winter oilseed rape, weed control methods, organic farming system, bio-preparations, soil humus layer

1. Introduction

The development of organic farming was prompted by the environmental concerns, health-related issues, and the search for solutions to social problems.

In Lithuania, organic farms account for more than 5% of the area under cultivation and are in line with the EU average. According to the popularity of organic farming, Lithuania surpasses the neighboring Poland but is far behind the other Baltic States. Although organic farms represent a small proportion of the total num-ber of farms, their number has increased rapidly over the past decade. In 2017, 2448 organic farms were certified in Lithuania; they cover about 244,000 hectares of the agricultural land [1]. Most of the organic production farms are 10–30 ha in size.

Oilseed rape is one of the world's most important oil crops [2, 3]. The cultivation of oilseed rape on organically managed farms was encouraged by a search for healthy, high-quality, and safe food. According to the data from the public organization "Ekoagros," in 2017, the total area devoted to oilseed rape production on organically managed farms in Lithuania amounted to 3962.2 ha, including 3250.98 ha of winter oilseed rape and 711.22 ha of spring oilseed rape. The main reasons why winter oilseed rape production area on organically managed farms is not increasing are the problems associated with plant nutrition [4, 5], weed, disease and pest control, and unstable plant overwinter survival, and all these factors result in low rapeseed yields [6, 7]. Many organic farms in Lithuania are located on infertile soils, and the erosion and productivity problems are relevant there. Organic crop production farms are prevalent in this region; therefore the problem of crop rotation, nutrient, and humus balance is highly relevant. In the organic production farms, the inclusion of oilseed rape in the crop rotation is very important because it is characterized by phytosanitary properties, is a good pre-crop for other crops, and improves soil properties [8].

In the organic production farms, in the absence of the possibility of controlling weeds with herbicides, a great deal of attention is paid to nonchemical methods of weed control—mechanical, thermal, and natural crop-weed competition/self-regulation. Weed control by using wet water steam has not been extensively studied in the world. More comprehensive studies on the thermal weed control by water steam have been carried out by Lithuanian scientists [9–13].

2. Material and methods

Field experiments were conducted in 2014–2017 at the Experimental Station (54°53′ N, 23°50′ E) of Aleksandras Stulginskis University on an *Endocalcaric Endogleyic Luvisol* (*LV-can.gln*) according to the WRB 2014. Agrochemical properties of the experimental soil with a regular humus layer (averaged data of 2014, 2015, and 2016) were as follows: pH, 7.30; humus, 1.79%; contents of available nutrients in the soil: P_2O_5, 199.0 mg kg^{-1}; K_2O, 97.7 mg kg^{-1}; total nitrogen, 0.079%. Agrochemical properties of the experimental soil with a thickened humus layer were as follows: pH, 7.20; humus, 2.19%; contents of available nutrients in the soil: P_2O_5, 277.7 mg kg^{-1}; K_2O, 123.0 mg kg^{-1}; total nitrogen, 0.115%.

Two-factor field experiments were established using a split-plot design. Winter oilseed rape was grown in the soil with a regular humus layer (23–25 cm) (experiment I) and in the soil with a thickened humus layer (45–50 cm) (in 1988, a thickened humus-rich layer was artificially formed using fertile soil delivered from elsewhere) (experiment II). The length of initial plots was 14 m, width 6 m, and the area 84 m^2. The length of the harvested plots was 10 m, width 2 m, and the area 20 m^2. The experiments included four replications. The winter oilseed rape crop was preceded by black fallow.

The study object was winter oilseed rape (*Brassica napus* L. spp. *oleifera biennis* Metzg.) agrocenosis.

Experimental treatments:
Factor A: nonchemical weed control methods:

1. Thermal (water steam)

2. Mechanical (interrow cultivation)

3. Self-regulation/smothering (natural weed-crop competition, sowing with narrow interrows)

Factor B: biological preparations (bio-preparations):

1. Without bio-preparations

2. With bio-preparations

A winter rape cultivar "Cult" (Sweden, SW Seed) was grown in the experiment. The crop was sown at a seed rate of with a Multidrill M300 sowing machine. In 2014, winter rape was sown on September 1, in 2015 on August 27, and in 2016 on August 29. In the thermal weed control treatment, the oilseed rape crop was grown with interrow spacing of 48 cm, and weeds were killed using a tractor-mounted wet water steam unit at a 3–4 leaf growth stage (BBCH 13–14) of winter rape. The thermal power of the device is 90 kW, with a capacity of 120 kg h^{-1} steam; the device is run on liquefied gas. The temperature of steam is 99°C, with thermal treatment time of 2 s (Sirvydas, Kerpauskas, 2012).

The principal scheme of the tractor-mounted thermal weed control unit using wet water steam is presented in **Figure 1**.

2.1 Description of operation of the mobile thermal weed control unit

Liquefied gas is fed through tube 6 into the combustion chamber 4 of demountable steam boiler 8. There the burning gas heats water present in the steam boiler 8. The wet water steam which has formed in steam boiler 8 gets into the steam separator, in which steam dampness is reduced. Then the wet water steam which has passed through the steam overheater 11 is fed through the tube 5 into steam diffusers 2, which spread/distribute steam in the environment of target weeds.

The height of the steam diffusers is adjusted by the height adjustment mechanism 3. To prevent the liquefied gas from cooling, the gas cylinder 9 is placed into the heating tank 10. Hot water from steam boiler 8 is fed into the gas cylinder's heating tank 10 through tube 7. The mobile thermal weed control unit is mounted on a tractor 12 with a mounting device 11.

In the mechanical weed control treatment, the interrows were loosened with an interrow cultivator (KOR-4.2-01, Ukraine) using two passes. In the weed control treatment involving weed smothering (self-regulation), the winter oilseed rape was

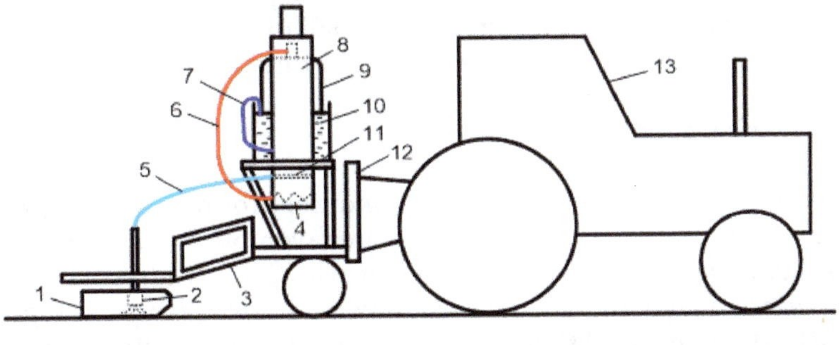

Figure 1.
The principal technological scheme of the tractor-mounted thermal weed control unit: (1) protectors of steam diffusers, (2) steam diffusers, (3) diffusers' height adjustment mechanism, (4) combustion chamber, (5) tube by which wet water steam is fed to steam diffusers and discharges, (6) tube by which gas is fed to combustion chamber, (7) tube by which hot water is fed to the heating tank of gas cylinder, (8) demountable steam boiler, (9) gas cylinder, (10) heating tank of gas cylinder, (11) steam overheater, (12) mounting device of thermal weed control unit, and (13) tractor.

grown with 12.0 cm interrow spacing. Winter rape was not fertilized, and no chemical plant protection products were applied.

In the treatments with the use of the bio-preparations, pre-sowing, the seeds of winter rape were coated with the bio-organic fertilizer Nagro (BioPlant) (0.5 l per ton of seeds and 10 l of water), and during the growing season, the winter rape crop was sprayed twice with the bio-preparations (in the autumn with Terra Sorb Foliar (BioIberica) (2 l ha^{-1}), in the spring with Terra Sorb Foliar (1 l ha^{-1}) and 0.3% Conflic (Atlantica Agricola)).

2.2 Assessment of weed incidence in the crop

The analysis of weed seedlings was carried out at winter rape 3–4 leaf growth stage (BBCH 13–14) in the autumn and after resumption of vegetation in the spring (BBCH 50) before the application of thermal and mechanical weed management methods. In each experimental plot, in four randomly selected 0.10 m^2 record plots, the number of weed seedlings and weed species composition were established. This analysis was done for the second time 5–7 days after application of the weed control methods in the marked record plots. The number of weed seedlings was recalculated as per m^2. The efficacy (E) of the weed management methods for the change in the number of weed seedlings was calculated according to the formula:

$$E = (S1 - S2) / S1 \times 100\%, \quad (1)$$

where S1 is the number of weed seedlings per m^2 before application of the weed control methods and S2 is the number of weed seedlings per m^2 after application of the weed control methods.

At winter rape green silique stage (BBCH 79), the number of weeds and weed species composition were determined in each plot in four 0.25 m^2 record plots; he weeds were dried in a drying chamber at 60°C temperature and weighed [14].

2.3 Statistical analysis

The significance of the differences between the means was estimated using the *t* criterion; the interplay between the traits was determined by the correlation regression methods. The statistical analysis of the experimental data was performed using software STAT from the software package SELEKCIJA [15]. The experimental data that did not fit the normal distribution law, prior to the statistical evaluation, were transformed using the function $y = \ln x + 1$.

2.4 Meteorological conditions

In 2014 autumn was warm and long, so conditions for rape growing were favorable. In winter, meteorological conditions were favorable for rape over-wintering. In 2015, autumn was warm and humid. During the first decade of January 2016 of very cold weather and the absence of snow, over-wintering of rape was not successful. In 2016, the conditions for rape preparation for wintering and for over-wintering were favorable. In 2017, rape vegetation renewed on March 31. April was cold and humid with 35.3 mm more rainfall than usual. As a result, some winter rape has not over-wintered.

3. Experimental results

3.1 Weed incidence in the winter rape crop in the autumn and spring growing seasons

The following annual weed species predominated in the winter rape crop: *Chenopodium album* L., *Tripleurospermum perforatum* (Merat) M. Lainz, *Stellaria media* (L.) Vill., *Viola arvensis* Murray, *Veronica arvensis* L., *Sinapis arvensis* L., *Capsella bursa-pastoris* (L.) Medik, and *Poa annua* L.

In the soil with both regular and thickened humus layers, in the treatments where winter rape was grown with wide interrow spacings (48 cm), when the light and moisture conditions were favorable, the number of emerged weeds was higher than in the treatments with narrow interrows (12 cm), except for the spring growing season of 2017. The application of the bio-preparations in most cases reduced the number of weed seedlings in the winter rape crop both during the autumn and spring growing seasons.

Experiment I: in the soil with a regular humus layer. The most effective weed control method in organic winter rape crop was mechanical: efficiency 26.7–71.5% without biological preparations and 54.2–71.7% with biological preparations (**Figure 2**). The efficiency of the thermal weed control method was lower than the mechanical one. In Ref. [16], it was stated that the efficiency of mechanical and thermal weed control was 50–100%. Biological preparations enhanced the effectiveness of thermal and mechanical weed control techniques only in 2014. The effectiveness of the self-regulation method for the change of weed sprouts was negative throughout the study years.

Experiment II: in the soil with a thickened humus layer. The most effective weed control method in rapeseed crop was mechanical: efficiency was 39.8–75.1% without biological preparations and 53.0–68.9% with biological preparations (**Figure 3**). The efficiency of the thermal weed control method was lower than the mechanical one. Bio-preparations enhanced the effectiveness of mechanical weed control only in 2014. The method of self-regulation did not reduce the number of weed sprouts.

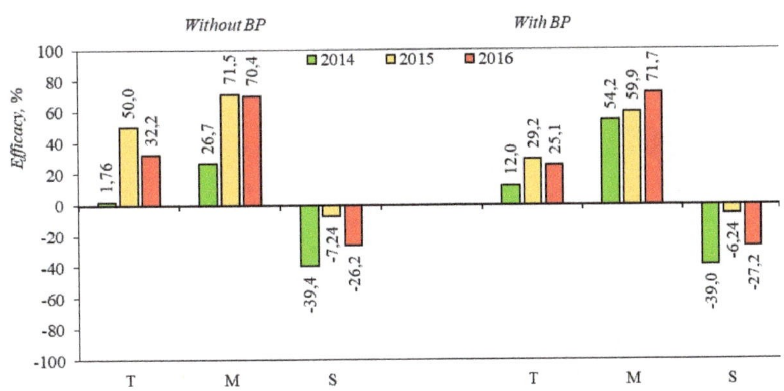

Figure 2.
The efficacy of the nonchemical weed control methods for the change in the number of weed seedlings in the winter oilseed rape crop, grown in the soil with a regular humus layer in the autumn (2014–2016). Note. T, thermal; M, mechanical; S, self-regulation; BP, bio-preparations.

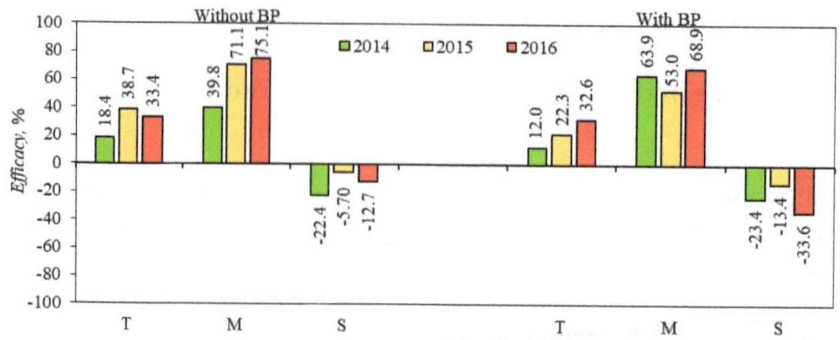

Figure 3.
The efficacy of the nonchemical weed control methods for the change in the number of weed seedlings in the winter oilseed rape crop, grown in the soil with a thickened humus layer in the autumn (2014–2016). Note. T, thermal; M, mechanical; S, self-regulation; BP, bio-preparations.

3.2 Weed incidence in the winter rape crop before harvesting

Experiment I: in the soil with a regular humus layer, the nonchemical weed management methods did not exert any pronounced effect on the weed number in the winter rape crop before harvesting throughout study years (**Figure 4**). Using bio-preparations in 2015 in experimental plots with mechanical weed control method, weed number was significantly 42.9% lower than in experimental plots with self-regulation. In 2016 and 2017, there were no significant differences in weed number between the different weed control methods.

The use of bio-preparations did not significantly affect the number of weeds in the oilseed rape crop during all study years compared to the treatment where they were not used.

There was no significant difference in weed dry matter mass between different weed control methods in plots with bio-preparations or without bio-preparations in 2015 (**Figure 5**). In 2016 and 2017, weed killing with interrow loosening and without biological preparations resulted in significantly lower weed dry mass than this in plots where thermal weed control and self-regulation methods were used, respectively, 2.0 and 2.5 times and 3.0 and 4.5 times. In 2017, the weed dry matter mass in plots with thermal weed control was significantly 32.3% lower than that of the self-regulation plots. In 2016 in plots with bio-preparations and mechanical

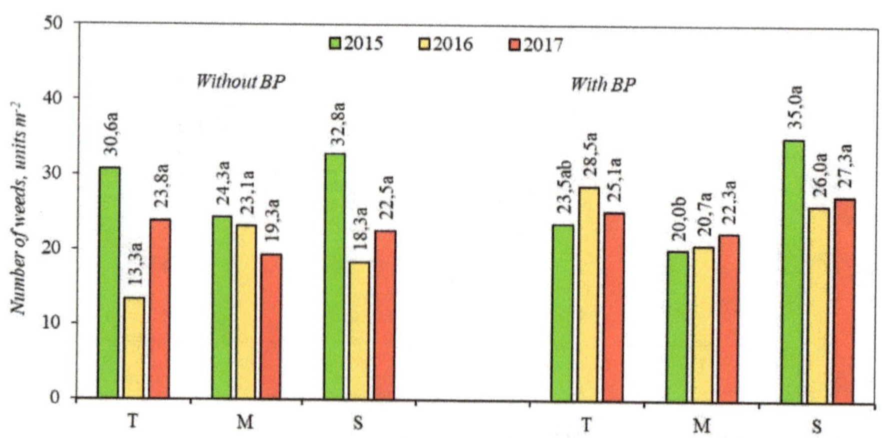

Figure 4.
The number of weeds in the winter oilseed crop, grown in the soil with a regular humus layer, before harvesting (2015–2017). Note. The differences between the averages of treatments of factor a, marked by not the same letter (a, b), are significant (P < 0.05). T, thermal; M, mechanical; S, self-regulation; BP, bio-preparations.

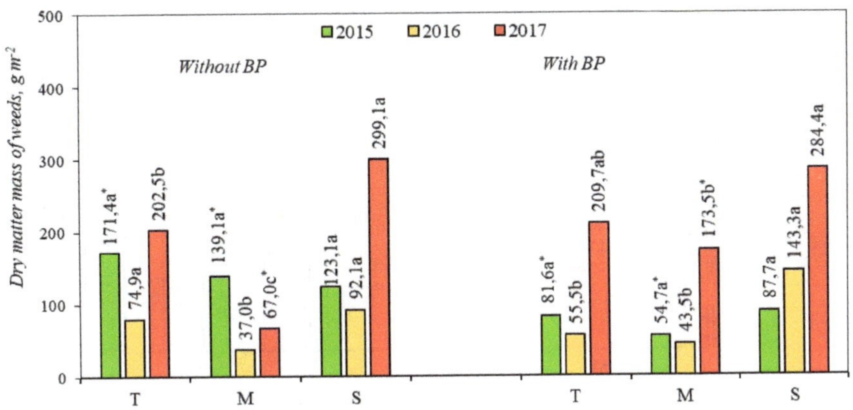

Figure 5.
The dry matter mass of weeds in the winter oilseed crop, grown in the soil with a regular humus layer, before harvesting (2015–2017). Note. The differences between the averages of treatments of factor a, marked by not the same letter (a, b, c), and between the averages of treatments of factor B, marked by an asterisk, are significant (P < 0.05). T, thermal; M, mechanical; S, self-regulation; BP, bio-preparations.

weed control method, the weed dry matter mass was significantly 2.6 and 3.3 times lower than that in the self-regulation weed control method plots. In 2017, mechanical weed control resulted in a significantly 39.0% lower dry mass of weeds than self-regulation method. It was found that thermal weed control resulted in a 44.0% reduction in dry mass of weeds using thermal weed control with water steam [16]. The similar results were obtained in the experiments in Lithuania—weed numbers were 3.2–4.4 times lower in plots with mechanical weed control method than this in plots with self-regulation weed control method and respectively weed dry mass by 2.2–3 times [17, 18].

With the use of biological preparations, the dry matter mass of weeds decreased significantly by 2.1 and 2.5 times in the fields of thermal and mechanical weed control methods only in the droughty 2015, as the more fertile rapeseed crop using biological preparations better suppressed weeds. K. Różyło and E. Pałys [19] found that as the assimilation leaf area of rape increased, the dry matter mass of weeds decreased. In 2017 with the use of biological agents, the dry weight of weeds increased significantly 2.6 times in the fields of mechanical weed control.

Experiment II: in 2015 in the soil with a thickened humus layer, interrow loosening without biological preparations significantly decreased weed number by 38.8 and 36.8% compared to thermal weed control and self-regulation (**Figure 6**). In 2016 and 2017, there were no significant differences in the number of weeds 0in the plots with various weed control methods without biological preparations. Using bio-preparations in 2015 in plots with mechanical weed control method, and in 2017 in plots with thermal weed control method, the weed numbers were significantly lower than in the plots of self-regulation weed control method by 42.9 and 34.7%, respectively. In 2016, there was no significant difference in the number of weeds in the plots with different weed control methods and with biological preparations.

Without biological preparations in 2015 and 2016, there were no significant differences in weed dry mass between the different weed control methods (**Figure 7**). Different weed control techniques used in plots with biological preparations in 2015 had no significant effect on the dry matter mass of the weeds. In 2016 in plots with mechanical weed control and with biological preparations, weed dry matter mass was significantly 4.1 and 5.1 times lower than that in plots with thermal weed control and self-regulation. In 2017 the dry matter mass of weeds was 6.7 and 5.7

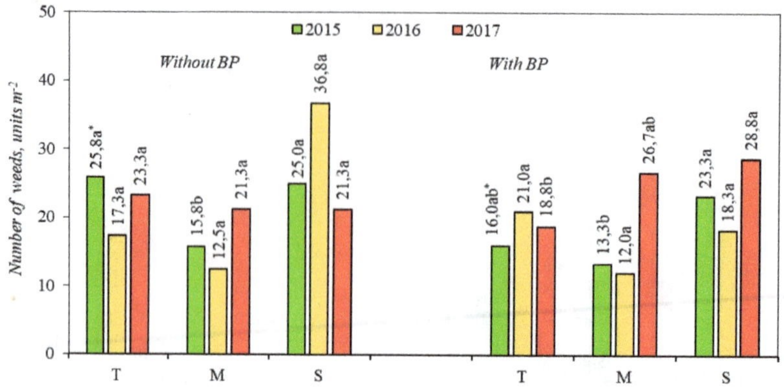

Figure 6.
The number of weeds in the winter oilseed crop, grown in the soil with a thickened humus layer, before harvesting (2015–2017). Note. The differences between the averages of treatments of factor a, marked by not the same letter (a, b), and between the averages of treatments of factor B, marked by an asterisk, are significant ($P < 0.05$). T, thermal; M, mechanical; S, self-regulation; BP, bio-preparations.

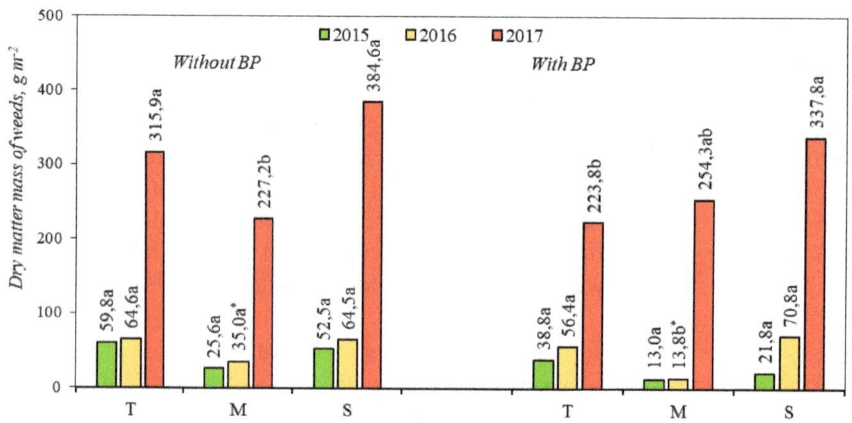

Figure 7.
The dry matter mass of weeds in the winter oilseed crop, grown in the soil with a thickened humus layer, before harvesting (2015–2017). Note. The differences between the averages of treatments of factor a, marked by not the same letter (a, b), and between the averages of treatments of factor B, marked by an asterisk, are significant ($P < 0.05$). T, thermal; M, mechanical; S, self-regulation; BP, bio-preparations.

times higher, respectively, without the use of biological preparations and 11.1 and 5.8 times higher with the use of biological preparations than in 2015 and 2016. Apparently, the higher weed dry matter content was caused by the lower rape crop and the humid and cold weather during rape vegetation. The mechanical weed control without biological preparations were found to have a significantly lower mass of weed dry mass than the thermal weed control and self-regulation, by 28.1 and 40.9% lower. The use of biological preparations and thermal weeds to control using wet water vapor resulted in a significantly by 33.7% lower dry mass of weeds than that in the self-regulation plots.

The use of biological agents significantly reduced weed dry matter in mechanical weed control fields only by 2016.

In 2017, negative, very strong, significant correlations were determined between the winter rape plant population density and weed dry matter mass: in the soil with a regular humus layer $r = -0.95$, $P < 0.01$; in the soil with a thickened humus layer $r = -0.91$, $P < 0.05$. [20]. R. Kosteckas (2011) [20] also found that the dry matter mass of weeds correlates with the density of rape crop.

4. Conclusions

Annual weeds predominated in the winter oilseed rape crop: *Chenopodium album* L., *Tripleurospermum perforatum* (Merat) M. Lainz, *Stellaria media* (L.) Vill., *Viola arvensis* Murray, *Veronica arvensis* L., *Sinapis arvensis* L., *Capsella bursa-pastoris* (L.) Medik., and *Poa annua* L.

In the soil with both humus layers, regular and thickened, the most efficient weed control method was mechanical weed management both during the autumn (efficacy 26.7–75.1%) and spring (efficacy 37.1–76.7%) growing seasons.

Thermal and mechanical weed control in combination with the bio-preparations in droughty years significantly reduced the number of weed seedlings. Dry matter mass of weeds most markedly decreased through the application of the mechanical weed management method.

Author details

Aušra Marcinkevičienė[1], Marina Keidan[2], Rita Pupalienė[1*], Rimantas Velička[1], Zita Kriaučiūnienė[1], Lina Marija Butkevičienė[1] and Robertas Kosteckas[1]

1 Vytautas Magnus University, Kaunas, Lithuania

2 Public Institution "Ekoagros"

*Address all correspondence to: rita.pupaliene@vdu.lt

References

[1] Melnikienė R, editor. Agriculture and Food Sector in Lithuania. Vol. 2019. Vilnius: Lietuvos Agrarinės Ekonomikos Institutas/Lithuanian Institute of Agrarian Ecomnomics; 2018. p. 216

[2] Crnobarac J, Marinković B, Jeromela-Marjanović A, Balalić I, Jaćimović G, Latković D. The effect of variety and sowing date on oilseed rape yield and quality. Agriculture and Food. 2015;**3**:241-245

[3] Vincze E. The effect of sowing date and plant density on yield elements of different winter oilseed rape (Brassica napus var. napus f. biennis L.) genotypes. Columella. 2017;**4**(1):21-25. DOI: 10.18380/SZIE.COLUM.2017.4.1.suppl

[4] Alaru M, Talgre L, Eremeev V, Tein B, Luik A, Nemvalts A, et al. Crop yields and supply of nitrogen compared in conventional and organic farming systems. Agricultural and Food Science. 2014;**23**:317-326. DOI: https://doi.org/10.23986/afsci.46422

[5] Engstörm L, Stenberg M, Wallenhammar AC, Stahl P, Gruvaeus I. Organic winter oilseed rape response to N fertilisation and preceding agroecosystem. Field Crops Research. 2014;**67**:94-101. DOI: 10.1016/j.fcr.2014.07.01

[6] Valantin-Morison M, Meynard JM. Diagnosis of limiting factors of organic oilseed rape yield. A survey of farmers' fields. Agronomy for Sustainable Development. 2008;**28**(4):527-539. DOI: 10.1051/agro:2008026

[7] Zihlmann U, Scherrer C, Krebs H, Oberholzer HR, Vögeli GA, Nemecek T, et al. Integrierter und biologischer Anbau im Vergleich. ART-Bericht. 2010;**722**:1-16

[8] Velička R, Rimkevičienė M, Marcinkevičienė A. Changes in the properties of a loamy gleyic cambisol as related to the saturation of crop rotations with rape. Eurasian Soil Science. 2006;**39**(9):1002-2293

[9] Sirvydas A, Kerpauskas P, Čekanauskas S, Čingienė R. Vandens garo sklaida profiliuotame dirvos paviršiuje termiškai naikinant piktžoles. Vandens ūkio inžinerija. LŽŪU mokslo darbai (in Lithuanian). 2006;**30**(50):28-35

[10] Čekanauskas S. Aukštatemperatūrės aplinkos veiksniai termiškai naikinant piktžoles/High Temperature Environment Factors in Weed Thermal Control [thesis]. Akademija: Lithuanian University of Agriculture; Čekanauskas; 2007

[11] Staniulienė R. Aukštatem-peratūrinės aplinkos poveikis sunkiai termiškai sunaikinamoms piktžolėms [thesis]. Akademija: Lithuanian University of Agriculture; 2010

[12] Sirvydas K. Terminis piktžolių Naikinimas/Thermal Weed Control. Monograph. Akademija: Lithuanian University of Agriculture; 2012. p. 327

[13] Mockevičienė R. Necheminių piktžolių kontrolės priemonių ir biologinių prepratų įtaka vasarinių rapsų Agrocenozei / the Influence of Non-chemical Weed Control and Biopreparations on the Agrocenosis of Spring Oilseed Rape [Thesis]. Akademija: Aleksandras Stulginskis University; 2017

[14] Stancevičius A. Piktžolių apskaita ir laukų piktžolėtumo kartografavimas/ Weed accounting and field weed mapping. Akademija; 1979

[15] Raudonius S. Application of statistics in plant and crop research: Important issues. Žemdirbystė = Agriculture. 2017;**104**(4):377-382

[16] Raffaelli M, Fontanelli M, Frasconi C, Sorelli F, Ginanni M, Peruzzi A. Physical weed control in processing tomatoes in Central Italy. Renewable Agriculture and Food Systems. 2011;**26**(2):95-103. DOI: 10.1017/S1742170510000578

[17] Kerpauskas P, Sirvydas AP, Lazauskas P, Vasinauskienė R, Tamošiūnas A. Possibilities of weed control by water steam. Agronomy Research. 2006;**4**:221-225

[18] Velička R, Mockevičienė R, Marcinkevičienė A, Pupalienė R, Butkevičienė LM, Kriaučiūnienė Z, et al. Necheminių piktžolių kontrolės būdų efektyvumo palyginimas vasarinių rapsų pasėlyje ekologinės žemdirbystės sąlygomis/ the comparison of non-chemical weed control methods efficiency in spring oilseed rape crop under the conditions of organic farming system. Žemės ūkio mokslai. Agricultural Sciences. 2015;**22**(44):189-197. DOI: 10.6001/zemesukiomokslai.v22i4.321

[19] Różyło K, Pałys E. Impact of plant density on the canopy area index and weed infestation depending on different cultivars of winter oilseed rape (*Brassica napus* L., ssp. oleifera Metzg.). Annales Universitatis Mariae Curie-Skłodowska Lublin–Polonia. 2014;**LXIX**(2):44-55

[20] Kosteckas R. Skirtingu Intensyvumu tręštų vasarinių rapsų (*Brassica napus* L.) Biopotencialo Formavimosi dėsningumai įvairaus Tankumo pasėliuose/the Regularities of Biopotential Formation of Spring Oilseed Rape (*Brassica napus* L.) Fertilized at Different Intensities in Different Crop Stand Densities [Thesis]. Akademija: Lithuanian University of Agriculture; 2011

12

Role of Organic Farming in Agriculture

Muthuraman Yuvaraj, Peyandi Paraman Mahendran and Eman Tawfik Hussien

Abstract

Organic farming could be an all-encompassing generation administration framework that empowers and improves agroecosystem wellbeing, counting biodiversity, natural cycles, and soil biological activity. It stresses the role of management activities in preference to the use of off-farm data, considering that regional conditions require locally adapted systems. This can be achieved using agronomic, biological, and mechanical methods, in equal share to synthetic materials, to carry out any specific role inside the organization. Organic farming is still only a small industry, which represents only 2% of global food sales. However, it is growing in importance in the world. It is hard to get information due to lack of official statistics and the level of confidentiality of systems of organic produce. Soil practices such as crop rotations, organic fertilizers, symbiotic associations, cover crops, inter-cropping, and minimum tillage are central to organic practices. The static arrangements of soil are achieved by soil fauna and vegetation. Besides, cycling of nutrients and energy is enhanced by increasing the retentive abilities of the soil for nutrients and water.

Keywords: organic farming, vermicompost, soil health

1. Introduction

Organic farming is defined as a production system that avoids or largely eliminates the usage of synthetic compounded fertilizers, growth regulators, pesticides, and farm animal feed additives. To maximize feasible extent, the organic farming system depends on crop rotation, green manures, legumes, animal manures, crop residues, off-farming natural squanders, and aspect of biological pest control to preserve soil fertility and productivity to sustain plant and thereby curbing pests, diseases, and unwanted weeds. Organic farming methods are internationally regulated and legally executed by many countries, based in great part on the standards set by the International Federation of Organic Agriculture Movements (IFOAM), an international umbrella organization for organics established in 1972 [1].

After the launch of green revolution in India, substantial growth in the output of food grains was achieved. This was achieved through the utilization of improved crop varieties and higher levels of inputs of plant foods and plant protection chemicals. Merely there has been a rise in production, which was accomplished by the monetary value of soil health. The organic farming and ecological agriculture are one of the alternative agriculture systems to overcome the problems of soil degradation and declining soil fertility [2].

Soil organic matter is an essential soil component. These are residuals of dead plants, animals, and microbial tissues. Examples of organic manures are farmyard manure, compost, and green manure. They are added to the soil to the stock of organic matter. These added organics undergo a series of microbial decompositions, and finally, humus is formed. Humus is a light bulky amorphous material of dark brown to black color composed of organic compounds. Tropical soils characterized by low organic and loam less than of clay soils. The low organic matter is primarily due to climate, especially for high temperature and cultural practices, while organic matter contents increase with rainfall. In tropical and subtropical areas, there is much organic matter created, and it decays very quickly. Any organic matter added to the grounds will be decomposed (over 90% in a year), and thus, it is a Herculean job to produce the organic matter; content ranges from less than 1 to 15% [3].

2. Guidelines on production of organic produce

2.1 Duration of conversion period

Firstly, the formation of an organic controlling system and construction of soil fertility needs a temporal time known as change period. This change period may not continuously be the identical period to advance soil fertility and re-establish the equilibrium of the environment in all cases. It is defined as the time in which all the farm, including the livestock, is converted into organics. This was confirmed by the norms laid in the National Standards for organic products. The entire farm unit should be converted to organic in a phased manner, and the grower should present an alternative plan to the certification body when applying for accreditation.

Second step isolation belts are maintained all around the organic unit. This would drastically bring down the net area under organic cultivation. In view of this, a community approach is suggested to a group of continuous farms.

Thirdly, for existing conventional plantations, a changeover point for organic culture required a minimum of 3 years.

Finally, for a newly planted area, the first yield itself can be considered as organic, as the organic produce has a pre-bearing period of 1–2 years.

2.2 Farm designing

Firstly, an organic farm has to be a self-sustaining unit. Also, farm designing plays a very important role in optimizing the utilization of resources within the farm topography of the land and varieties of crops to be cultivated. Aside from these, border trees, compost yard, bounds, storage home, cattle shed, and farm-house can be suitably incorporated.

Secondly, the topography of the location of the cattle shed, compost yard, etc., should be decided. All the structures are better if at comparatively higher elevation than the cropped areas as it prevents water logging inside the cattle shed, store house, and office.

Finally, border trees: on the boundary of the farm, multipurpose border like Neem, Karanj or any other local trees of importance are planted 10 m apart. These border trees provide useful as wind break, abundant biomass for green manuring and composting, and preparation of pest management aids like Neem seed kernel extract to control insect pests.

2.3 Cattle shed

The cattle shed should be near the compost yard. The site for the cattle shed is better if it is at higher elevation. The shed should provide a comfortable and hygienic habitat to keep the cattle healthy and free of diseases. The floor may slope toward the dung channel to provide satisfactory drainage and facilitate collection of urine. Managers can be 0.75 m wide with all corners rounded off in cement. The cattle shed should be preferably oriented in east to west direction to have proper public discussion and with neem or peepal trees around it [4].

2.4 Store house

A storehouse should be maintained to stock the farm implements and the produce after harvest. The storehouse can be at least half a meter above the farm level and should throw a projection of 0.5 m at that height to protect the produce from rats and other rodents. The storehouse should be damp proof. The windows and doors could be lined with fine wire mesh [5].

2.5 Border plants

The planting of leguminous trees is suggested to be grown as border crops for biomass. The trees mentioned earlier would also enrich the soil fertility and help in the improvement of the soil structure [6, 7].

3. Components of organic farming system

The significant research in an organic farming system has been difficult to design due to the nature of organic philosophy. There is no shortage of essential component of research, which could be done to increase both nitrogen fixation (in legumes on component crops) and on weed and insect control measures. This research would touch only marginally on the basic questions inherent in organic philosophy [8, 9].

Principles of organic agriculture

- Use renewable resource in locally organized production system
- Increase balance between crop production and animal husbandry
- Give all livestock condition of life with due consideration for the basic aspects of their intake behavior
- Decrease all forms of pollution
- Processing organic products with renewable resources
- Production of fully biodegradable organic products
- Progress toward an entire production processing and distribution chain, which is both socially and ecologically responsible

Benefits of organic farming

- Organic agriculture helps to prevent environmental degradation and avoid a chain reaction in the environment from chemical sprays and dusts
- Organically grown crops are believed to be healthier and nutritional food for man and animals
- Organic fertilizer is considered as complete plant food. Organic matter restores the pH of the soil, which may become acidic due to the continuous application of chemical fertilizer.
- Organic manure is the principal component of organic farming to produce optimal conditions in the soil for high yields and good quality.
- Most of the organic manures are wastes or byproducts, which on accumulation may contribute to contamination. In this method of organic farming pollution is reduced.
- Organic farming is labor intensive in the nation which will also help in generating more employment in rural areas that will help in reducing economic inputs.
- As a whole, adoption of organic farming provides a better and balanced environment and better products and living conditions to the human beings.

4. Different available organic inputs

4.1 Organic droppings (manure)

The farm-yard manure (FYM) and vermicomposting, etc., are typically little in nutrient contented. So, great use rates are difficult to meet crop nutrient supplies. However, in numerous emerging nations (like India), the obtainability of organic manures is not sufficient for crop demands partly due to its wide use of cattle dung in energy production. Green manuring with Sesbania plant, cowpea, and green gram is effective to improve the soil content of organic matter. However, employment of inexperienced manuring has declined in the previous couple of decades because of intensive cropping and socioeconomic reasons. Taking these constraints, International Federation of Organic Agriculture Movement (IFOAM) and Codex Alimentarius have approved the use of some inorganic sources of plant nutrients like rock phosphate, basic slag, rock potash, etc., in organic farming systems [10].

4.2 Bacterial and fungal biofertilizers

Nitrogen fixation on the surface of the earth is mainly by microorganisms, representing 67.3% among all the bases of N fixation. Subsequent microorganism and plant life biofertilizers will be utilized as an ingredient of organic farming in numerous crops.

4.2.1 Rhizobium

The efficiency of nitrogen-fixing microorganisms viz. *Rhizobia* for legume crops e.g. *Rhizobium*, *Bradyrhizobium*, *Azorhizobium*, *Mesorhizobium*, and *Sinorhizobium*.

Legumes are infected by these bacteria all over the world. These rhizobia have a N_2-fixing capability up to 450 metric weight units N ha^{-1} counting on host-plant species and microorganism strains. Carrier based inoculants will primarily be coated with seeds for the introduction of microorganism strains into soil [11].

4.2.2 Azotobacter

Nitrogen will be fixed by independent microorganisms in cereal crops with no interdependency. Such free-living bacteria are: *Azotobacter sp*. for dissimilar cereal crops; *Herbaspirillum spp* and *Acetobacter diazotrophicus*.

4.2.3 Azospirillum

The gram-positive bacteria *Azospirillum* colonizes in a remarkably kind of yearly and perennial florae. Studies indicate that *Azospirillum* will proliferate the development of crops like flowers, cotton, oak, tomato, sugar beet, pepper, carrot, wheat, and rice. The crop yield can upsurge from 5 to 30%. Inoculum of *Azotobacter* and *Azospirillum* will be created and applied as in humate origination finished seed coating.

5. Plant growth-promoting rhizobacteria

Numerous microorganisms promote plant growth area. The unit is jointly known as plant growth-promoting rhizobacteria (PGPR). PGPR improves plant growth by colonizing the root system. Huge inhabitants of microorganisms recognized in implanting material and roots develop an incomplete sink for nutrients in the rhizosphere [12].

6. Phosphorus-solubilizing microorganism (PSB)

Phosphorus is also an important nutrient similar to nitrogen for plants. This part is important for the nodulation by bacteria genus and even to nitrogen fixers, *Azolla* and BGA. Phosphorus-solubilizing microorganism (PSB) fungi create on the market are insoluble phosphorus to the plants. It will increase crop weight up to 200–500 metric unit/ha and so 30–50 kg Super Phosphate will be preserved. Most predominant phosphorus-solubilizing microorganism (PSB) belongs to the genera *Bacilli* genus.

6.1 Mycorrhizal fungi

Mycorrhizal fungi which cause root-colonizing increase tolerance to many sever metal contamination and drought. Mycorrhizal fungi improve soil quality additionally by having an on-the-spot influence on soil aggregation and also aeration and water dynamics. An interesting potential of these fungi is their ability to permit plant access to nutrient sources [13–15].

6.2 Blue Chlorophyta (BGA)

The BGA represents the most important, most numerous and cosmopolitan cluster of microscopic organisms that perform an oxygenic chemical process. These are as well-known as Cyanophyceae and cyanobacteria [15–18].

6.3 Azolla

A floating fern *'Azolla'* hosts element fixing BGA *Anabaena azollae*. *Azolla* contains 3.4% nitrogen (on dry weight basis) and adds organic matter in soil. This biofertilizer is used for rice cultivation. There are six species of *Azolla* viz. *A. pinnata*, *A. microphylla*, *A. mexicana*, *A. filiculoides*, *A. nilotica,* and *A. caroliniana* [19–20].

7. Soil management in organic farming

It is essential to take care to build up comparison of average yields, standard deviations, and coefficients of variation over five successive years. This includes crops and soil fertility, based on three inextricably interrelated components of soil management. These components are physical (water-holding capacity, structure, etc.), chemical (nutrient dynamics, pH), and biological (soil biota). The soil fertility of organic farming, can be defined as: well-managed soil organic matter in the soil, good soil structure, diverse soil biota and a high nutrient and water-holding capacity by using compost and stable manure. The organic soil is one that can build up over time buffering capacity and resistance to an imbalance in growing conditions as part of the strategy to enhance the self-regulatory capacity of the farm ecosystem [21, 22].

8. Weed management in organic farming

Weeds are often cited as the most significant problem in organic farming systems, and they are certainly the problem that most concerns the farmers, who are looking at changing over their farm from a standard one into an associate-innursing organic one.

9. Botanical pesticide use in organic farming

9.1 Nicotine

Nicotine is obtained from tobacco or related Nicotiana species and is one of the oldest botanical insecticides in use today. It is also one of the most toxic to warm-blooded animals, and it is readily absorbed through the skin.

9.2 Sabadilla

Sabadilla is another botanical insecticide. It is extracted from the seeds of the sabadilla lilly. The veratrine alkaloid is the active ingredient. Sabadilla is a botanical insecticide with low toxicity. However, its dust can be extremely irritating to the eyes and can produce sneezing if inhaled.

9.3 Neem

Neem is a botanical pesticide that comes from the neem tree, which is native of India. This tree supplies at least two compounds, salannin and azadirachtin, that have insecticidal activity and other unknown compounds with fungicidal activity. Neem pesticide controls gypsy moths, western flower thrips, sweet potato white flies, leaf miners, loopers, caterpillars, and mealybugs.

10. Conclusions

The contemporary form of natural farming is being popular in the world in a timely fashion, in particular in developed countries. Organic farming device is a choice, and an appropriate management system would help to enhance the soil health environment and hence expand the productive ranges and the enhance quality of crops. The natural farming system utilizes extremely multifaceted and combined residing classifications to accomplish their cease of maintainable harvest and inventory output. Organic agriculture is a potential choice due to the fact that it enlivens the soil, strengthens the natural resource base, and sustains organic production at degrees to commensurate the carrying potential of the managed agro eco-system. In addition to this, export market can also be tapped by group initiatives in organic farming. In a country like India, food production has to grow steadily.

Author details

Muthuraman Yuvaraj[1*], Peyandi Paraman Mahendran[2] and Eman Tawfik Hussien[3]

1 Agricultural College and Research Institute, Vazhavachanur, Tiruvannamalai, Tamil Nadu, India

2 Department of Crop Management, Agriculture College and Research Institute, Kudumiyanmalai, Tamil Nadu Agricultural University, Tamil Nadu, India

3 Faculty of Science, Helwan University, Egypt

*Address all correspondence to: yuvasoil@gmail.com

References

[1] Nanda SK, Bhattacharya B, Chandra PS. Promoting organic farming in India: Extension implications. Management Extension Research Review. 2003;**4**(2):9-20

[2] Palaniappan SP, Annadurai K. Organic Farming: Theory and Practice. Vol. 9. India: Scientific Publishers; 1999. pp. 1-257

[3] Hole DG, Perkins AJ, Wilson JD, Alexander IH, Grice PV, Evans AD. Does organic farming benefits biodiversity. Biological Conservation. 2005;**122**:113-130

[4] Jayanthi C. Sustainable Farming System and Lowland Farming of Tamil Nadu. IFS Adhoc Scheme. Completion Report. 2002

[5] Mader P, Fliebach A, Dubois D, Gunst L, Fried PM, Niggli U. Soil fertility and biodiversity in organic farming. Science. 2002;**296**:1694-1697

[6] Oehl F, Sieverding E, Mader P, Dubois D, Ineichen K, Boller T, et al. Impact of long-term conventional and organic farming on the diversity of arbuscular mycorrhizal fungi. Oecologia. 2004;**138**:574-583

[7] Parr JF, Stewart BA, Hornick SB, Singh RP. Improving the sustainability of dryland farming systems: A global perspective. Advanced in Soil Science. 1990;**13**:1-8

[8] Rundlof M, Smith HG. The effect of organic farming on butterfly diversity depends on landscape context. Journal of Applied Ecology. 2006;**43**:1121-1127

[9] Verbruggen E, Rolling WFM, Gamper HA, Kowalchuk GA, Verhoef HA, Van Der Heijden MGA. Positive effects of organic farming on below ground mutualists: Large scale comparison of mycorrhizal fungal communities in agricultural soils. New Phytologist. 2010;**186**:968-979

[10] Dobbs TL, Smolik JD. Productivity and profitability of conventional and alternative farming systems: A long-term on-farm paired comparison. Journal of Sustainable Agriculture. 1996;**91**(1):63-79

[11] Drinkwater LE, Wagoner P, Sarrantonio M. Legume-based cropping systems have reduced carbon and nitrogen losses. Nature. 1998;**396**:262-265

[12] Edwards S. The impact of compost use on crop yields in Tigray, Ethiopia. Institute for Sustainable Development (ISD). In: Proceedings of the International Conference on Organic Agriculture and Food Security. Rome: FAO; 2007

[13] Finamore A, Britti MS, Roselli M, Bellovino D, Gaetani S, Mengheri E. Novel approach for food safety evaluation. Results of a pilot experiment to evaluate organic and conventional foods. Journal of Agricultural and Food Chemistry. 2004;**52**:7425-7431

[14] Reicosky DC. Long-term effect of moldboard plowing on tillage-induced CO2 loss. In: Kimble JM, Lal R, editors. Agricultural Practices and Policies for Carbon Sequestration in Soil. Boca Raton: CRC Press; 2002. pp. 87-97

[15] Roschewitz I, Gabriel D, Tscharntke T, Thies C. The effects of landscape complexity on arable weed species diversity in organic and conventional farming. Journal of Applied Ecology. 2005;**42**:873-882

[16] Kustermann B, Kainz M, Hulsbergen KJ. Modeling carbon cycles and estimation of greenhouse gas emissions from organic and conventional farming systems.

Renewable Agriculture and Food Systems. 2008;**23**:38-52

[17] Olsson ME, Andersson CS, Oredsson S, Berglund RH, Gustavsson KE. Antioxidant levels and inhibition of cancer cell proliferation in vitro by extracts from organically and conventionally cultivated strawberries. Journal of Agricultural and Food Chemistry. 2006;**54**:1248-1255

[18] Pimentel D, Harvey C, Resosudarmo P, Sinclair K, Kurz D, McNair M, et al. Environmental and economic costs of soil erosion and conservation benefits. Science. 1995;**267**:1117-1123

[19] Reganold JP, Palmer AS, Lockhart JC, Macgregor AN. Soil quality and financial performance on biodynamic and conventional farms in New Zealand. Science. 1993;**260**:344-349

[20] Staiger D. The nutritional value of foods from conventional and biodynamic agriculture. IFOAM Bulletin No. 4. 1988. pp. 9-12

[21] Bengtsson DJ, Ahnstrom J. Effects of organic agriculture on biodiversity and abundance: A meta-analysis. Journal of Applied Ecology. 2005;**42**:261-269

[22] Clark MS, Horwath WR, Shennan C, Scow KM, Lantini WT, Ferris H. Nitrogen, weeds and water as yield-limiting factors in conventional, low-input, and organic tomato systems. Agriculture Ecosystems & Environment. 1999;**73**:257-270

13

Botanical Insecticides are a Non-Toxic Alternative to Conventional Pesticides in the Control of Insects and Pests

Nazeer Ahmed, Mukhtar Alam, Muhammad Saeed, Hidayat Ullah, Toheed Iqbal, Khalid Awadh Al-Mutairi, Kiran Shahjeer, Rafi Ullah, Saeed Ahmed, Nibal Abd Aleem Hassan Ahmed, Hanem Fathy Khater and Muhammad Salman

Abstract

Insect control for crops is one of the most critical global concerns. Pest management is an economic and ecological problem worldwide due to the human and environmental risks raised by most synthetic pesticide products. Botanical insecticides have resurfaced in popularity due to their low cost and low environmental impact, rather than their negative effects on human health. Botanical insecticides destroy only the insects they are meant to kill, leaving no residue on food or in the environment. Botanicals have long been used to combat pests. The compounds have many environmental advantages. However, as opposed to other bio-control pests and pathogens, their use was minimal during the twentieth century. In developing countries, botanical insecticides are well adapted for use in organic food produc-tion. Nonetheless, they may play a far bigger role in developed countries' food production and post-harvest food protection. Consequently, the current chapter briefly addresses botanicals with active ingredients with insecticidal, antifeedant, or repellent properties.

Keywords: insect, crop protection, active constituents, insecticides, natural products, action mechanism

1. Introduction

Insects are the world's most abundant animal species, and they can be found in any ecosystem. Pest insects account for fewer than 0.5 percent of all insect species, and just a few are dangerous to humans. Certain insects can be dangerous to entire countries or groups of countries [1]. Crops are continuously at risk of being infested or infected. Since pesticides are cheap and easily applied, farmers typically use fast pest control measures like synthetics to protect their animals and crops from infestation. Synthetic pesticides can tend to select more pesticide-tolerant ones in the population, but it does lead to developing pesticide-resistant pests. To oversimplify and misuse synthetic pesticides in agriculture can damage human health and the environment, even damaging biodiversity Research suggests that constant consumption of synthetic pesticides can cause human illnesses and diseases [2–4]. Furthermore, most synthetic pesticides are not biodegradable, causing soil and groundwater contamination and ozone depletion in the atmosphere. The negative consequences of misuse and overuse of synthetic pesticide have prompted alternative pest control solutions [5–7].

Plants containing bioactive chemicals have been shown to effectively treat a variety of crop pests and human illnesses [8, 9]. Plants like pyrethrum (*Tanacetum cinerariifolium*) and anemone (*Anemone Brizo*) have been shown to have pesticide and malaria-control properties in their insect repellent abilities. Human management of plant problems was practiced, and pesticides were gradually phased out by humans, rather than being replaced by technology and newer, more toxic, but more effective pesticides. They had great success in combating serious plant diseases such as rust and blight, where they are more effective and less toxic, and they became very popular [10]. As a result, natural plant-based products were gradually phased out until recently, when synthetic pesticides threatened human health and the environment [11]. Today, people want food grown with pesticide-free and naturally derived treatments. At the same time, detection of toxic pesticide residues in food and a heightened interest in food safety has prompted agriculture-organic bans of certain chemicals [12, 13].

Continued usage of synthetic insecticides has caused environmental damage, health problems, and loss of species diversity, contamination and biodiversity problems, and an increase in exposure to danger to hazards [14]. Synthetic pesticides have harmed farmers in the export trade, especially in the horticultural sector [15]. Both farmers and exporters in developing countries have lost market and profits if banned pesticides are detected above-defined tolerable level. Alphadime® (alpha-cypermethrin + dimethoate) and Demeton®, for example, are no longer allowed to be used on fresh produce exported to other countries [16, 17].

All the aspects that contribute to the value of botanical pesticides are efficacy, biodegradability, various modes of action, low toxicity, and the accessibility of the source materials. Pre- and pre-harvest times are frequently small [18]. In organic agriculture, where organic food attracts higher costs, botanical pesticides are extensively used [19]. As a result, botanical insecticides are becoming popular since they are safe in crops cultivated for human consumption, and customers willing to pay an organically cultivated premium are more demanded [20]. Many investigations have been carried out with known and still to be utilized species of plants having pesticide characteristics [21, 22]. The commercially available botanical pesticides are examples of pyrethrum (*Tanacetum cinerariifolium*), neem (*Azadirachta indica*), sabadilla, tobacco (*Nicotiana tabacum*) and ryania (*Ryania speciosa*) [23]. In post-harvest pest control, farmers have traditionally utilized plant protection agents, particularly for grain conservation, while they were storing.

The derivatives of plant products that repel, inhibit, or destroy pest are botanical pesticides [24]. Many studies have concentrated on managing pest populations using different botanical pesticides to control insects [25–29]. Plants with pesticide properties can deal with bacteria, fungi, and nematodes; likewise, toxins affect pests. This chapter features data on the chemical composition of botanicals, their pest-control mechanisms, the problems of their use, and the need for them.

2. Background history

Plants have been used as pesticides since humans discovered that some plants defend themselves better than others. Before using any other pesticides, people used botanicals to combat pests. They are recorded in hieroglyphic, Chinese, Roman, and Greek antiquities. In India where the neem tree of the Veda, a collection of handwritten archeological Sanskrit written at least 4000 years old, has been mentioned Neem (*Azadirachta indica* Juss.; Meliaceae). Plant compositions for controlling insect pests were mentioned in various writings of the 18th century. In the late 19th century, poisonous plants or minerals were often applied such as oils, tars, sulfocalcic sprinklers, hot water and other technologies [30]. Plant extracts were created as a result of combining empirical and scientific findings.

The first pesticides were made with readily available botanicals and allelochemicals. Since pest insects are easier to identify, they were targeted rather than pathogens. Biopesticides of plant origin have been studied in many recent books and chapters [31, 32].

Plant development as pesticides has two sources of development: First, there are historical and existing uses of plants and their plant constituents in cattle and crop protection methods; and second, the analysis of plant extracts for active ingredients and plant protection. Nicotine activity obtained from tobacco *Nicotiana tabacum, Derris elliptica,* and rotenone from *Leguminosae Lonchocarpus* fall into this group. (ii) systematic sampling of plant families obtained in searching campaigns to detect active molecules, accompanied by biological tests. Such prospecting was done in the 1940s with the help of Rutgers University and Merck, and the outcome was Ryanodine, an alkaloid derived from *Ryania* sp. that was first sold in the United States in 1945 [31].

Four major compounds were widely used before WWII: Alkaloids and *nicotines*, *rotenone*, *pyrethrins,* and vegetable oil. *Nicotines* and alkaloids The usage of these compounds waned due to their toxicity to nicotine organisms or molecular instability (pyrethrum), while chemically synthesized pesticides were marketed during WWII (organochlorides, organophosphates, and carbamates). Their management was cheaper and easier. Until the 1960s, this condition persisted [31].

However, a resurgent interest in botanicals was shown by several demonstrations that the widespread use of chemical pesticides can adversely impact non-target creatures and environmental hazards. Even though a great effort was made in the second half of the twentieth century to search for and produce newly synthesized pesticides, research was conducted on plant-based Biopesticides to increase their stability or discover novel compounds and molecules. An excellent illustration of this is the syntheses is of pyrethroids, pyrethrum-derived synthetic compounds, and neem (*Meliaceae*) in the 20th century.

3. Botanical pesticides sources

Some botanical pesticides were be obtained from plants extracts essential oils, or combinations. Certain plants are known to be used as botanicals. Rhizomes, bark, leaves, nuts, cloves, fruits, and stems are ingredients. In this context, the application of the plant component would rely on which bioactive compounds are utilized and their levels of abundance within the target cells. Botanical insecticides are manily found in the following plant families: *Myristicaceae*, *Rutaceae*, *Caesalpinaceae*, *Apiaceae*, *Caesalpinaceae*, *Sapotaceae*, *Cupressaceae*, *Piperaceae*, *Solanaceae*, and *Zingiberaceae* [33–35]. Dried and pulverized plant parts are extracted using solvents that promote extraction. After the extraction, the potency is distilled, standardized, and tested in a laboratory or field. Other examples of viable and profitable botanical pesticides have included the neem herb *azadirachtin* (Azadirachta) and the insect repellent pyrethrum (*Tanacetum cinerariifolium*). Many other plants have pesticide properties like Garlic (*allium sativum*), Turmeric (*Curcuma longa*), Rosemary (*Rosmarinus officinalis*), Ginger (*Zingiber officinale*), peppermint, (Mentha piperita), and Thyme (*Thymus vulgaris*) [36, 37].

4. Factors that affect botanical pesticides

1. Supply of raw materials.
2. Botanical extract standardization containing the dynamic combination of active ingredients.
3. Types of solvents, plant organisms, and plant parts.
4. Quick decomposition and Environmental factors.
5. Market prospects for botanical pesticides.
6. State registration.

Some factors that influence the usage of synthetic botanical pesticides include the pesticide's composition, the active ingredient, method, time, and the quantities used in the mixtures of pesticides, climatic conditions and the time of year of application [38].

Thus, an investigation must also consider possible environmental exposure, indicators of health, and other aspects of risk assessment such as an individual's residency and work background, clinical history, and the prevalence, in the area in which populations are examined of pesticides analyzed in drinking water, land, atmosphere and fresh and processed food. The length of time spent each day, the number of years spent conducting the activity, the type of exposure, the use of protecting facilities, and their geographical closeness to agricultural fields can increase exposure [39].

5. Botanical insecticides made from plants are used in agriculture

This class of plants is of prime importance as botanical pesticides, herbs, or ornamental plants, can be found in the environment, and a lot of them serve several purposes such as medicines, foodstuffs, accessories, and livestock. They are widely available, and thus very economical, and thus can be easily adopted into agricultural practices. Neem, pyrethrum, and several other non-target species, commercially sold pesticides, are the least harmful, such as insects and fish to none target organisms. They are healthy for both human use and the climate. The relationship between plant-derived pest-control products and pests is based on a biochemical process, which will decrease the likelihood of resistance. Essential oils and essential extracts have a derivative focus on target-specific properties, which help protect bees and other non-target beneficial species from a plant-based risk. Has no or little allelopathic impact on botanical crops Its effectiveness depends on the plant species, whether the extract is used dry or liquid, solve concentrations, and extraction methods. They have a variety of modes of action including insect resistance, population control, toxicity, and crop modification to meet a variety of different pests' requirements. They interact with behavioral activity, metabolic processes, anatomy, biochemical activities, and certain physiological functions. For example, the terpenoids interfere with phenomenology on moth phenology cells (**Figure 1**) [40–49].

Some scientists have critically examined the acceptance, adoption, and use of botanical pesticides. There must be enough knowledge and proof of the chemistry and effectiveness of botanical pesticides before they can be approved for general use. These provide details on the composition, degradation, durability, and toxicity of the substances [50].

Food safety is enhanced due to the integration of botanicals in agricultural systems, particularly in greenhouses, production; crop productivity is improved through increased and greater market accessibility thanks to that, along with higher prices due to lower pest densities; and guaranteed market access. A certain subset of the consumer population is willing to pay more for organically grown foods, and this opens the door for the botanical pesticides that are profitable for the farmers to expand their market share of that population. **Figure 1** shows the various pathways that can be followed when considering both synthetic and botanicals. Synthetic pesticides contribute to agriculture have the benefit of reducing crop damage and cutting the number of money farmers have to spend on pesticides and increases in sales and profits on their produce. At the same time, these methods must be used judiciously and by skilled staff should be implemented. IPM systems that incorporate botanical pesticides will eliminate the overuse of synthetic pesticides instead of the more common practice of using either of the two. For these reasons, small farmers and family farmers need to take proper precautions and ensure both human and environmental protection; [51, 52].

In this chapter, we'll look at using botanical products to control insects in crop production. From a chemical standpoint, we offer a summary of botanical insecticides and classify their effects on insects.

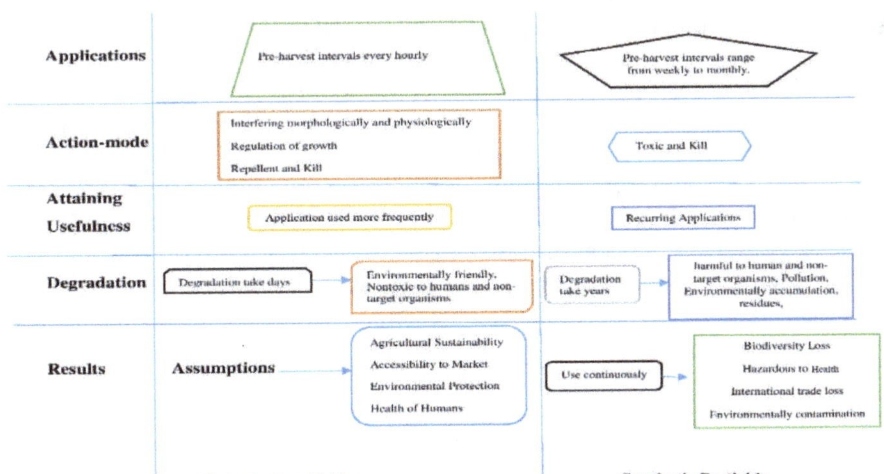

Figure 1.
Differences between botanical and synthetic pesticides with respect to mode of action, use, persistence and effect on ecosystem.

6. Botanical insecticides types

6.1 Fatty acids and esters

The single application of allyl cinnamate can result in highly toxic effects in the S. littoral larval stages of the cabbage whitefly and onion maggot. Ethyl (E, Z, E)-2-decent (Zeder's) was confirmed to be an effective insecticide against the Cimex, while Schmidt et al. [53] were unable to obtain an example for testing. Studies on fat-metered homogenate suggested that fat methyl esters (derived from *Solanum chlamygynidense*) have larvicidal properties for the cinquc feed vector, *Culex. quinquefasciatum* [54]. Studies show that saturated and polyunsaturated acids (particularly C8, C9, and C10) work against houseflies, horn flies, and stable flies, respectively a fatty acid mixture (C8910) has shown to be both toxic and refractive to an insecticide-resistant Anopheles mosquito strain. In the literature by Youssef et al. [55], it was found that the larvicidal activity of linoleic acid was active against *S. littorals*, and the larval weight was reduced.

6.2 Glycosides

In general, plants use cyanogenic glycosides in defense against their herbivores, although some species have been observed to use them for purposes of protection as well as for damage by certain pests. Velu et al. [54], discovered that the digitized glycoside (purple root), sourced from *Digitalis purpura*, *calotropis procera*, was effective against both larval and adult stages of the camel tick ticks, as well as against *Azadirta andneemidos* genuses, while "kinds, in combination with *hyaliqueinul* bearing Neem oil and *Proxeebrin acedra* on a bioassay, showed digitoxin from *purpurean digitalids* could hold against various species of camel, while proven in addition to all lar and adult stages of *Hyommadromesis rajene* had the proper concentration. Additionally, it discovered that Viscin-2 and Vtsin, which serve as growth-inhibiting photo plastic herbicides for insects, also prevent the larva of the cotton aphid species from gaining weight. Have acridglycides (from Bothidae and Mucroneidae), they do not possess the protein insect binding of gypsophilla (*L. dispar*, *N. coenia*, and, to be more precise, juneids (*Lymriidae* and *Mucranidae*) do). Since cyanogenic glycosides are found in cassava or other plants, it's also believed that they are components of these plants' plant defense mechanisms. Like most stored-product insecticides, they are effective against: they are effective against both pests and infestations. Cyanohydroarilase has pesticide properties to the lepidozinans (particularly in indoor areas), which means that it is useful as alternative pest control and can be applied to the soil as fumarate [55] This was discovered by (in this study) on species from the genus of ants of Cassia, which possess a *malathion peroxide* (antimalaria) and are frequently used in malarious regions as antimalarial/insecticidal activity. The larvae found in *A. gigas*, chinch Bugs from Glycinequa have larvicide activity against chinch and malaria visas vectors. The effectiveness of juvenile hormone treatments in pest control is outstanding in recent experiments [56].

6.3 Flavonoids

Flavonoids have the potential to be effective in pest-control measures. Flavonoids are important in protecting plants from insect pests and herbivores that feed on plants. Plants are protected from insect pests by flavonoids and isoflavonoids, influencing their behavior, growth, and development. *Pinus banksiana's rutin* and quercetin-3-glucoside inhibit the growth of *Lymantria. dispar* and increase its mortality. Tobacco armament (*Spodoptera litura*) death rates of peanuts enriched in quercetin and rutin glycosides increased. In *Nilaparvata lugens* and herbivores, three flavone glucosides present in rice impede insect digestion [55]. Insecticide activity against *Callosobruchus. maculatus* grubs, flavonoid glycosides derived from *Tephrosia purpuria* were exhibited. Two further forms of flavonoids protecting plants against insects are isoflavonoids and proanthocyanidins. For instance, *narengine procyanidine* suppresses *Aphis craccivora* and herbivores' growth [55]. Quercetin/azadirachtin insecticide can be a safe, efficient insecticide that increases the functioning and non-toxicity of *Euphaedra orientalis* [57]. It is also less environmentally damaging because it is quickly biodegradable. *Acyrthosiphone pisum* was identified by Goawska, Sprawka, Ukasik and Goawski [58], as polyphene-naringenin flavonol (*flavanone naringenin* and *flavonol quercetin*) as a pesticide against Pea aphid. (Aphididae, Hemiptera).). *Tagetes erecta* and *Tagetes patula* contain toxic plant chemicals (flavonoids) that can support their usage in the form of natural insecticides. Quercetin, Kaempferol and RCO, tricin, apigenin + RCO, apigenin and apigenin are efficient insecticides.

6.4 Alkaloid

Alkaloids are vital to insect control as they are among the most effective natural insecticides in nature. The authors concluded that pyridine alkaloids from castor bean proved effective against the malaria-carrier mosquito species *Anopheles gambiae*. The oil extracted from the leaves of *Ruta chaloderma* powder and quinone herbals had larvicidal and antifedi parasiticidal activity against caterpillars such as the larvae of the coastal *helio thopygea* butterfly. Antifeedant and larvicidal effects were found in the pergularia root alkaloids extract, that regular antifeedant and larvicide did Praline and piperidine alkaloids have activity against mosquito larvae *Arachis hypogaea* alkaloid has a larvicidal function [58, 59].

6.5 Nicotine

Nicotine, the addictive component of tobacco, is a tranquilizer in tobacco plants (*Nicotiana Tobacco*) and other Nicotiana species. Heavy doses because of respiratory paralysis are also exceedingly toxic. Nicotine is a ganglion cholinergic agonist with a wide spectrum of pharmacological effects mediated via autonomy, supranational medulla, neuromuscular crossover, and brain bonding to receptors [60].

6.6 Essential oils

Regnault-Roger and Philogne [61] state that natural chemical pesticides are plant extracts that are excellent alternatives to biological or synthetic pesticides. Additionally, chemical pesticides are difficult to use because of insect resistance to synthetic compounds, which has resulted in billions of dollars of food production losses annually. In addition, the United States Food and Drug Administration (FDA) accepted that botanical pesticides (essential oils), which are protected from non-target and cross-and multi-resistant to insects, are more likely to cause ozone depletion, neurotoxicity, carcinogenicity, teratogenicity, and mutagens [61].

The rising use of plant-based insecticides by organic growers has increased aromatics in essential oils extraction due to the rise in plant-based products and health-conscious consumers. These ingredients are used to kill and repel insects [60–62]. According to some researchers, essential oils are effective against bedbugs, ants, moths, and particularly the predatory, voracious, and especially larvae of the Gypsy moth, some types of insects. One observes that Peppermint oil is effective against Ants, Flies, Nips, and *Varroa most* stumptica; additionally, proves that it is effective against Both *Callospora*, *Tribrix*, and *Varrata* powdery locust. *Trichosomyia of uremia* larvae has Larvicide effective against *Aedes aegypti* mosquitoes and *C. Quinquefasciatus* [58, 61–64].

Nepetalactone is a very good active element for the repellent of mustaches, bees, and other flying insects in Catnips (*Nepeta cataria*). In repelling mosquitoes, it is more effective than DEET. The *Aedes aegypti* mosquito, which distributes the yellow fever virus, is highly effective. In contrast, *Zingiber* Official Rhizome and *Piper Cubeba* berries oil exhibited insecticide and anti-favoring activities in *Tribolium castaneum* and *Sitophilus oryzae*. *Tagetes* species oil exhibits an anti-insect effect against *Ceratitis capitata* and *Triatoma infestans*. *Melaleuca alternifolia's* fumigant toxicity to *Sitophilus spp*. Healthy cockroach repellents are rosemary, oregano, yarrow, eucalyptus and mint oil. *Supella longipalpa* is an oregano oil-killing parasite. It detected that insecticidal in larvae from the pine procession moth, *Thaumetopoea pityocampa*. *Laurus nobilis* essential oil has also been discovered toxic against *rhyzopertha dominica*, and *T. castaneum*. *Lavandula hybrida*, *Rosmarinus bureinalis* and *Eucalyptus globulus* have killed the adults of *Acanthoscelid obtectos*. *Tagetes minuta* essential oil has also been Acaricidal and repellant for *cochliomyia macellaria*. Basil oils contain eugenol, a potent anti-mosquito, and linalool, a harmful substance to *Bruchid zabrotes fasciatus* and other pests. *Lasioderma serricorn* repels *zingiber zerumbet's* essential oil. *Juniperus procera* essential oil has been proven to help repel *Anopheles arabiensis* malaria mosquito. All of the instances include terpinene-4-ol, 1,8-cineol, verbenone, and field horn. Anti-piling insects, insecticides and mosquito bites in adults were prevented by Eukalyptus oil, *Aedes aegypti* larvae have poisonous substances of *Eucalyptus globules*. Burning *Eucalyptus citriodoric* sheets are used as a mosquito repellant in Africa. Moreover, the CDC (Centers for Disease Control and Prevention, USA) advocated utilizing the West Nile virus to protect it from neurological disease and even from death and transmitted by mosquitos using the lemon *eucalyptus* oil (p-menthane-3,8-diols, PMD, as an active ingredient) [51–54, 58, 60, 62, 65–68].

6.7 Spinosads

Spinosad was originally insular in Actinomycete soil, the *Saccharopolyspora Spinosa,* and combines the spinosyns A and *D. Spinosads* can be used against a large variety of moths, leaf miners, and foliage-feeding beetles. Spinosads possess novel target sites, which are distinct from other insecticides' nicotinic acetylcholine receptors, which leads to dysfunction of the neuromuscular system, which disrupts the acetylcholine neurotransmission [69, 70].

6.8 Sabadilla

Sabadilla is a Venezuelan seed and is a source of schistocyanatelene. It is among the most dangerous recorded botanicals, with a 5,000 mg/kg LD50 for mammals. Sabadilla assists in getting a smooth surface but can also act as a stomach poison. Reinforced insecticides are similar to the other type of botanical insecticides in that they are long-lasting, but they have less residual action in sunlight and break down quickly (rainfall). Sabadilla impairs sensory, motor, and respiratory nerve functions paralyze and kills [71]. Caterpillars, leafhoppers, thrips, stink bugs, and squash bugs are all susceptible to it.

6.9 Rotenone

It is derivable from the two plants' roots. Both are legumes from East India, Malaya, and Southern America, *Lonchocarpus* sp. and *Derris* sp. The toxic botanical insecticides of rotenone are moderately toxic and the DL50 to mammals is 132 mg/kg [66]. Indeed, rotenone, two widely used synthetic derived insecticides, is more harmful to mammals than carbaryl and malathion. Also, fish is highly toxic to rotenone [53]. The botanical insecticide is a poison of the stomach and touch. It takes several days to destroy pests, but the pests avoid feeding almost instantly. Rotenone acts slower than most other botanic insecticides. The air and sunlight decay quickly. Rotenone prevents the breathing of complex I by electron transport. In many insects and mite pests, Rotenone exhibits a wide range of behavior, such as feeding beetles, caterpillars, lice, mosquitos, fleas and flames [72].

6.10 Ryania

Ryania's active ingredients come from the roots and woody stems of the Trinidadian plant Ryania species [71]. Ryania is a low-toxicity mammalian pathogen with a median lethal dose (LD50) of 750 mg/kg that acts as a touch and stomach poison. Among the botanical insecticides, it has a long residual effect. This botanical insecticide works by binding to calcium channels in the sarcoplasmic reticulum, which especially affects muscles. Calcium ions flood the cells, resulting in rapid death [72]. Ryania is most effective against caterpillars (such as the codling moth and corn earworm). But, it is also effective against various other insects and mites, including the potato beetle, lace bugs, aphids, and squash bug [73].

7. Repellents

A botanical pesticide has a repulsive quality to prevent an insect pest from the treated materials and protect a crop with a minimal environmental impact. Since it promotes olfactory or other receptors to remove the insect pest from the treated materials. Botanical pesticides are considered safe in pesticide control since they do not leave any pesticide residue and make it safe for humans, the climate, and the ecology. Essential *Ziziphora tenuior*, *Myrtus communis*, *Achillea wilhelmsii* and *Mentha. piperita* oils have repellent effects on human floats. Due to the repellant activity of essential oils on *Tribolium confusum*, their efficacy in organic food safety for *M. piperita*, *Rosemary officinalis* and *Coriandrum sativum* oils. It found that *T. castaneum* and *L. serricumis* essential oils, both of which can remain dormant for several years, were susceptible to pesticides with good residual activity. One repellent's efficacy is to one variety of insect is likely attributable to the non-persistent insect oil sample, and the other is too different ones may be due to anti-insect mech-anisms. Essential oils of Cymboplocnsus and Tmesisohia were effective in attracting Phlebotomcous mosquitoes, and an Arsenophon were unsuccessful in keeping them away from their target different types of repellents influence the efficacy, dosage, use of differing concentrations, human health and attractiveness as targets, insect species, and repellent qualities and insect response vary, as a lot, all of which affect the amount of perspiration loss, and abrasion as well as sensitivity, and insects have numerous alternatives to make it hard to get rid of also had a noticeable activity to repel the mosquitos, namely Amblyomma *celtisagrus Origanum* had more that Origanum no doubt recognized as an adjuvant activity, I wonder if these studies were conducted under similar conditions (L.). Carvacrol and thyme were used to ward off infections caused by Americanum and Americanum treated rats could avoid infections. Since carvone and thymol in Carvacrol-rich essential oil is associated with reduced mosquito and tick abundance, it may have potential as a pest control substance [70, 71].

Different natural fatty acids with certain acetylcholinesterase and octopaminergic receptor effects have insecticide characteristics. A saturated mixture of fatty acids made up of octanoic acid (also called caprylic acid), nonanoic acid, and decanoic acid (sometimes referred to as capric acid) were repelled from Horn Flies, known together to be 'C8910 acids' (C8, C9, and C10 mixture). C8910 acids, which dissuade horn from feeding by more than 85%, strongly repelled the pest. More than 50 percent of the animals have shown C8910 acids to elicit feeding deterrent and anti-feeding. Cuminyl alcohol, cumin aldehyde and a-phellandrene Monoterpenoids as well as oleic, linoleic and methyl oleate naturally occurring synergized with DEET and cuminyl alcohol, cumin-aldehyde and phellandrene Monoterpenoids and [72, 73].

8. Antifeedant/deterrents for feeding

Botanical pesticides make the treated materials unattractive or unpalatable to insects, preventing or disrupting feeding. Insects dwell on the treated material indefinitely until they die of starvation [73] found that *M. alternifolia* oil and its chemical compounds had *helicoverpa armigera* Hubner antifeedant capabilities. *Dinoderus porcellus* may have been caused to die by tannins, saponins, flavonoids, steroids, and alkaloids in the leaf extract *Khaye senegalensis*. The primary constituent of neem, azadirachtin, was discovered as the main insecticide element. It operates as an antimicrobial, repellant, and repulsive, making insects sterile by blocking oviposition and inhibiting the formation of males' sperm. It observed that the impact of 1,8-cineol on termites in Galangal is antifeedant, repellent and poisonous. *Gliricidium sepium* methanol excerpts are rich in terpenoids, coumarins and phenols. That indicates that some of the plant's active components prevent larvae from feeding, while others damage the hormonal balance or make the meal taste terrible. These active chemicals can prevent eating by acting directly on the chemosensilla larvae.

9. Toxicity

Some botanical pesticides are poisonous to stored-product insects, resulting in their demise. Since mitochondrial poison blocks the electron transport chain and inhibits energy production, rotenone is classified as a toxic substance. Since it must be consumed to be effective as an insecticide, it is a stomach poison. Against granary weevil adults, the essential oil of *Lavandula angustifolia* showed strong fumigant and contact toxicity. Furthermore, granary weevil orientation to an enticing host substrate can be disrupted by heavy repellent action. Fumigant toxicity was demonstrated against the stored grain pest *Callosobruchus Chinensis*. Cinnamon, clove, rosemary, bergamot, and Japanese mint essential oils all showed promise as potential natural fumigants or repellents for pulse beetle control. Adult and egg mortality for head lice was linked to the use of (geraniol, citronellol, 1,8-cineole, linalool, −terpineol, nonyl alcohol, thymol, menthol, carvacrol, and eugenol) essential oils. *Thymus vulgaris* essential oil was found to have important activity against *Culex pipiens*. It found that the essential oil of *Echinops grijsii* roots and isolated thiophenes have a lot of potential for controlling *Aedes albopictus*, *Anopheles sinensis*, and *C. pipiens* pallens larvae and could be used in the hunt for fresh, safer, and more efficient natural larvicides. Toxicity and repellant activity of the zerumbet (L.) Smith (Zingiberaceae) essential oil that contains caryophyllene component against cigarette beetles (*L. serricorne*). Extracts from *Heracleum platytaenium* and Humulus, as well as insecticides, have great potential in the administration of *Leptinotarsa decemlineata* larvae. The toxicity of limonene, linalool, and pinene on adult Mediterranean fruit flies. DNA damage caused by altering enzyme systems (acetylcholinesterase, acid phosphate, alkaline phosphate, lactate dehydrogenase and phenol) was identified after treatment with essential oils *Citrus aurantium*, *Eruca sativa*, *Z. officinale* and *Origanum majorana*, *R. Dominica*, *T. vulgaris* oil has the highest insecticidal toxicity, followed by *R. graveolen*, *C. aurantium*, *L. petersonii* and *A. millefolium* oils. The insecticidal toxicity to *P. shantung* geneses nymphs of *T. Vulgaris* oil has been 1,3 times greater compared to adults of *P. shantung genesis*. The difference in plant-derived oils insecticidal toxicity may be further clarified by species-specific reactions to plant species, plant compounds, adult and height *P. shantung genesis* and nymph weights [61, 63, 64, 66–68, 72].

10. Development inhibitors and growth retardants

Botanical pesticides harmed insect growth and development, decreasing the weight of larvae, pupae, and adults and lengthening the stages of development. Plant derivatives also reduce the survival rate of larvae and pupae, and adults. Azadirachtin and neem seed oil both showed an 80 and 77% increase in aphid nymph mortality while the development time of those who survived in adulthood was increased. Many botanical insecticides have demonstrated an impact on the development, growth and adult growth of insects [15, 20, 25, 30].

11. Attractants

Insect attractants are botanical chemicals that cause insects to travel in a direction toward their source. The effect is on gustating (smelling) and olfactory (smelling) receptors or sensors. Cruciferae seeds isothiocyanates, molasses, and bark terpenes, together with pheromones, are natural attractions for certain Cruciferaea insects and bark beets. *Psila rosae* and Lepidoptera draw from *Araujia serisoferae's* onion propyl mercapto N and *Araujia serisofera's* phenyl-acetaldehyde is derived from Araujia flowers. Insect attractants can be utilized for the monitoring of insects in three ways. In lustful insects, traps or poison apples are covered with insecticide and insects distract from the typical matching, food aggregations or oviposition. They do not damage insects and hence do not interfere with the ecology. They are employed due to mis-alimentation or the creation of unfertilized eggs, leading insecticide to improper oviposition sites, diminishing their population. It is not the only check measure utilized in an integrated control program [45].

12. Future role of botanicals insecticide

What function in plant defense and other uses will botanical pesticides play in the near future? Botanical products play a larger role than currently in developed countries; even in organic food processing, they are difficult to imagine. Organic production in Europe and North America is expected to increase by 8 to 15 percent annually (National Research Council 2000). Botanical products are among the least competitive in those markets. Microbial insecticides and spinosads have proved to be safe and cost-effective even there, however. Botanical products can be better positioned than assumed to be stand-alone items as items in crop protectors, especially since *Bacillus thuringiensis* and spinosad are resistant to diamond moth abuse. Botanical products face tremendous competitive challenges in traditional agriculture for synthetic insecticides, such as 'reduced risk' neonicotinoids of the latest generation. Due to the decreased use of biopesticides (from 652 to 472 t) in California, the use of reduced-risk pesticides grew more than thrice between 1998 and 2003 (from 138 t to 483 t).

Botanicals are also declining, representing less than 1 percent of California's biopesticide use. Overall, the botanicals are hard to assume that they are best applied in wealthy countries in public health (mosquitoes and cockroaches) and consumers (home and garden). In underdeveloped nations, where farmers cannot afford conventional pesticides and where the traditional use of plants and plant derivatives to store the safety of products is well established, the true usefulness of botanical pesticides is more acknowledged. Although traditional pesticides (for example, through government aid) are available to farmers, a lack of literacy and protection equipment leads to thousands of poisonings every year.

Traditional West African plants that provide postharvest insect protection have received more attention. Some of the most effective plants employed are widely known for their active substances (e.g., Tephrosia rotenoids, Nicotiana nicotine, Securidaca methyl salicylate, and Ocimum eugenol), while others are volatile, which are a natural spray that destroys adult plagues and their descendants. Those materials are relatively stable in their current form, according to at least one evaluation.

Certain plants can effectively preserve grain against storage pests in developing nations. Many of these plants have a tropical spectrum and are possibly cultivable in underdeveloped nations. Pesticide efficacy in plant adoption is, however, only one element. The logistics of the processing, preparation, and consumption of botanical products will reduce their use [72]. Maybe, rather than screening new plants and insulate new bioactive compounds that pick up our interests, are not likely to be useful, this scientific community needs to focus its efforts on growing and applying existing botanicals.

13. Conclusion

The natural environment offers a multitude of different plant species which have helped develop cures for human, animal, and plant sicknesses. The use of synthetic pesticides is often questioned on environmental health, strict regulation of their use, and strict control on pesticide residues in agricultural produce demand are all required precautions to ensure that must be taken. Pesticides produced synthetically are still hazardous to both environmental health, animals, and human beings subject to toxic or otherwise hazardous chemicals that remain on the ground or in the atmosphere after their use. Concerning their regenerative nature and contribu-tion to human and environmental protection, botanicals must be reconsidered and their effectiveness in controlling crop pests.

Large-scale agriculture could be practiced in marginal lands where food is not in abundance to escape the competition with source plant extracts. The development of such crops in semiarid areas could benefit communities. Rhizomes and herbaceous plants may be grown in areas under a tree canopy of shortness but with minimum disturbance to the trees. Biochemical compounds that have pesticide properties in plants are produced through biotechnological collaborations.

The natural presence of insect-based plant compounds, as a precious alternative to synthetic or chemical pesticides, are botanical insecticides used for the protection of crops from negative or side effects in conventional insecticides. The chemical features of botanical pesticides, notably repellents, feeding dissuasive agents, toxicants, growth retardants and chemosterilants and attributes (essentials, flavonoids, alkaloids, glycoside, ester, and fatty acids), and their impact on insects in various forms. Instead of synthetic insecticides, botanical insecticides must be used, and organic cultivators in developed countries accept certain botanical insecticides. We, therefore, advocate the use and encouragement of botanical insecticidal products and research into new sources of botanical insecticides are being conducted.

Acknowledgements

The author is grateful to Research Scientist Hewa Lunuwilage Chamila Darshanee (Sri Lanka) for reviewing this chapter in the early stage. The authors would like to thank the Science and Technology Development (STDF), Egypt entitled: "Eco-friendly Pesticides against Pests of Medical, Veterinary, and Agricultural Importance" ID: 41608.

Notes/thanks/other declarations

We are thankfull to staff of department of agriculture for their support and encouragement.

Author details

Nazeer Ahmed[1*], Mukhtar Alam[1], Muhammad Saeed[1], Hidayat Ullah[1], Toheed Iqbal[2], Khalid Awadh Al-Mutairi[3], Kiran Shahjeer[4], Rafi Ullah[1], Saeed Ahmed[5], Nibal Abd Aleem Hassan Ahmed[6], Hanem Fathy Khater[7] and Muhammad Salman[8]

1 Department of Agriculture, University of Swabi, Anbar, Khyber Pakhtunkhwa, Pakistan

2 Faculty of Plant Protection, Department of Entomology, The University of Agriculture, Peshawar, Khyber Pakhtunkhwa, Pakistan

3 Faculty of Science, Dpartment of Biology, University of Tabuk, Tabuk, Saudi Arabia

4 Department of Zoology, Abdulwali Khan University, Mardan, Khyber Pakhtunkhwa, Pakistan

5 Agricultural Research Center, Londrina State University, Londrina, Brazil

6 Faculty of Sciences in Kurma, Department of Science, Taif University, Mecca, Saudi Arabia

7 Faculty of Veterinary Medicine, Department of Parasitology, Benha University, Toukh, Egypt

8 Department of Entomology, Faculty of Crop Protection, The University of Agriculture, Peshawar, KP-Pakistan

*Address all correspondence to: dr.nazeer@uoswabi.edu.pk

References

[1] Wafaa MH, Rowida SB, Hussein AHS. Botanical insecticide as simple extractives for pest control, Cogent Biology. **2017,** 3:1, 1404274. https://doi.org/10.1080/23312025.2017.1404274.

[2] Damalas CA, Koutroubas SD, Farmers' exposure to pesticides: toxicity types and ways of prevention, Toxics 1 (2015) 1-10. **2014,** 3; 150-151.

[3] Sande D, Mullen J, Wetzstein M, Houston J. Environmental impacts from pesticide use: a case study of soil fumigation in Florida tomato production. Int J Environ Res Public Health. **2011,** 12; 4649-4661.

[4] Mahmood I, Imadi SR, Shazadi K, Gul A, Hakeem KR. Effects of pesticides on environment, in: Plant, Soil and Microbes, Springer, Cham, **2016,** pp. 253-269.

[5] Shabana YM, Abdalla ME, Shahin AA, El-Sawy MM, Draz IS, Youssif AW. Efficacy of plant extracts in controlling wheat leaf rust disease caused by *Puccinia triticina*, Egypt. J Basic Appl Sci. **2017,** 1; 67-73.

[6] Wimalawansa SA, Wimalawansa SJ. Agrochemical-related environmental pollution: effects on human health, Glob. J Biol Agric Health Sci. **2014,** 3 72-83.

[7] Karpagam T, Devaraj A. Studies on the efficacy of Aloe vera on antimicrobial activity, Int. J. Res. Ayurveda and Pharm. **2011,** 4; 1286-1289.

[8] Thiruppathi S, Ramasubramanian V, Sivakumar T, Thirumalai AV. Antimicrobial activity of *Aloe vera* (L.) Burm. f. against pathogenic microorganisms. J Biosci Res. **2010, 4;** 251-258.

[9] Raja N. Botanicals: Sources for eco-friendly biopesticides. J. Biofertil Biopestic. **2014,** 5; e122.

[10] Nikkhah M, Hashemi M, Mohammad B, Habibi N, Farhoosh R. Synergistic effects of some essential oils against fungal spoilage on pear fruit. Int J Food Microbiol. **2017,** 285-294.

[11] Karaca G, Bilginturan M, Olgunsoy P. Effects of some plant essential oils against fungi on wheat seeds. Indian J Pharm Educ Res. 2017.

[12] Mishra RK, Bohra A, Kamaal N, Kumar K, Gandhi K, Sujayanand GK, Mishra M. Utilization of biopesticides as sustainable solutions for management of pests in legume crops: achievements and prospects, Egypt. J. Biol. Pest Control. **2018,** 28 (1); 3.

[13] Shabana YM, Abdalla ME, Shahin AA, El-Sawy MM, Draz IS, Youssif AW. Efficacy of plant extracts in controlling wheat leaf rust disease caused by *Puccinia triticina*, Egypt. J Basic Appl Sci. **2017,** 1; 67-73.

[14] Nashwa SMA, Abo-Elyousr AMK. Evaluation of various plant extracts against the early blight disease of tomato plants under green house and field conditions. Plant Prot Sci. **2012, 2;** 74-79.

[15] Business Daily. Chemical ban hits vegetable exports to the EU Market. 14 February 2013.

[16] Business Daily. (2014). Regulator suspends use of pesticides on vegetables. 13 October 2014.

[17] Neeraj GS, Kumar A, Ram S, Kumar V. Evaluation of nematicidal activity of ethanolic extracts of medicinal plants to *Meloidogyne incognita* (kofoid and white) chit wood under lab conditions, Int J Pure Appl Biosci. **2017,** 1; 827-831.

[18] Srijita D. Biopesticides: an eco-friendly approach for pest control,

World J. Pharm. Pharm. Sci. **2015**, 6; 250-265.

[19] Misra HP. Role of botanicals, biopesticides and bioagents in integrated pest management, Odisha Rev. **2014,** 62-67

[20] Erenso TF, Berhe DH. Effect of neem leaf and seed powders against adult maize weevil (*Sitophilus zeamais* Motschulsky) mortality. Agric Res. **2016**, 2; 90-94.

[21] Jawalkar N, Zambare S, Zanke S. Insecticidal property of *Datura stramonium* L. seed extracts against *Sitophilus oryzae* L. (Coleoptera: Curculionidae) in stored wheat grains. J Entomol Zool Stud. **2016,** 4; 92-96.

[22] Asif M, Tariq M, Khan A, Siddiqui MA. Biocidal and antinemic properties of aqueous extracts of *Ageratum* and *Coccinia* against root-knot nematode, *Meloidogyne incognita in vitro*. J Agric Sci. **2017**, 2; 108-122.

[23] Hikal WM, Baeshen RS, Said-Al Ahl HA. Botanical insecticide as simple extractives for pest control, Cogent Biol. **2017**, 3 (1); 1404274.

[24] S. Ali, S.M.H. Muhammad, A. Muneer, H. Faisal, F. Muhammad, H. Dilbar, S. Muhammad, G. Abdul, Insecticidal activity of turmeric (*Curcuma longa*) and garlic (*Allium sativum*) extracts against red flour beetle, *Tribolium castaneum*: a safe alternative to insecticides in stored commodities, J. Entomol. Zool. Stud. 3 (2014) 201-205.

[25] Isman MB. Bridging the gap: moving botanical insecticides from the laboratory to the farm, Ind. Crops Prod. **2017**, 110; 10-14.

[26] Isman MB, Grieneisen ML. Botanical insecticide research: many publications, limited useful data, Trends Plant Sci. **2014**, 19 (3); 140-145.

[27] *Mkenda PA, Stevenson PC, Ndakidemi P, Farman DI, Belmain SR. Contact and fumigant toxicity of five pesticidal plants against Callosobruchus maculatus (Coleoptera: Chrysomelidae) in stored cowpea (Vigna unguiculata),* Int J Trop Insect Sci. *2015, 35 (4); 172-184.*

[28] Stevenson PC, Isman MB, Belmain SR. Pesticidal plants in Africa: a global vision of new biological control products from local uses. Ind Crops Prod. **2017,** 110; 2-9.

[29] Philogène BJR, Regnault-Roger C, Vincent C. Botanicals: yesterday's and today's promises. In: Regnault-Roger C, Philogène BJR, Vincent C (eds) Biopesticides of plant origin. Lavoisier and Andover, **2005,** UK, pp 1-15.

[30] Copping LG. The biopesticide manual, 2nd edn. British Crop Protection Council, Farnham, 2001, p 528.

[31] Regnault-Roger C, Philogène BJR, Vincent C (eds). Biopesticides of plant origin. Lavoisier, Paris, **2005,** p 313.

[32] Ahmad W, Shilpa S, Sanjay K. Phytochemical Screening and antimicrobial study of *Euphorbia hirta* extracts. J Med Plant Stud. **2017,** 2; 183-186.

[33] Gakuubi MM, Wanzala W, Wagacha JM, Dossaji SF. Bioactive properties of *Tagetes minuta* L. (Asteraceae) essential oils. a review, Am J Essent Oil Nat Prod. **2016,** 2; 27-36.

[34] Vidyasagar GM, Tabassum N. Antifungal investigations on plant essential oils. a review, Int J Pharm Pharm Sci. **2013,** 2; 19-28.

[35] Chougule PM, Andoji YS. Antifungal activity of some common medicinal plant extracts against soil borne phytopathogenic fungi *Fusarium oxysporum* causing wilt of tomato, Int. J. Dev. Res. **2016,** 6 (3); 7030-7033.

[36] *Zarubova L, Lenka K, Pavel N, Miloslav Z, Ondrej D, Skuhrovec J. in: Botanical Pesticides and Their Human Health Safety on the Example of Citrus Sinensis Essential oil and Oulema Melanopus Under Laboratory Conditions, Mendel Net.* **2014**, *pp. 330-336.*

[37] Dosemeci M, Alavanja MC, Rowland AS, Mage D, Zahm SH, Rothman N, Lubin JH, Hoppin JA, Sandler DP, Blair A. A quantitative approach for estimating exposure to pesticides in the agricultural health study. Ann Occup Hyp. **2002,** 46: 245-260.

[38] Arcury TA, Quandt SA, Bart DB, Hoppin JA, Mc auley, Grzy-wacs JG, Robson MG. Farmworker exposure to pesticides: methodology issues for the collection of comparable data. Environ Health Persp. **2006,** 11:923-928.

[39] Srijita D. Biopesticides: an eco-friendly approach for pest control, World J. Pharm. Pharm. Sci. **2015**, 6; 250-265.

[40] Castillo-Sánchez LE, Jiménez-Osornio JJ, Delgado-Herrera MA, Candelaria-Martínez B, Sandoval-Gío JJ. Effects of the hexanic extract of neem *Azadirachta indica* against adult whitefly *Bemisia tabaci*, J. Entomol Zool Stud. **2015,** 5; 95-99.

[41] Joseph B, Sujatha S. Insight of botanical based biopesticides against economically important pest, Int. J. Pharm. Life Sci. **2012,** 11; 2138-2214.

[42] Gaikwad RS, Kakde RB, Kulkarni AU, Gaikwad DR, Pancha VH. *In vitro* antimicrobial activity of crude extracts of *Jatropha* species. Curr. Bot. **2012**, 3; 09-15.

[43] Hernandez-Moreno D, Casa-Resino I, Lopez-Beceiro A, Fidalgo LE, Soler F, Perez-Lopez M. Secondary poisoning of non-target animals in an Or-nithological Zoo in Galicia (NW Spain) with anticoagulant rodenticides: a case report, Vet. Med. (Praha) **2013,** 10; 553-559.

[44] Nawaz M, Juma M, Hongxia H. Current status and advancement of biopesticides: microbial and botanical pesticides. J Entomol Zool Stud. 2016, 2; 241-246.

[45] Arafat Y, Shahida K, Wenxiong L, Changxun F, Sehrish S, Niaz A, Saadia A. Allelopathic evaluation of selected plants extract against broad and narrow leaves weeds and their associated crops, Acad. J. Agric. Res. **2015,** 10; 226-234.

[46] Sarkar E, Samarendra NC, Chakraborty P. Allelopathic effect of *Cassia tora* on seed germination and growth of mustard, Turk. J. Bot. **2012,** 36; 488-494.

[47] Gurjar MS, Shahid A, Akhtar M, Kangabam SS. Efficacy of plant extracts in plant disease management, Agric. Sci. **2012,** 3; 425-433.

[48] Kushram T, Yadu YK, Sahu MK, Kulmitra AK, Kumar R. Bio efficacy of botanical insecticides against defoliators pests on soybean, Int. J. Curr. Microbiol. Appl. Sci. **2017**, 3; 2196-2204.

[49] Kareru P, Rotich ZK, Maina EW. Use of Botanicals and Safer Insecticides Designed in Controlling Insects: The African Case, In Tech, **2013.**

[50] Nefzi A, Abdallah BAR, Jabnoun-Khiareddine H, Saidiana-Medimagh S, Haouala R, Danmi-Remadi M. Antifungal activity of aqueous and organic extracts from *Withania somnifera* L. against *Fusarium oxysporum* f.sp. *radicis-lycopersici*, J. Microbial Biochem. Technol. **2016**, 3; 144-150.

[51] Damalas CA. Safe food production with minimum and judicious use of

pesticides, in: Food Safety, Springer, Cham, 2016, pp. 43-55.

[52] Giner M, Avilla J, Balcells M, Caccia S, Smagghe G. Toxicity of allyl esters in insect cell lines and in *Spodoptera littoralis* larvae. Archives of Insect Biochemistry and Physiology, **2012,** *79*(1), 18-30. https://doi.org/10.1002/arch.2012.79.issue-1.

[53] Schmidt S, Tomasi C, Pasqualini E, Ioriatti C. The biological efficacy of pear ester on the activity of Granulosis virus for codling moth. Journal of Pest Science. 2008, 81, 29. https://doi.org/10.1007/s10340-007-0181-x

[54] Velu K, Elumalai D, Hemalatha P, Babu M, Janaki A, Kaleena PK. Phytochemical screening and larvicidal activity of peel extracts of *Arachis hypogaea* against chikungunya and malarial vectors. International Journal of Mosquito Research. 2015, *2*(1); 01-08.

[55] Yousef H, EL-Lakwah SF, EL Sayed, Y. A. (2013). Insecticidal activity of linoleic acid against Spodoptera littoralis (BOISD.). Egyptian Journal of Agricultural Research, 91(2), 573.

[56] Acheuk F, Doumandji-Mitiche B.. Insecticidal activity of alkaloids extract of *Pergularia tomentosa* (Asclepiadaceae) against fifth instar larvae of *Locusta migratoria* cinerascens (Fabricius 1781) (Orthoptera: Acrididae). International Journal of Science and Advanced Technology. 2013; *3*(6), 8-13

[57] Kumar P, Bhadauria T, Mishra J. Impact of application of insecticide quercetin/azadirachtin and chlorpyrifos on earthworm activities in experimental soils in Uttar Pradesh India. Science Postprint. **2015,** *1*(2), e00044.

[58] Goławska S, Sprawka I, Łukasik I, Goławski A. Are naringenin and quercetin useful chemicals in pest-management strategies? Journal of Pest Science. **2014,** *87*; 173-180. https://doi.org/10.1007/s10340-013-0535-5

[59] Velu K, Elumalai D, Hemalatha P, Babu M, Janaki A, Kaleena PK. Phytochemical screening and larvicidal activity of peel extracts of Arachis hypogaea against chikungunya and malarial vectors. International Journal of Mosquito Research, (2015). 2(1), 01-08.

[60] Acheuk F, Doumandji-Mitiche B. Insecticidal activity of alkaloids extract of *Pergularia tomentosa* (Asclepiadaceae) against fifth instar larvae of *Locusta migratoria* cinerascens (Fabricius 1781) (Orthoptera: Acrididae). International Journal of Science and Advanced Technology. **2013,** *3*(6); 8-13.

[61] Regnault-Roger C, Philogène BJR. Past and current prospects for the use of botanicals and plant allelochemicals in integrated pest management. Pharmaceutical Biology. **2008,** *46*; 41-52. https://doi.org/10.1080/13880200701729794.

[62] Dadang EDF, Djoko P. Effectiveness of two botanical insecticide formulations to two major cabbage insect pests on field application, Int. Soc. Southeast Asian Agric. Sci. **2009,** 1; 42-51.

[63] Regnault-Roger C, Vincent C, Arnason JT. Essential oils in insect control: Low-risk products in a high-stakes world. *Annual Review of Entomology.* **2012,** *57*; 405-424. https://doi.org/10.1146/ annurev-ento-120710-100554.

[64] Sithisut D, Fields PG, Chandrapathya A. Contact toxicity, feeding reduction and repellency of essential oils from three plants from the ginger family (Zingiberaceae) and their major components against *Sitophiluszeamais* and *Tribolium castaneum*. J Stor Prod. **2011,** *104*; 1445-1454.

[65] Tripathi AK, Prajapati V, Khanuja SPS, Kumar S. Effect of

d-limonene on three stored-product beetles. Journal of Economic Entomology. **2003,** *96*; 990-995. https://doi.org/10.1093/jee/96.3.990.

[66] Yamamoto-Ribeiro MMG, Renata G, Cássia YK, Flavio DF, Simone AGM, Expedito LS, Benicio AAF, Mikcha JMG, Machinski JM. Effect of *Zingiber officinale* essential oil on *Fusarium verticillioides* and fumonisin production, Food Chem. **2013,** 141; 3147-3152.

[67] Stevenson PC, Isman MB, Belmain SR. Pesticidal plants in Africa: a global vision of new biological control products from local uses, Ind. Crops Prod. **2017,** 110; 2-9.

[68] Yousef H, EL-Lakwah SF, EL Sayed YA. Insecticidal activity of linoleic acid against *Spodoptera littoralis* (BOISD.). *Egyptian Journal of Agricultural Research.* **2013,** *91*(2), 573.

[69] Lin QS, Chen SH, Hu MY, Haq MU, Yang L, Li H. Biodegradation of cypermethrin by a newly isolated actinomycetes HU-S-01 from wastewater sludge, Int. J. Environ. Sci. Technol. 2011, 1; 45-56.

[70] IPPC F. ISPM No 5: glossary of phytosanitary terms, Int. Stand. Phytosanit. Meas. (1) (2006) 57-79.

[71] Stanojkovic T,´ Kolundžija B, Ciric B,´´ Sokovic M,´ Nikolic D, ´Kundakovic T.´ Cytotoxicity and antimicrobial activity of *Satureja kitaibelii* (Wierzb. ex Heuff) (Lamiaceae), Digest J. Nanomater. Biostruct. **2013,** 2; 845-854.

[72] Morse S, Ward A, McNamara N, Denholm I (2002) Exploring the factors that influence the uptake of botanical insecticides by farmers: a case study of tobacco-based products in Nigeria. Exp Agric 38:469-479

[73] Liao, M., Xiao, J. J., Zhou, L. J., Yao, X., Tang, F., Hua, R. M., ... Cao, H. Q. (2017). Chemical composition, insecticidal and biochemical effects of Melaleuca alternifolia essential oil on the Helicoverpa armigera. Journal of Applied Entomology, 2017, 1-8

14

Nano Pesticides Application in Agriculture and their Impact on Environment

Malik Asif, Shayesta Islam, Mushtaq A. Malik, Zaffar Mahdi Dar, Amjad Masood, Saima Shafi, Bisma Rashid and Showkat Sidique

Abstract

Environmental contamination and the tolerance developed by the pests, pathogens are some of the environmental issues related to the aimless utilization of chemical pesticides. It has became matter of serious concern for environment, food quality and soil health. Nanotechnology, envisaged as a swiftly emerging field has capability to reform food systems in agriculture. Nanotechnology provides an imperishable solution to these problems by the establishment of nano-pesticides. The functional components or the conveyor molecules used are of nano size. The performance of these nano sized particles is much better the traditional pesticides, as the smaller size aids in proper spreading on the pest surface. Amelioration in solubility of operational components, betterment in stability of formulation, gradual liberation of operational components and enhancement in mobility are some of the paramount advantages of nano particles attributed to the minute size of particles and greater surface area. Thus, nano particles have strengthened activity against target pests in comparison to bulk materials. Furthermore, nano-formulations sustain productive use in agriculture by offering systemic properties, uniform leaf coverage and enhanced soil properties. Despite all the positive aspects, it might have certain negative effects as well, like exposure of humans through distinct routes Viz, exposure to nano pesticides either directly or indirectly like adsorption through skin, or inhalation while breathing air or transfer from one energy level to another by taking contaminated food and water.

Keywords: nanopesticide, formulations, environmental risk, agriculture, future aspect

1. Introduction

Pest management is still greatly dependent on the utilization of pesticides, through the application of organic chemical originated components on crops, products and urban habitat, regardless of divergent alternative methods available. Most of the registered pesticides intrude with the nervous system of insects through their prime method of action, that is they are neurotoxic, and are considered to enhance the threat of expansion of various neurodegenerative diseases, such as

Parkinson's disease [1]. In order to lessen the utilization of neurotoxic compounds, various novel products have been launched during recent times in the market, which include insect killing agents like, destroying agents of adenosine triphosphate (ATP) or controlling agents of insect growth, in spite of this environmental affect is a matter of great concern. In this connection, the matter has been correlated to the lethality caused to mammals, environmental pollution, and biomagnifications in the food chain. Along with amplified tolerance rate developed by insects species to currently used products, the aforesaid concerns have became fundamental challenges in agriculture, and may significantly restrict the beneficial active compounds. To tackle this trouble, there is need of novel pest control strategies for pest management through the initiation of novel ideas related to pest tolerance and advancement in technologies. Nanotechnology has come up with as considerably alluring field of research to achieve these aims, by furnishing advanced methodologies for designing novel operational components with microscopic measurements, formation procedures and dissemination, which are collectively referred to as "nanopesticides." Nanopesticide research, is the introduction of nanotechnology to protect crops. This field encompasses extensive research aspects viz.,

- Learning the basic interrelation notion of nano-scale materials.
- Preparation of nanoemulsions from active components and their dispersal using pesticides.
- Utilizing nanomaterials as active agents to develop novel nanopesticide formulations.
- Using these nanomaterials as nanocarriers for pesticide delivery [2–4].

This nanopesticide research is comprehensive and is believed to address the major drawbacks of the existing pest management strategies and comes up with new advanced nano-based formulations that stay stable and active, that is not influenced by sun, heat, and rain in the target environment, enter the target organism (insect), resist defense of the pest, remain amiable to plants and mammals, formulation and manufacture is lucrative, and ideally possess a new-fangled approach of action [2, 5, 6]. NPs in diverse fashion and chemical makeup including metal, metal oxides, semiconductor quantum dots (QDs), carbon, ceramics, silicates, lipids, polymers, proteins, dendrimers, and emulsions were made by utilizing a wide range of materials synthesized or used from natural materials [7, 8]. Some familiar advantages of NP-based pesticide formulations include:

- Solubility of water-insoluble operative constituents is enhanced.
- Stability of the formulation is elevated.
- Eradicates dangerous organic solvents.
- Ability to liberate operative constituents slowly.
- Early deterioration is averted by improvement in stability.
- High mobility and insecticidal activity is attributed to nano size
- Endurance is believed to get increased due to large surface area [9].

The ideas related to advancement in the manufacturing of nano based formulations are

1. The advancement in traditional pesticide formulations.

2. The establishment of delivery systems.

3. Utility of NPs as nano carriers, and use of solid NPs as operative agents.

2. Nano pesticide formulations

Utilization of technologies such as encapsulation and restrained liberation methods for the use of pesticides is of remarkable concern. Various researchers have laid emphasis on the importance of Nanocapsules and nanoparticles for agricultural purposes [10]. A number of companies have prepared different formulas that comprise of NPs with the dimensions ranging from 100 to 250 nm. A few make use of Nanoemulsions, which are prepared by dissolving nanoparticles pesticides evenly either in water or oil with dimensions ranging from 200 to 400. The manifold applications can be achieved like protective measures, treatment, or preventing reaped product by easily integrating these emulsions into gels, creams, liquids. One of the modern desired restricted liberation of agricultural chemicals is the utilization of materials incorporated with silica. Porous hollow silica NPs (PHSN) was used as pesticide carrier to record the restricted discharge pattern of pesticide namely avermectin [11]. It was revealed that PHSNs can be utilized to restricting pesticide dissemination applications, as these carriers noticeably deferred the release of the pesticide. NPs can imbibe and bind with other compounds effortlessly because of large surface areas, flow without difficulty in lepidopteran systems and certainly be utilized for pesticide development [12]. It has been reported that most terpene compounds have antifeedant action and are extremely volatile. Insecticidal activity and life span of the formulations significantly enhanced by integrating certain plant extracts with nano-silica [13]. Formulations made by integrating apinene and linalool with nano-silica improved biological activity of the plant pure chemicals, stability of the formation with superior zeta potential, restricted liberation of the plant derived components compound, and prolonged life span of the plant isolates [13]. These formulations exhibit great antifeedant action against *Spodoptera litura* and *Achaea Janata* (L. These nanoformulations smoothly disseminate, which was made clear by the dissemination dispersion studies. It was concluded that life span analysis of nano formulations with terpenes have no impact on dissemination, dimensions, zeta potential, or biological activity of the nanoformulations up to period of 6 months after harvesting [3, 13].

3. Solid nanoparticles as nanopesticides

A number of nanoparticles have been recommended as potential candidates both as nano carriers as well as operational pesticide agencies or biological pesticides, because of intrinsic antibacterial/pesticidal properties.

3.1 Inert dusts as insecticides

The mechanism of activity of stable dusts such as silica, alumina, and clay is due to the destruction of the waxy coating on the insect's epidermis by surface assimilation and scraping. This destruction trigger insects to dehydrate, dry out and die. Because the mechanism of action is physical, it is difficult to acquire insect resistance [14].

3.1.1 Silica

The utility of silica based nanoparticles (SNPs) for controlling pests in agricultural is reasonably recommended because silicon has the capacity to escalate the resistance of plants to abiotic and biological stresses [12]. The deadly effects of unstructured water loving, and oil loving SNPs to the rice weevil *Sitophilus oryzae* were studied by [15] and found that they were much more helpful and efficient than bulk silica. No new infections were detected after depositing for period of 60 days. Silica particles are physically absorbed by lenticular lipids that destroy the protective cuticle and destroy insects purely by physical means with a similar mode of action observed on diatoms particles used for protection of stored grain [12, 16]. In addition, modified hydrophobic surface-charged silica (3-5 nm) have been fruitfully used in veterinary medicine to control a wide range of essential agricultural and animal ectoparasites [17]. The fungal growth was decreased and germination of seeds was promoted by effective application of surface charged silica as a thin film of seeds [18]. The metabolic process of certain agricultural crop groups and horticultural plants does not get affected by the superficial application of NPs to the leaves and stem. They do not encourage any modification in gene expression in insect organs, so they deserve approval as nano pesticides. Additionally, World Health Organization (WHO) declared the utilization of unstructured silica as a nanopesticide risk free for human consumption. In traditional pesticides these SNPs also serve as carriers. Comparison to small amounts of chlorfenapyr, the biological effectiveness was enhanced by loading the insecticide chlorfenapyr into dispersed SNPs [19]. This is assigned to the nanoscale size and intrinsic insecticidal activity of SNPs. In addition, it has been reported that porous hollow SNPs can safeguard the fleeting insecticide avermectin from UV degradation and liberate it for a long time. SNP carriers have been observed to promote the sustained release of avermectin for approximately 30 days [20].

3.1.2 Diatomaceous earth (DE)

DE is comprised predominantly of unstructured silica originated by decomposing small sized plants under high temperature and pressure. These are employed to safeguard stored grain from pest insects. It has deleterious effects on grain properties especially on bulk density because it is required in high doses. Researchers are engaged to integrate DEs with other insecticides to attain high efficiency, so as to make it effective in low doses. For example, combination of DE and plant decoction bitterbarkomycin was used to control grain pest *Rhyzopertha dominica*, and was recorded to be efficient at low dose concentration of 150 ppm [21].

3.2 Alumina

Nanostructured alumina (NSA) was recognized as a successful insecticide in case of grain pests, *S. oryzae* (L) and *R. dominica* (F) by [22]. They also collated NSA formulations against a foremost traditional DE commodity and declared their formulation to be efficient. NSA dusts manufactured using a modified glycine-nitrate combustion process were reported to be more effective against, *S. oryzae* (L) and *R. dominica* (F) in the recent research. The results revealed that, key factors determining insecticidal efficacy are dimension of particle, surface area, and external features. However, these are not believed to be only factors responsible, and to attain superior outcomes for specific species, there is need to improve preparation pathways [23].

3.3 Clays

Nanoclays are fine layers of silicate minerals with following dimensions (thickness 1 nm and width 70–150 nm). Source of nanoclays is montmorillonite clays, which are mostly found in volcanic ash, formed by size reduction and surface modification and are biocompatible and have less hazardous associated. Among these living materials anionic clays are found to be most promising. They have fruitfully served as conveyor for the α-naphthalene acetate, which is regulator for growth in plants and for the restrained liberation of the 2,4-dichlorophenoxyacetate. Natural antibiotic cinnamate, which is utilized for pest management, is susceptible to quick deterioration in soil and is required in significant concentration. It was declared that by loading it on double layered hydroxide low pace liberation of the antibiotic and prolonged longevity in the soil is achieved [24]. This indicates the outstanding potential of nano-clays to be used for low pace/target specific dissemination of pesticides and fungicides.

4. Preparation of nano pesticides by encapsulating pesticide nano particles

Encapsulation is a process of encompassing or enveloping one substance inside another fabric on a very little scale or may be characterized as the method by which a material is encompassed and safe guarded from external conditions that would break it down, yielding capsules extending from less than one micrometer to several hundred micrometers in size [25]. Encapsulation can be obtained by few strategies with distinctive purposes in mind. Encapsulation of materials may take place deliberately that the central material be restricted inside capsule walls for a particular period of time. On the other hand, central substances can be incorporated so that the core material will be released either progressively through the capsule walls, known as controlled release or diffusion, or when outside conditions activate the capsule walls to break, melt, or dissolve [26]. The different areas in which encapsulated materials have numerous applications include agriculture, pharmaceuticals, foods, cosmetics and fragrances, textiles, paper, paints, coatings and adhesives, printing applications, and many other industries. Encapsulation permits the farmers to apply pesticides less frequently instead of requiring profound concentration and toxic initial applications to be released over time, allowing farmers to apply the pesticides less often rather than requiring very highly concentrated and perhaps toxic initial applications followed by repeated applications to fight the losses due to leaching, evaporation, and degradation [27]. Protection of pesticides from full contact with the components reduces the risk to the environment and those that might be exposed to the chemicals and provides a more efficient technique for pest control [28].

5. Toxicity and environmental impacts

The vulnerability and environmental consequences of nanoscale-based formulations is of great concern, which needs to be addressed. The amount of nanoformulations in soil, surface water, and groundwater, and effects impacts on non-target organisms have not been predicted much. The fate of nano formulations depends on number of chemical parameters, like pH, ionic strength, and dissolved molecules in the media [29]. Emphasis is laid on certain fundamental concerns mentioned under following sections.

5.1 Direct toxicity on humans and environment

The harmfulness of numerous pesticides utilized in a nano-scale formulation has been determined by specialists, the probable negative impact and the probable impacts of nano dimensions must be taken in to consideration during nano-scale formulations. More inquiry is needed to affirm the harmfulness of nano-scale formulations and the factors contributing to harmfulness of nano-particles like size, charge, shape, and chemistry.

5.2 Durability and persistence

The fundamental criteria for analyzing dangers associated with the application of nano pesticides is their longevity in the environment. The non target organisms get exposed when there is suspended liberation or release for a prolonged period of time [30].

5.3 Bioavailability

The nanocarriers promote transfer of some of the immobilized operational components, which is considered one of the likely detrimental effects. Thus increase the availability to organisms [31]. Enhancement in absorption by target organisms have been reported by certain nanoformulations. It must be ensured that no damage caused to non-target organisms.

5.4 Release profile/degradation

The two important factors which govern the liberation of operational components involve class of nanocarrier and the dissemination of active components in matrix. Research suggested that environmental affect and virulence of operational ingredients is contributed by set of factors viz., surface desorption, dissemination via the polymer matrix, and deterioration of the polymer [31].

5.5 Fate of the carrier

The centre of attention of research now are carriers established using natural biologically degradable polymers like, polysaccharides, or lipids, which deteriorate into by-products of least concern [13].

6. Nanopesticides interaction with other environmental contaminants

The evaluation of pesticides is done on a product-by-product basis, in a similar fashion to other contaminants. In the real environment, mixed interactions are possible as they cooccur with other substances. A nanopesticide may have collaborative interaction with other contaminants through the Trojan-horse effect, so hazard analysis of environmental mixtures is generally not needed for product hazard estimation [24]. In this effect the substance are carried to a tissue of organism or to an organism by interaction of substance with ENP, due to which the interior exposure to contaminants is enhanced, and might not get accumulated. The need of hour is to integrate these interrelations into authoritative hazard analyzing schemes.

7. Nano-pesticides for greener agriculture

Nano-pesticides are defined as preparations that deliberately introduce elements in the nm size range and the characteristics related with this small size range are new, these nanopesticides have earlier been launched in the market. Nanopesticides cannot be treated as a single category but have been blended in a numerous products. The various classifications of Nanopesticides include organic components viz., A.I., polymers and inorganic components like oxides of metals in diverse configurations like particles and micelles [32]. One of the most economical and multifaceted method of containing insect pests is the residues remaining on the surface after application. Microencapsulation; a nanotechnological approach can be employed to modify the insecticidal value by protecting the operational components from environmental conditions and by promoting persistence. The usage of engineered nanoparticles (ENPs) is gaining considerable attention in the pesticide sector due to the establishment of a wide varieties of plant protection products termed as "nanopesticides". Nanopesticides "involve either very small particles of a pesticide active ingredient (A.I) or other small engineered structures with useful pesticidal properties". The advantages offered by this emerging technology involves amplified efficacy, durability, and a cutting down the amount of active ingredient required. Nanoemulsions, nanocapsules like polymer and commodities containing pristine engineered nanoparticles, such as metals and their oxides, and nano-clays have been recommended. These products can be exploited to intensify the efficiency of prevailing pesticide operational components or results in improvement of environmental safety profiles or both [30].

8. Nano pesticide: future possibilities

Recent investigations have shown that nano-pesticides can reduce the deleterious effects of chemical based pesticides and furnish target-specific control of pests, and help develop intelligent nano-systems for minimizing problems like environmental imbalance, and negative effects on food security, and crop productivity [33]. They are effective for long term utility and provide solution to environmental related problems like nutrient richness in water bodies and accumulation non-biodegradable components in the food chain due to restricted liberation of operational ingredients. Furthermore, nano-pesticides show efficient pest control property due to amplified solubility and stabilities of operational components [34]. Still, there is necessity to modify the techniques in order to have remarkable benefits to agriculture. A few aspects for nano-scale pesticide delivery platforms discussed by [33] in their review include:

- The effectiveness of nanopesticide development is enhanced by use of green chemistry and environmentally viable principles [35].
- Upgrading traditional utilization of nanopesticide.
- Realistic utilization should be estimated at field level by comparing it with traditionally used products.

- The vulnerability of nanopesticides is estimated by Environmental impact assessment.
- Modification in the policy for application of nano material in agriculture.
- Agrochemical industries will come up with many solutions by the launching of smart nanopesticide, that is solubility of the operational components, stability, restricted liberation, and dissemination of active components on specific organisms, but lot of research need to be done to get acquainted with the fate of nanopesticide in the environment.

Author details

Malik Asif[1*], Shayesta Islam[2], Mushtaq A. Malik[1], Zaffar Mahdi Dar[1], Amjad Masood[3], Saima Shafi[1], Bisma Rashid[1] and Showkat Sidique[4]

1 FOA-Wadura, Division of Basic Sciences and Humanities, SKUAST-Kashmir, India

2 FOH-Shalimar, Division of Environmental Sciences, SKUAST-Kashmir, India

3 FOA-Wadura, Division of Agronomy, SKUAST-Kashmir, India

4 FOA, Wadura, Division of Agriculture Statistics and Economics, SKUAST-Kashmir, India

*Address all correspondence to: drasif_skuast@yahoo.com

References

[1] Ascherio A, Chen H, Weisskopf MG, Reilly EO, McCullough ML, Calle EE, et al. Pesticide exposure and risk of Parkinson's disease. Annals of Neurology. 2006;**60**:197-203

[2] Smith K, Evans DA, El-Hiti GA. Role of modern chemistry in sustainable arable crop protection. Philosophical Transactions of the Royal Society. 2008;**363**:623-637

[3] Usha RP, Madhusudhanamurthy J, Sreedhar B. Dynamic adsorption of a-pinene and linalool on silica nanoparticles for enhanced antifeedant activity against agricultural pests. Journal of Pest Science. 2014;**87**:191-200

[4] Benelli G, Lukehart CM. Special issue: Applications of greensynthesized nanoparticles in pharmacology, parasitology and entomology. Journal of Cluster Science. 2017;**28**(1):1-2

[5] Benelli G. Plant-mediated biosynthesis of nanoparticles as an emerging tool against mosquitoes of medical and veterinary importance: A review. Parasitology Research. 2016a;**115**(1):23-34

[6] Benelli G. Green synthesized nanoparticles in the fight against mosquito-borne diseases and cancer—a brief review. Enzyme and Microbial Technology. 2016b;**95**:58-68

[7] Niemeyer CM, Doz P. Nanoparticles, proteins, and nucleic acids: biotechnology meets materials science. Angewandte Chemie International Edition.2001;**40**:4128-4158

[8] Oskam G. Metal oxide nanoparticles: Synthesis, characterization and application. Journal of Sol-Gel Science and Technology. 2006;7:161-164

[9] Sasson Y, Levy-Ruso G, Toledano O, Ishaaya I. Nanosuspensions: Emerging novel agrochemical formulations. In: Ishaaya I, Nauen R, Horowitz AR, editors. Insecticides Design Using Advanced Technologies. Springer: Berlin; 2007. pp. 1-32

[10] Pavel A, Creanga DE. Chromosomal aberrations in plants under magnetic fluid influence. Journal of Magnetism and Magnetic Materials. 2005;**289**: 469-472

[11] Wen LX, Li ZZ, Zou HK, Liu AQ, Chen JF. Controlled release of avermectin from porous hollow silica nanoparticles. Pest Management Science. 2005;**61**:583-590

[12] Barik TK, Sahu B, Swain V. Nanosilica—from medicine to pest control. Parasitology Research. 2008;**103**:253-258

[13] Madhusudhanamurthy J, Usha Rani P, Sambasiva RKRS. Organic–inorganic hybrids of nano silica and certain botanical compounds for their improved bioactivity against agricultural pests. Current Trends in Biotechnology and Pharmacy. 2013;**7**:615-624

[14] Shah MA, Khan AA. Use of diatomaceous earth for the management of stored-product pests. International Journal of Pest Management. 2014; **60**(2):100-113

[15] Debnath N, Das S, Seth D, Chandra R, Bhattacharya S, Goswami A. Entomotoxic effect of silica nanoparticles against Sitophilus oryzae (L). Journal of Pest Science. 2011;**84**:99-105

[16] Korunić Z. Diatomaceous earth, a group of natural insecticide. Journal of Stored Products Research. 1998;**34**:87-97. DOI: 10.1016/S0022-474X(97)00039-8

[17] Ulrichs C, Mewis I, Goswami A. Crop Diversification Aiming Nutritional

Security in West Bengal: Biotechnology of Stinging Capsules in Nature's Water-Blooms. Ann Arbor: Ann Tech Issue of State Agri Technologists Service Assoc; 2005. pp. 1-18

[18] Robinson D, Salejova-Zadrazilova G. Observatory nano. In: Nano Tachnologies for Nutrient and Biocide Delivery in Agricultural Production. Working Paper Version. 2010

[19] Song MR, Cui SM, Gao F, Liu YR, Fan CL. Dispersible silica nanoparticles as carrier for enhanced bioactivity of chlorfenapyr. Journal of Pesticide Science. 2012;**37**(3):258-260

[20] Li ZZ, Chen JF, Liu F, Liu AQ, Wang Q, Sun HY, et al. Study of UV-shielding properties of novel porous hollow silica nanoparticle carriers for avermectin. Pest Management Science. 2007;**63**(3):241-246

[21] Athanassiou CG, Korunic Z. Evaluation of two new diatomaceous earth formulations enhanced with abamectin and bitterbarkomycin, against four stored-grain beetle species. Journal of Stored Products Research. 2007;**43**(4):468-473

[22] Stadler T, Buteler M, Weaver DK, Sofie S. Comparative toxicity of nano structured alumina and a commercial inert dust for Sitophilus oryzae (L.) and Rhyzopertha dominica (F.) at varying ambient humidity levels. Journal of Stored Products Research. 2012;**48**:81-90

[23] Buteler M, Sofie SW, Weaver DK, Driscoll D, Muretta J, Stadler T. Development of nanoalumina dust as insecticide against Sitophilus oryzae and Rhyzopertha dominica. International Journal of Pest Management. 2015;**61**(1):80-89

[24] Park EJ, Yi J, Kim Y, Choi K, Park K. Silver nanoparticles induce cytotoxicity by a Trojan-horse type mechanism. Toxicology In Vitro. 2010;**24**(3):872-878

[25] Del Carmen Gimenez-López M, Moro F, Torre AL, Gómez-García CJ, Brown PD, Slageren JV, et al. Encapsulation of single-molecule magnets in carbon nanotubes. Nature Communications. 2011;**2**:407

[26] Doane WM. Encapsulation of pesticides in starch by extrusion. Industrial Crops and Products. 1992;**1**:83-87

[27] Lewis DH, Cowsar DR. Principles of controlled release pesticides. In: Scher HB, editor. Controlled Release Pesticides. Washington, DC: American Chemical Society; 1977. pp. 1-16

[28] Ulbricht H, Hertel T. Dynamics of C60 encapsulation into single-wall carbon nanotubes. The Journal of Physical Chemistry. B. 2003;**107**:14185-14190

[29] Kookana RS, Boxall ABA, Reeves PT, Ashauer R, Beulke S, Chaudary Q, et al. Nanopesticides: Guiding principles for regulatory evaluation of environmental risks. Journal of Agricultural and Food Chemistry. 2014;**62**:4227-4240

[30] Kah M, Beulke S, Tiede K, Hofmann T. Nano-pesticides: State of knowledge, environmental fate and exposure modelling. Critical Reviews in Environmental Science and Technology. 2013;**43**(16):1823-1867

[31] Kah M, Hoffmann T. Nano pesticide research. Current trends and future priorities. Environment International. 2014;**63**:224-235

[32] Ragaei M, Sabry AH. Nanotechnology for insect pest control. International Journal of Science, Environment and Technology. 2014;**3**:528-545

[33] Nuruzzaman M, Rahman MM, Liu Y, Naidu R. Nanoencapsulation, nano-guard for pesticides: A new

window for safe application. Journal of Agricultural and Food Chemistry. 2016;**64**(7):1447-1483

[34] Venugopal NVS, Sainadh NVS. Novel Polymeric Nanoformulation of Mancozeb – An Eco-Friendly Nanomaterial. International Journal of Nanoscience [Internet]. World Scientific Pub Co Pte Lt; Aug 2016;**15**(04):1650016. Available from: http://dx.doi.org/10.1142/s0219581x16500162

[35] Ocsoy I, Paret ML, Ocsoy MA, Kunwar S, Chen T. Nanotechnology in plant disease management: DNA-directed silver nanoparticles on graphene oxide as an antibacterial against Xanthomonasperforans. ACS Nano. 2013;**7**(10):8972-8980

Permissions

All chapters in this book were first published by InTech Open; hereby published with permission under the Creative Commons Attribution License or equivalent. Every chapter published in this book has been scrutinized by our experts. Their significance has been extensively debated. The topics covered herein carry significant findings which will fuel the growth of the discipline. They may even be implemented as practical applications or may be referred to as a beginning point for another development.

The contributors of this book come from diverse backgrounds, making this book a truly international effort. This book will bring forth new frontiers with its revolutionizing research information and detailed analysis of the nascent developments around the world.

We would like to thank all the contributing authors for lending their expertise to make the book truly unique. They have played a crucial role in the development of this book. Without their invaluable contributions this book wouldn't have been possible. They have made vital efforts to compile up to date information on the varied aspects of this subject to make this book a valuable addition to the collection of many professionals and students.

This book was conceptualized with the vision of imparting up-to-date information and advanced data in this field. To ensure the same, a matchless editorial board was set up. Every individual on the board went through rigorous rounds of assessment to prove their worth. After which they invested a large part of their time researching and compiling the most relevant data for our readers.

The editorial board has been involved in producing this book since its inception. They have spent rigorous hours researching and exploring the diverse topics which have resulted in the successful publishing of this book. They have passed on their knowledge of decades through this book. To expedite this challenging task, the publisher supported the team at every step. A small team of assistant editors was also appointed to further simplify the editing procedure and attain best results for the readers.

Apart from the editorial board, the designing team has also invested a significant amount of their time in understanding the subject and creating the most relevant covers. They scrutinized every image to scout for the most suitable representation of the subject and create an appropriate cover for the book.

The publishing team has been an ardent support to the editorial, designing and production team. Their endless efforts to recruit the best for this project, has resulted in the accomplishment of this book. They are a veteran in the field of academics and their pool of knowledge is as vast as their experience in printing. Their expertise and guidance has proved useful at every step. Their uncompromising quality standards have made this book an exceptional effort. Their encouragement from time to time has been an inspiration for everyone.

The publisher and the editorial board hope that this book will prove to be a valuable piece of knowledge for researchers, students, practitioners and scholars across the globe.

List of Contributors

Charity N. Atoma
Department of Agricultural Extension and Management, Delta State Polytechnic, Ozoro, Nigeria

Olufemi M. Adesope and Linda C. Familusi
Department of Agricultural Economics and Extension, Faculty of Agriculture, University of Port Harcourt, Nigeria

Ezekiel O. Agbalagba and Hannah O. Agbalagba
Federal University of Petroleum Resources, Nigeria

Terrence Thomas
North Carolina A&T State University, Greensboro, NC, USA

Cihat Gunden
Ege University, Izmir, Turkey

Befikadu Legesse
Smart-Eco Consulting, Silver Spring, MD, USA

Nader Heydari
Iranian Agricultural Engineering Research Institute (AERI), Agricultural Research, Education and Extension Organization (AREEO), Karaj, Iran

Anton Perpar and Andrej Udovč
Department of Agronomy, Biotechnical Faculty, University of Ljubljana, Slovenia

Mazón-Suástegui José Manuel, Ojeda-Silvera Carlos Michel, Batista-Sánchez Daulemys, Tovar-Ramírez Dariel, Arcos-Ortega Fabiola, Murillo-Amador Bernardo, Nieto-Garibay Alejandra and Morelos-Castro Rosa María
Centro de Investigaciones Biológicas del Noroeste S.C. (CIBNOR), La Paz, B.C.S., Mexico

García-Bernal Milagro and Díaz-Díaz Miriam
Universidad Central de Las Villas (CBQ), Santa Clara, Villa Clara, Cuba

Avilés-Quevedo María Araceli
Freelance Services for R&D in Marine Aquaculture, La Paz, B.C.S., Mexico

Abasolo-Pacheco Fernando, Ferrer-Sánchez Yarelys, Alvarado-Mendoza Alex and Bonilla-Montalvan Boris
Universidad Técnica Estatal de Quevedo (FCA), Quevedo, Ecuador

Hamadttu Abdel Farag El-Shafie
Date Palm Research Center of Excellence, King Faisal University, Saudi Arabia

Kassio Ferreira Mendes
Department of Agronomy, Federal University of Viçosa, Viçosa, MG, Brazil

Rodrigo Nogueira de Sousa
Department of Soil Science, "Luiz de Queiroz" College of Agriculture, University of São Paulo, Piracicaba, SP, Brazil

Ana Flávia Souza Laube
Department of Chemistry, Federal University of Viçosa, Viçosa, MG, Brazil

Mauro Wagner de Oliveira and Terezinha Bezerra Albino Oliveira
Federal University of Alagoas, Rio Largo, Brazil

Luís Cláudio Inácio da Silveira and Márcio Henrique Pereira Barbosa
Federal University of Viçosa, Viçosa, Brazil

Alexandre Bosco de Oliveira
Federal University of Ceará, Fortaleza, Brazil

Manoel Gomes Pereira
Research and Development Sector, Triunfo Mill, Boca da Mata, Brazil

Helena Saraiva Koenow Pinheiro, Pedro Armentano Mudado Xavier and Lúcia Helena Cunha dos Anjos
Soil's Department, Agronomy Institute, Federal Rural University of Rio de Janeiro (UFRRJ), Seropédica, RJ, Brazil

Cesar da Silva Chagas and Waldir de Carvalho Júnior
Embrapa Solos, Rio de Janeiro, RJ, Brazil

Aušra Marcinkevičienė, Rita Pupalienė, Rimantas Velička, Zita Kriaučiūnienė, Lina Marija Butkevičienė and Robertas Kosteckas
Vytautas Magnus University, Kaunas, Lithuania

Marina Keidan
Public Institution "Ekoagros", Kaunas, Lithuania

Muthuraman Yuvaraj
Agricultural College and Research Institute, Vazhavachanur, Tiruvannamalai, Tamil Nadu, India

Peyandi Paraman Mahendran
Department of Crop Management, Agriculture College and Research Institute, Kudumiyanmalai, Tamil Nadu Agricultural University, Tamil Nadu, India

Eman Tawfik Hussien
Faculty of Science, Helwan University, Egypt

Nazeer Ahmed, Mukhtar Alam, Muhammad Saeed, Hidayat Ullah and Rafi Ullah
Department of Agriculture, University of Swabi, Anbar, Khyber Pakhtunkhwa, Pakistan

Toheed Iqbal
Faculty of Plant Protection, Department of Entomology, The University of Agriculture, Peshawar, Khyber Pakhtunkhwa, Pakistan

Khalid Awadh Al-Mutairi
Faculty of Science, Dpartment of Biology, University of Tabuk, Tabuk, Saudi Arabia

Kiran Shahjeer
Department of Zoology, Abdulwali Khan University, Mardan, Khyber Pakhtunkhwa, Pakistan

Saeed Ahmed
Agricultural Research Center, Londrina State University, Londrina, Brazil

Nibal Abd Aleem Hassan Ahmed
Faculty of Sciences in Kurma, Department of Science, Taif University, Mecca, Saudi Arabia

Hanem Fathy Khater
Faculty of Veterinary Medicine, Department of Parasitology, Benha University, Toukh, Egypt

Muhammad Salman
Department of Entomology, Faculty of Crop Protection, The University of Agriculture, Peshawar, KP-Pakistan

Malik Asif, Mushtaq A. Malik, Zaffar Mahdi Dar, Saima Shafi and Bisma Rashid
FOA-Wadura, Division of Basic Sciences and Humanities, SKUAST-Kashmir, India

Shayesta Islam
FOH-Shalimar, Division of Environmental Sciences, SKUAST-Kashmir, India

Amjad Masood
FOA-Wadura, Division of Agronomy, SKUAST-Kashmir, India

Showkat Sidique
FOA, Wadura, Division of Agriculture Statistics and Economics, SKUAST-Kashmir, India

Index

Printed in the USA
CPSIA information can be obtained
at www.ICGtesting.com
LVHW080033211023
761596LV00010BA/1739